ORGANIC CHEMISTRY

8

PROGRESS IN ORGANIC CHEMISTRY
8

Joint Editors

W. CARRUTHERS, Ph.D., D.Sc.
Senior Lecturer in Chemistry
University of Exeter

and

J. K. SUTHERLAND, Ph.D.
Professor of Organic Chemistry
University of Manchester

LONDON
BUTTERWORTHS

THE BUTTERWORTH GROUP

ENGLAND
BUTTERWORTH & CO (PUBLISHERS) LTD
London: 88 Kingsway, WC2B 6AB

AUSTRALIA
BUTTERWORTHS PTY LTD
Sydney: 586 Pacific Highway, NSW 2067
Melbourne: 343 Little Collins Street, 3000
Brisbane: 240 Queen Street, 4000

CANADA
BUTTERWORTH & CO (CANADA) LTD
Toronto: 14 Curity Avenue, 374

NEW ZEALAND
BUTTERWORTHS OF NEW ZEALAND LTD
Wellington: 26–28 Waring Taylor Street, 1

SOUTH AFRICA
BUTTERWORTH & CO (SOUTH AFRICA) (PTY) LTD
Durban: 152–154 Gale Street

First published 1973

© Butterworth & Co. (Publishers) Ltd., 1973

ISBN 0 408 70382 2
Suggested U.D.C. No.

Typeset and Printed in England by
Chapel River Press, Andover, Hants

FOREWORD

EIGHT THEMES covering a range of topics of current interest to organic chemists are discussed in the present volume, following the pattern laid down by the original editor of the series, Professor J. W. Cook, and successfully followed in earlier volumes. Four of the chapters are concerned directly with aspects of the chemistry of natural products and a fifth discusses the synthesis of a group of natural products by routes which simulate the supposed biosynthetic pathways. A sixth chapter reviews a fascinating series of chemiluminescent reactions, while the seventh and eighth chapters discuss practical and mechanistic aspects of two important reactions in organic chemistry. As in earlier volumes of the series all the authors are specialists who have themselves contributed to the fields of work which they have reviewed.

In the first chapter, Dr. Clarkson reviews recent work on the synthesis of compounds of the prostaglandin series. The chemical structures of these compounds only became generally known about ten years ago, but since then they have aroused a great deal of interest on account of their powerful physiological action and potential therapeutic usefulness. Because of the difficulties encountered in obtaining sufficient material from natural sources attention has turned to the chemical synthesis of the compounds, and Dr. Clarkson gives an illuminating account of the ingenious methods which have been used to overcome the formidable difficulties involved.

The variety of complex structures found among the terpenoid class of natural products has provided a constant challenge to the synthetic organic chemist, as well as to those interested in biosynthesis. With a view to providing experimental evidence in support of biosynthetic proposals and speculations, and also, possibly, of developing more direct synthetic routes to the compounds concerned, increasing attention has been paid in recent years to the synthesis of natural products by routes which simulate the supposed biosynthetic pathway. In Chapter 2 Dr. Money's stimulating article provides a valuable survey of recent work in this direction in the terpenoid field, and draws attention to the remarkable successes which have already been achieved.

Two other classes of natural products of great current interest are the cannabinoids, which include the active components of *Cannabis* preparations, and the penicillins. The latter have recently acquired added interest through the discovery of the antibacterial activity of the related cephalosporins, and this work has given rise to ideas about structure—activity relationships in the series which are leading to attempts at chemical synthesis of analogues of the natural products. Among the cannabinoids, considerable progress has been made in elucidating the structures of many of the naturally occurring compounds and in synthesising some of them. The fascinating chemistry involved in both these general fields is ably reviewed in Chapters 3 and 4 by Dr. Razdan and by Dr. Stoodley.

In Chapter 5 Dr. Hardy and Dr. Ridge give a clear and systematic account of recent work on the chemistry of cyclic peptides, an important group of compounds which includes a number of naturally occurring substances with powerful biological activity. The activity of many of the latter is dependent not only on their structure but also on their conformation, and the chemical factors which control this and the methods used for determining the conformations are among the topics discussed in this article.

One of the most intriguing natural phenomena is bioluminescence, the emission of visible light by living organisms. It is now known that this light emission is due to chemical reaction, but it is only in the last ten years that some insight has been gained into the mechanism of the phenomenon. The different types of chemical reaction which can give rise to chemiluminescence, the mechanism of light emission and the relation of the naturally occurring processes to the laboratory reactions are considered by Dr. McCapra in his most interesting and lucid account in Chapter 6.

The penultimate chapter of the book is concerned with nitration. Although this reaction has been used by organic chemists for a very long time and has been much studied it still holds some secrets. In a stimulating article Dr. Hartshorn and Dr. Schofield discuss some recent work on nitrating systems and some results which have been obtained from nitration which bear on the problem of aromatic reactivity.

Finally, Dr. McQuillin reviews catalytic hydrogenation under homogenous conditions with soluble transition metal catalysts. Because of the greatly increased selectivity which can often be obtained, these reactions have been receiving increased attention in recent years, and Dr. McQuillin gives a valuable survey of the scope of these reactions and discusses present knowledge about the reaction mechanisms.

W. CARRUTHERS
J. K. SUTHERLAND

CONTENTS

	PAGE
FOREWORD	v

1 THE SYNTHESIS OF PROSTAGLANDINS 1
R. CLARKSON, Ph.D., *Imperial Chemical Industries, Limited, Alderley Park, Cheshire*

2 BIOGENETIC-TYPE SYNTHESIS OF TERPENES 29
T. MONEY, Ph.D., A.R.I.C., *Associate Professor of Chemistry, University of British Columbia, Vancouver, Canada*

3 RECENT ADVANCES IN THE CHEMISTRY OF CANNABINOIDS 78
R. K. RAZDAN, Ph.D., *J. C. Sheehan Institute for Research, Cambridge, Massachusetts*

4 RECENT PENICILLIN CHEMISTRY 102
R. J. STOODLEY, M.Sc., Ph.D., *Lecturer in Chemistry, University of Newcastle upon Tyne*

5 RECENT ADVANCES IN THE CHEMISTRY OF CYCLIC PEPTIDES 129
P. M. HARDY, Ph.D. and B. RIDGE, Ph.D., *Lecturers in Chemistry, University of Exeter*

6 CHEMILUMINESCENCE OF ORGANIC COMPOUNDS 231
F. McCAPRA, Ph.D., D.I.C., *Reader in Chemistry, University of Sussex*

7 SOME ASPECTS OF RECENT WORK ON NITRATION 278
S. R. HARTSHORN, Ph.D., *Demonstrator in Chemistry, University of Exeter*
K. SCHOFIELD, Ph.D., D.Sc., F.R.I.C., *Reader in Organic Chemistry, University of Exeter*

8 HOMOGENEOUSLY CATALYSED HYDROGENATION 314
F. J. McQUILLIN, M.A., D.Phil., D.Sc., *Reader in Organic Chemistry, University of Newcastle upon Tyne*

INDEX 339

1
THE SYNTHESIS OF PROSTAGLANDINS
R. Clarkson

INTRODUCTION	1
Chemistry	1
Biological Activity	2
Synthetic Aims	3
Choice of Synthetic Target	4
SYNTHESES OF E AND F PROSTAGLANDINS	5
Syntheses Incorporating Cyclopentane Ring Formation	5
Syntheses Using Bicyclic Intermediates	10
The Bicyclo [3, 1, 0] hexane Route	10
The Bicyclo [2, 2, 1] heptane Route	14
The Indane Route	19
Other Approaches	21
PROSTAGLANDINS FROM A NON-MAMMALIAN SOURCE	22
SYNTHESES OF 13, 14-DIHYDROPROSTAGLANDINS	23
SYNTHESES OF 11-DESOXY-PROSTAGLANDINS	25

INTRODUCTION

THE NAME prostaglandin was given by von Euler in 1935[1] to that part of the lipid fraction of human seminal plasma which showed powerful smooth muscle stimulatory properties. It was some 28 years later, however, before the chemical structures of the compounds responsible became generally known[2]. Since this initial phase, the growth of prostaglandin research has been almost logarithmic and the compounds have been shown to be involved in a wide range of physiological processes and to have potential therapeutic utility in several areas[3].

Chemistry

Over 20 different natural prostaglandins have now been isolated and the structures of the more important compounds are given in *Figure 1.1*. All are derived from the C-20 cyclopentane acid, prostanoic acid, which has the absolute stereochemistry shown. In prostaglandin-E_1 (PGE_1) the cyclopentane ring carries a β-ketol function and the alkyl chain contains a *trans*-allylic alcohol system. Prostaglandins of the second series (e.g. PGE_2) have in addition a *cis* double bond in the 5,6-position, while those of the third series (e.g. PGE_3) have a further *cis*-double bond at the 17,18-position. Replacement of the 9-keto group of either prostaglandin E_1, E_2 or E_3 by an α-hydroxyl group affords prostaglandin $F_{1\alpha}$, $F_{2\alpha}$ and $F_{3\alpha}$ respectively. Prostaglandins of the A-series (A_1, A_2, A_3) have the 9-keto-Δ^{10} system and are readily prepared by mild acid-catalysed dehydration of the corresponding prostaglandin E. Treatment of either a prostaglandin A or E with strong base leads to the

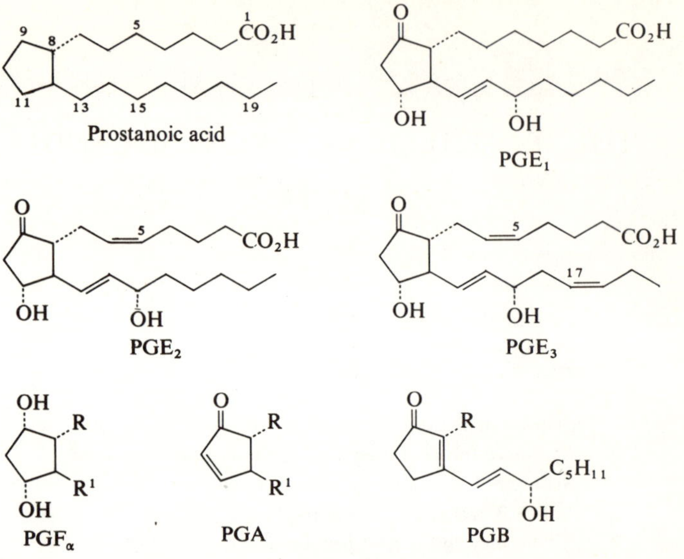

Figure 1.1

B-series which shows the dienone chromophore (λ_{max} 278 nm). This is frequently used as a means of estimating prostaglandins A or E. The six compounds (E, Fα, of the 1st, 2nd and 3rd series) are the primary prostaglandins[4] whose synthesis will be considered below. For details of their isolation, structure elucidation, biosynthesis and basic chemistry the reader is referred to the several excellent reviews which are now available, and in particular to those of Bergström[4] and of Samuelsson[5,6].

These primary compounds together with those of the A and B series are not the only naturally occurring prostaglandins. 19-Hydroxylated compounds are fairly common[7]; 8-iso-PGE$_1$ has been isolated[8]; a range of metabolites of primary prostaglandins have now been identified[9]; modified biosyntheses have produced homologues[10] and some rather unusual cyclic ethers[11]; and derivatives of 15-epi-A$_2$ have been isolated from a species of gorgonia[12]. The last finding is particularly interesting in that it could provide a supply of raw material for synthetic work (see below). However, these compounds are not themselves synthetic targets and the list is included merely to maintain perspective.

Biological Activity

This is a somewhat strange heading in a review concerned with synthesis but it is important that the main driving force behind the synthetic effort should be mentioned, however briefly.

The prostaglandins are fascinating not only on account of the wide range of biological response they produce but also because of their sheer potency. Concentrations as low as 1 part in 10^9 can readily be detected by biological assays. However, it is the demonstration, in the past few years, that prostaglandins have clear and important therapeutic utilities which has intensified the volume of research and has attracted almost all the large drug companies into the field.

Among the clinical studies recently carried out, the effects of prostaglandins on reproduction have been most widely studied and have shown the effectiveness of E_2 or $F_{2\alpha}$ as agents for inducing normal labour at term and for inducing therapeutic abortions. Their possible utility as contraceptives has also been demonstrated. Limited clinical investigation has been carried out into the use of prostaglandins for the relief of asthmatic bronchospasm and as agents to control blood pressure. Other potential clinical uses are in the treatment of thrombosis, of obesity and of certain types of stomach ulcers.

Of the many detailed reviews on the physiological actions of prostaglandins, those of Horton[13] and of Speroff and Ramwell[14] are particularly recommended. Shorter, more general accounts and speculations on the clinical future of prostaglandins are also available[15].

The prostaglandins have now been isolated from a wide variety of body tissues and fluids. They probably act as local regulatory hormones, being synthesised on demand at the site of action and then rapidly deactivated to limit their sphere of influence. This rapid metabolic deactivation of prostaglandins (c. 95% of a dose of E_1 is deactivated by one pass through the circulation[16]) is one of the problems which must be overcome before they can be generally used in therapy. Another serious drawback is that prostaglandins have too wide a range of activity and ways must be found of selecting the particular type of action required if undesirable side effects are to be avoided. This latter problem becomes more acute in the case of metabolically stable prostaglandins.

Synthetic Aims

The main target of the synthetic chemists has been to make the primary prostaglandins available for detailed evaluation in both the laboratory and the clinic.

Until recently PGE_1 and PGE_2 were biosynthesised (*see Figure 1.2*) from the essential fatty acids bishomo-γ-linolenic (Ia) and arachidonic (Ib) acids, respectively, using an enzyme preparation obtained from ram seminal

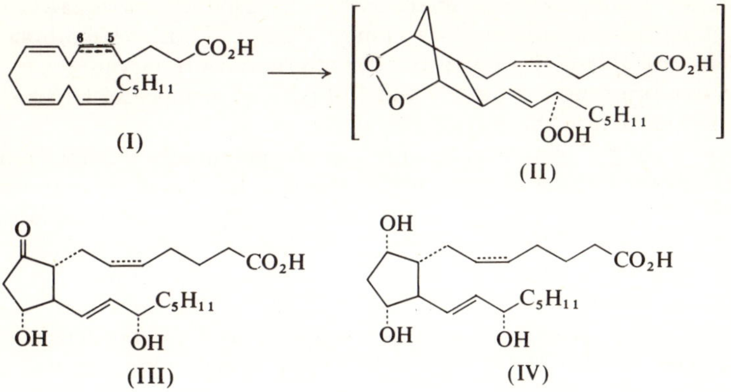

Figure 1.2. (a) 5,6-saturated; (b) 5,6-cis double bond

vesicles[17,18]. It is generally assumed that a peroxide such as (II) is an intermediate which on rearrangement and/or reduction affords either E (III) or F_a (IV) prostaglandins. This particular preparation produces mainly PGE which, on borohydride reduction, affords both PGF_a and its 9-epimer (PGF_β) in roughly equal amounts[19]. The limitations of such a biosynthesis are apparent when one considers that an enzyme preparation from 1 kg of seminal vesicle tissue is needed to transform one gramme of essential fatty acid to give, at best 250–300 mg of PGE. The seminal vesicles from one ram weight approximately 20 g.

The first criterion then is that a chemical synthesis should compete effectively with biosynthesis. The second is that the synthesis should be easily modified in order to prepare analogues with improved selectivity and stability to metabolic degradation. Metabolism appears to be confined to the side-chains—β-oxidation of the acid side-chain to produce the essentially inactive *bis* and *tetra*-nor-acids[20], and oxidation of the 15-hydroxyl group to give the inactive α,β-unsaturated ketone[21]. An effective synthesis should therefore allow variation of the substitution around these areas of the molecule.

Choice of Synthetic Target

The motivation for, and aims of, the intensive effort on prostaglandin synthesis have now been established, but before turning to the synthetic problem itself, the first consideration must be which of the prostaglandins is the most useful and practical synthetic target.

It is important to be able to prepare compounds of the E-, F_a- and A-series since each shows a different range of biological activity. (Prostaglandins of the B-series are virtually inactive.) Since F_a- and A-prostaglandins can be readily prepared from the corresponding E, the latter would seem to be the most useful primary synthetic target. Prostaglandins of the 1, 2 and 3-series in general differ only quantitatively in their pharmacological properties. Thus it is not surprising that most of the syntheses which will be discussed are aimed at PGE_1.

There are, however, two difficulties. Firstly, prostaglandins of the E-series are the most labile (*see* below), and secondly, the reduction of PGE_1 to give PGF_{1a}, for example, is not stereospecific and produces an equal amount of the comparatively uninteresting epimer $PGF_{1\beta}$. Thus the synthesis of the much more stable PGF_a is an attractive alternative primary target, especially if selective protection of the 11- and 15-hydroxyl groups can be arranged to permit oxidation to the corresponding E.

Even though A-prostaglandins show potentially useful biological properties, their direct total synthesis appears to have received little attention, probably because they were regarded as less satisfactory intermediates to the primary prostaglandins. However, the discovery of a non-mammalian source of a PGA_2 derivative[12] has stimulated research into methods for converting this into PGF_{2a} and PGE_2 and this is discussed later.

Of the other prostanoic acid derivatives that have attracted synthetic effort, only 13,14-dihydro- and 11-desoxy-prostaglandins are discussed in this review. This somewhat arbitrary choice is dictated partly by considerations of space, partly by reasons of chemical interest and partly by the biological activity of the compounds themselves. For details of the other syntheses such as those of PGB-compounds and oxa-prostaglandins, the reader is referred to earlier reviews[22].

SYNTHESES OF E AND F$_\alpha$ PROSTAGLANDINS

There are two main problems to be overcome during the synthesis of PGE$_1$. Firstly, since the molecule has four asymmetric centres and a *trans*-double bond, effective control of the relative stereochemistry is important. Three of the four asymmetric centres (8, 11 and 12) are contiguous on the five-membered ring and are mutually *trans*. This is the more stable arrangement and therefore relatively easy to establish. The same is true for the *trans*-double bond. However, stereochemical control of the fourth asymmetric centre at C-15 is unlikely because of its remote position. Secondly, prostaglandins of the E-series are labile due to the β-ketol system which dehydrates at a reasonable rate at values of pH outside the range 3–9.

In the synthesis of an F$_\alpha$-prostaglandin the extra asymmetric centre complicates the stereochemical problem but this is perhaps more than compensated for by the marked increase in the molecule's stability (PGFs are moderately labile at very low values of pH presumably due to the ene–diol system at C-11 to C-15 which is both allylic and homo-allylic).

It is interesting to compare the various strategies used to meet these problems, and for this purpose it is convenient to divide the types of syntheses into two classes: those in which the cyclopentane ring is closed in the synthesis and the stereochemical problems tackled subsequently, and those in which a bicyclic system is used to effect some degree of stereochemical control.

Syntheses Incorporating Cyclopentane Ring Formation

Of the three routes included in this section, only one has been successfully taken through to a primary prostaglandin. They clearly illustrate the limitations of this general approach; viz. the difficulties in controlling stereochemistry and the problem of selective protection of functional groups.

(a) In Corey's first synthesis of PGE$_1$[23] his main concern was to generate the β-ketol system under the mildest possible conditions and as late as possible in the scheme. The plan was to synthesise the 9-amine and convert this into the 9-ketone via the corresponding imine. Ingenious new methods were developed for oxidising model primary amines[24] though in the actual synthesis the known Ruschig method (illustrated below) was the only successful one.

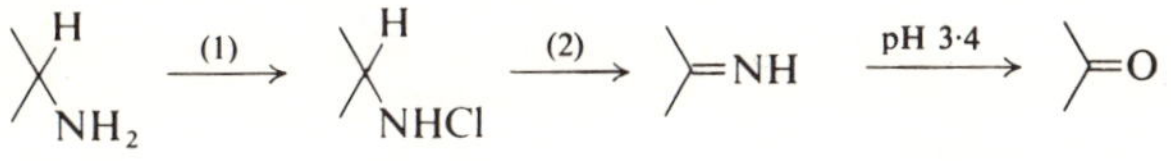

$$\underset{NH_2}{\overset{H}{\diagup\!\!\!\diagdown}} \xrightarrow{(1)} \underset{NHCl}{\overset{H}{\diagup\!\!\!\diagdown}} \xrightarrow{(2)} {=}NH \xrightarrow{pH\ 3\cdot 4} {=}O$$

Reagents: (1) *N*-chlorosuccinimide; (2) MeO$^-$

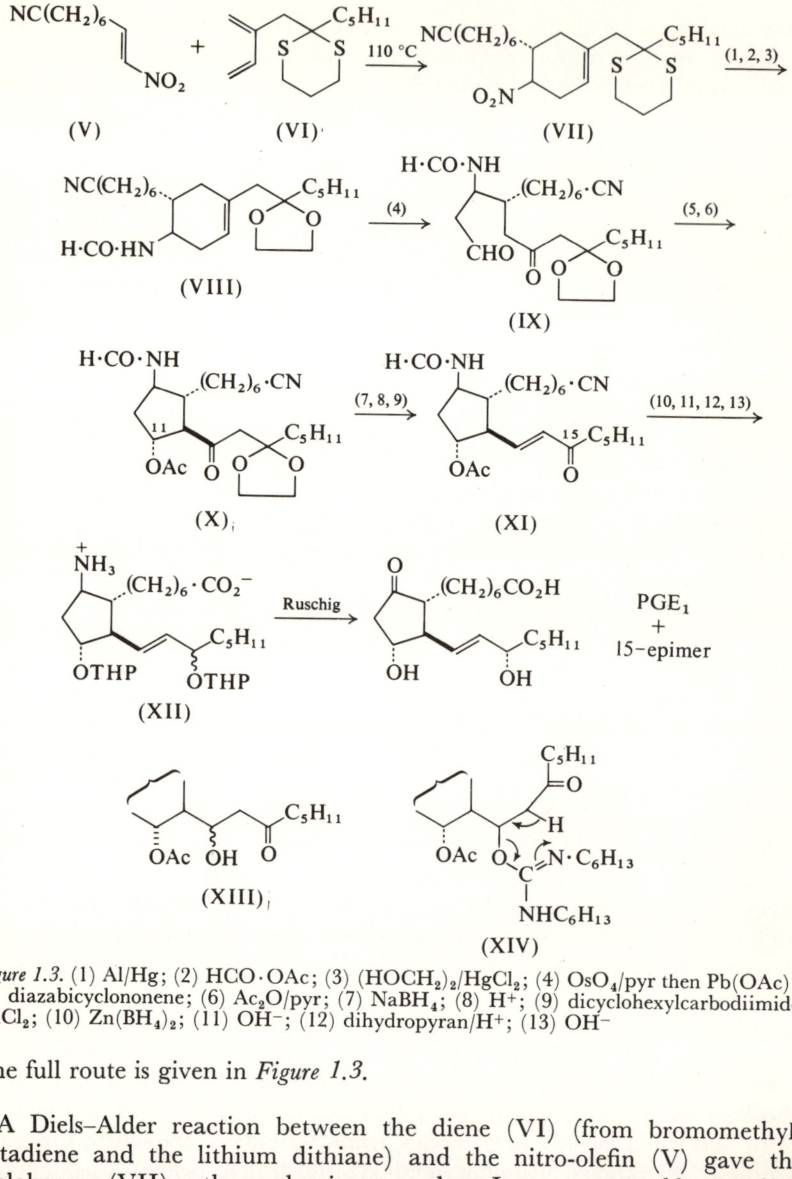

Figure 1.3. (1) Al/Hg; (2) HCO·OAc; (3) (HOCH$_2$)$_2$/HgCl$_2$; (4) OsO$_4$/pyr then Pb(OAc)$_4$; (5) diazabicyclononene; (6) Ac$_2$O/pyr; (7) NaBH$_4$; (8) H$^+$; (9) dicyclohexylcarbodiimide/CuCl$_2$; (10) Zn(BH$_4$)$_2$; (11) OH$^-$; (12) dihydropyran/H$^+$; (13) OH$^-$

The full route is given in *Figure 1.3*.

A Diels–Alder reaction between the diene (VI) (from bromomethylbutadiene and the lithium dithiane) and the nitro-olefin (V) gave the cyclohexene (VII) as the predominant product. It was converted by standard reactions into the formamido-ketal (VIII), the olefinic group of which was cleaved in two steps to the keto-aldehyde (IX). The cyclopentane ring was formed by an aldol condensation using 1,5-diazabicyclo[4,3,0]non-5-ene as base (this minimises dehydration of the aldol), and the hydroxy ketone was isolated as its acetate (X). As expected the alkyl chains were fixed in the *trans* position and the major product had the C-11 acetate in the 'natural' relative configuration (α). The configuration of the formamido group is of course of little consequence.

THE SYNTHESIS OF PROSTAGLANDINS

The sequence employed to convert (X) into the enone (XI) is particularly noteworthy. Reduction of the ketone with borohydride and cleavage of the ketal affords a β-hydroxy-δ-acetoxyketone (*see* partial structure XIII) which has to be dehydrated without concomitant elimination of the 11-acetoxy group to give the unwanted dienone. This was achieved using dicyclohexylcarbodiimide in the presence of cuprous ion. The iso-urea formed is assumed to eliminate by the cyclic mechanism shown in (XIV).

Reduction of the 15-ketone (XI) with zinc borohydride (used to minimise conjugate addition and base-catalysed elimination) gave a mixture of C-15 epimeric alcohols, which were taken to the amino-acid mixture (XII) in which the 11- and 15-hydroxyl groups were protected as tetrahydropyranyl ethers. Application of the Ruschig procedure then gave a mixture of PGE_1 and its 15-epimer which were separated by chromatography.

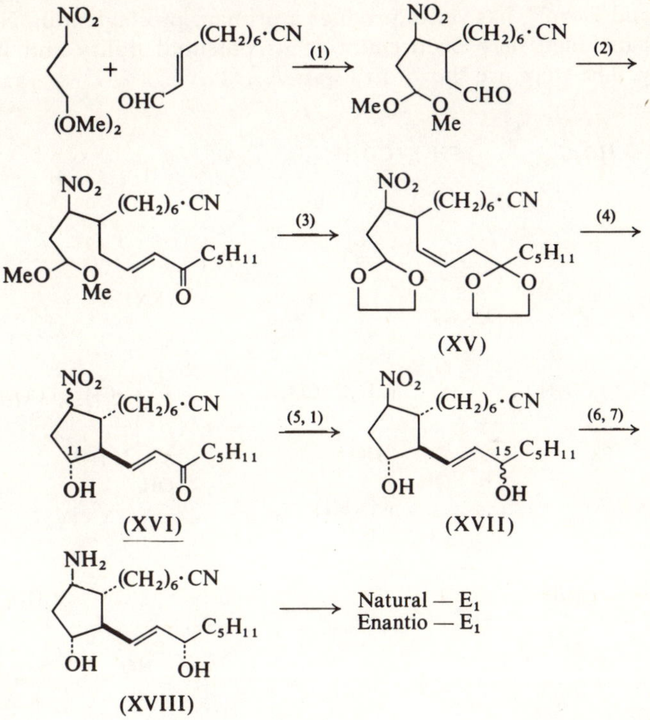

Figure 1.4. (1) base; (2) $(MeO)_2PO\ \bar{C}HCOC_5H_{11}$; (3) $(HOCH_2)_2/H^+$; (4) $SnCl_4$; (5) $Zn(BH_4)_2$; (6) chromatography; (7) Al/Hg

Improvements to the early steps of this route were described in subsequent papers[25,26] and are summarised in *Figure 1.4*. A fascinating reaction is the cyclisation of the bis-ketal (XV) (prepared as illustrated by a Michael reaction followed by ketalisation) which in the presence of anhydrous stannic chloride is almost completely stereospecific to give the nitrocyclopentane (XVI). If water was added to the reaction more of the 11β-hydroxy compound was obtained. The 11α-hydroxyketone was reduced to a mixture of diols (XVII) from which the required α-isomer could be separated by chromatography and the unwanted β-isomer recycled by oxidation to the enone (XVI). Reduction of the nitro-group of diol (XVII, one pure isomer at C-15) gave the amino-nitrile (XVIII) which could be resolved as the bromocamphorsulphonate. By completing the sequence as in *Figure 1.3*, both natural and enantiomeric isomers of PGE_1 were obtained. The various 15- and 11-epimers were also obtained and some rather interesting structure–activity relationships were demonstrated[27].

Although several of the stereochemical problems have been resolved, the generation of the β-ketol system from the β-hydroxyamine is experimentally very difficult, and thus the route is not suitable for either large-scale production or for the synthesis of analogues.

(b) The second synthesis to be considered under this heading, that of Miyano and Dorn[28], has yet to produce a primary prostaglandin. Nevertheless, it is included here on account of its potential utility and its direct simplicity. The steps are shown in *Figure 1.5*.

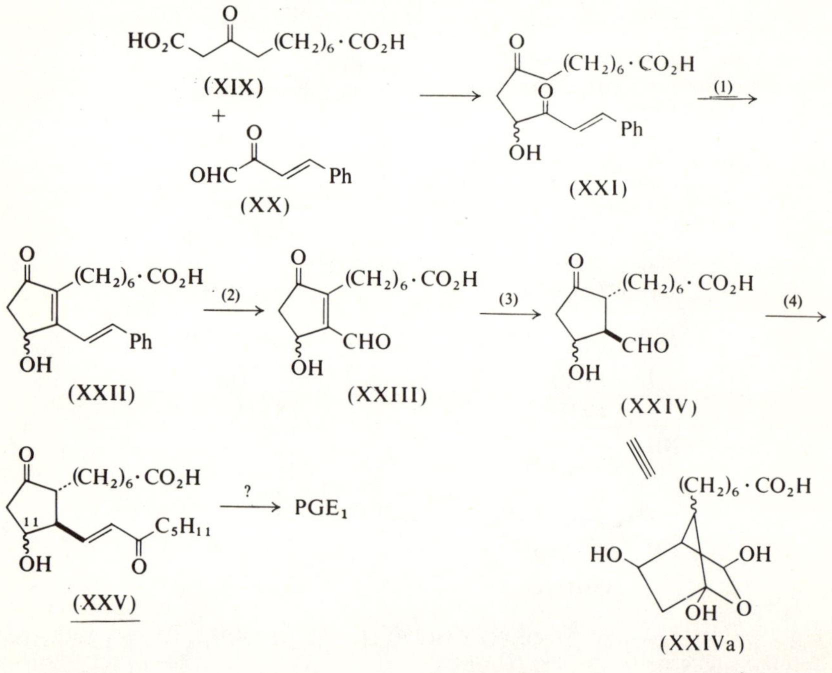

Figure 1.5. (1) base; (2) $OsO_4/NaIO_4$; (3) $Zn/AcOH$; (4) $Ph_3P=CHCOC_5H_{11}$

The cyclopentane system is formed by two aldol-type condensations which are induced to take place in the required direction by an ingenious choice of starting materials, namely the β-keto-acid (XIX) and styrylglyoxal (XX). The first condensation with concomitant decarboxylation gave the aldol (XXI) in quantitative yield which, with base, cyclised to the dienone (XXII). Selective cleavage of the styryl double bond in (XXII) afforded the aldehyde

(XXIII) which, because of its enedione system, could be reduced with zinc and acetic acid to give the crude saturated aldehyde (XXIV) which is assumed to exist as a mixture of cyclic hydrates, one of which is depicted by structure (XXIVa). Wittig condensation of this crude aldehyde gave a 20–25% yield of a mixture of 15-dehydro-PGE$_1$ and its 11-epimer, which were separated by chromatography.

β-Hydroxy-aldehydes such as (XXIV) are intermediates in several of the prostaglandin syntheses. They dehydrate readily and are usually reacted without attempted purification and under the mildest possible conditions. Thus it is a little surprising that Miyano and Dorn chose to use the feebly reactive triphenyl phosphorane (reagent 4) when much more reactive Wittig-type reagents are available (cf. the following synthesis and the one by Corey described in the next section). The low yield from this Wittig reaction could, however, reflect the purity of the starting aldehyde, suggesting poor steric control in the previous reduction.

The authors did not describe a method for selective reduction of the 15-keto group. However, prostaglandin 15-dehydrogenase is relatively easy to isolate[20] although the equilibrium set up by this enzyme is on the side of the 15-ketone. If a means could be found to shift this equilibrium position in favour of the 15-alcohol, this route could become the basis of a very attractive and direct synthesis.

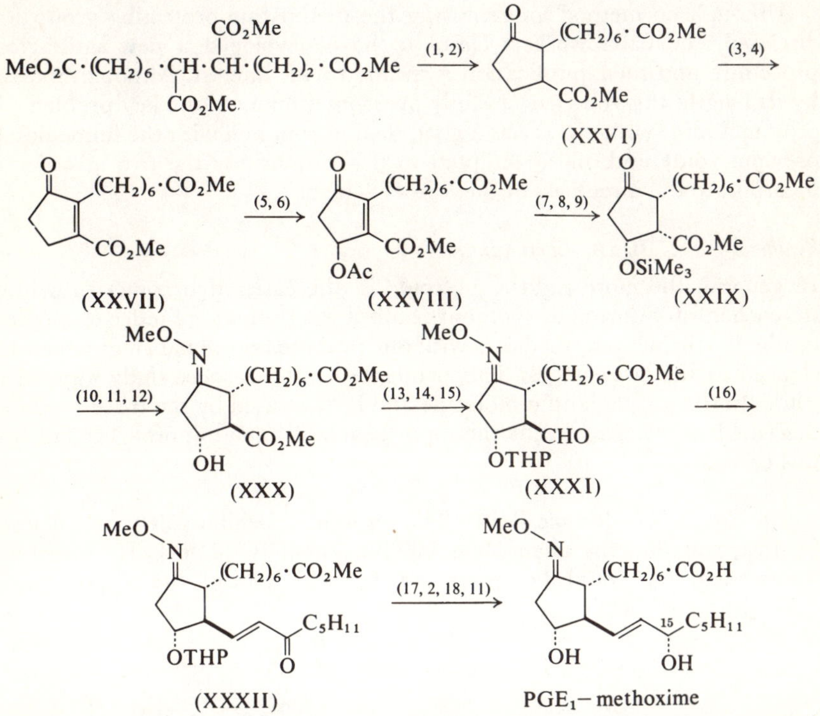

Figure 1.6. (1) NaH; (2) H$^+$; (3) Br$_2$; (4) Et$_3$N; (5) NBS; (6) AgOAc; (7) MeOH/H$^+$; (8) silylation; (9) H$_2$/Raney Ni; (10) MeONH$_2$·HCl/pyr; (11) K$_2$CO$_3$/MeOH; (12) esterification; (13) dihydropyran/H$^+$; (14) NaBH$_4$; (15) modified Moffat oxidation; (16) Bu$_3$P=CHCOC$_5$H$_{11}$; (17) NaBH$_4$; (18) p.l.c.

(c) Finch and Fitt's route[29] is the final one to be considered in this section (*see Figure 1.6*). Using the cyclopentenone (XXVII) they were able to introduce the 11-oxygen and elaborate the C_8 side-chain with excellent control of stereochemistry but, as in previous routes, effective protection of the β-ketol system turned out to be the main difficulty.

The enone (XXVII) (prepared as shown) was allylically acetoxylated then hydrogenated to the saturated ketone (XXIX). Note that the allylic oxygen function was protected from hydrogenolysis as the trimethylsilyl ether. The bulky silyloxy group also controls the direction of hydrogenation and leads to the all *cis*-configuration shown. The carbonyl of the β-ketol system was protected as the methoxime, and the secondary methoxycarbonyl group was epimerised with potassium carbonate in methanol to give, after re-esterification, compound (XXX) in which the required stereochemistry has now been established. It is somewhat surprising that since the hydroxyl group is in the β-position with respect to the ester function it does not eliminate under these conditions, to give the α,β-unsaturated ester.

The remainder of the synthesis is relatively straightforward. The aldehyde (XXXI) was very labile (it dehydrated to give the α,β-unsaturated aldehyde) and the reactive tributylphosphorane (reagent 16) was used in the subsequent Wittig reaction (cf. previous synthesis). It will be noted that again no control of C-15 stereochemistry is possible and the two isomers have to be separated.

Although no method for removing the methoxime protecting group was disclosed, the Ciba workers claim to have developed a new satisfactory procedure and their publication is awaited with interest. Although marred by its length, this route successfully overcomes many of the key problems in prostaglandin synthesis. It has a good deal in common with the immediately previous route and one is tempted to try to hybridise the two in order to incorporate the directness of the former scheme.

Syntheses Using Bicyclic Intermediates

In general, the more rigid a molecule is, the easier it becomes to control stereochemistry during its synthesis. Thus it is a common practice to use, in a synthetic scheme, intermediates with one or more additional rings which are cleaved in the later stages. This principle has been successfully applied in routes to the prostaglandin molecule and is illustrated by the three syntheses described below. At present the more generally useful approaches fall into this category.

The Bicyclo[3,1,0]hexane Route—The basic idea behind this route, devised by Just, was that the allylic–homoallylic system (C-11 to C-15) of correct stereo-chemistry could be generated by solvolysis of the appropriately

THE SYNTHESIS OF PROSTAGLANDINS

substituted cyclopropyl carbonium ion (derivable from the epoxide, for example) as shown on p. 10 (XXXIII → XXXIV).

An encouraging precedent was available from the work of Wiberg and Ashe[30], who had shown that solvolysis of the simple bicyclic tosylate (XXXV) gave mainly ring-opened products (XXXVI; R = Ts and Ac).

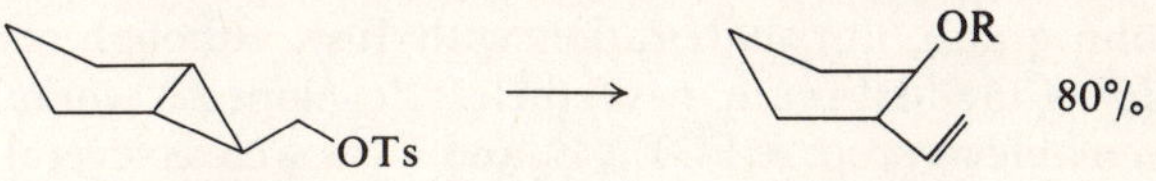

Details of their route were first published by Just and Simonovitch[31] in 1967 but they had neither resolved the fairly formidable stereochemical problems nor found solvolytic conditions which favoured cyclopropane ring opening, and it is unlikely that they isolated any natural prostaglandin although their

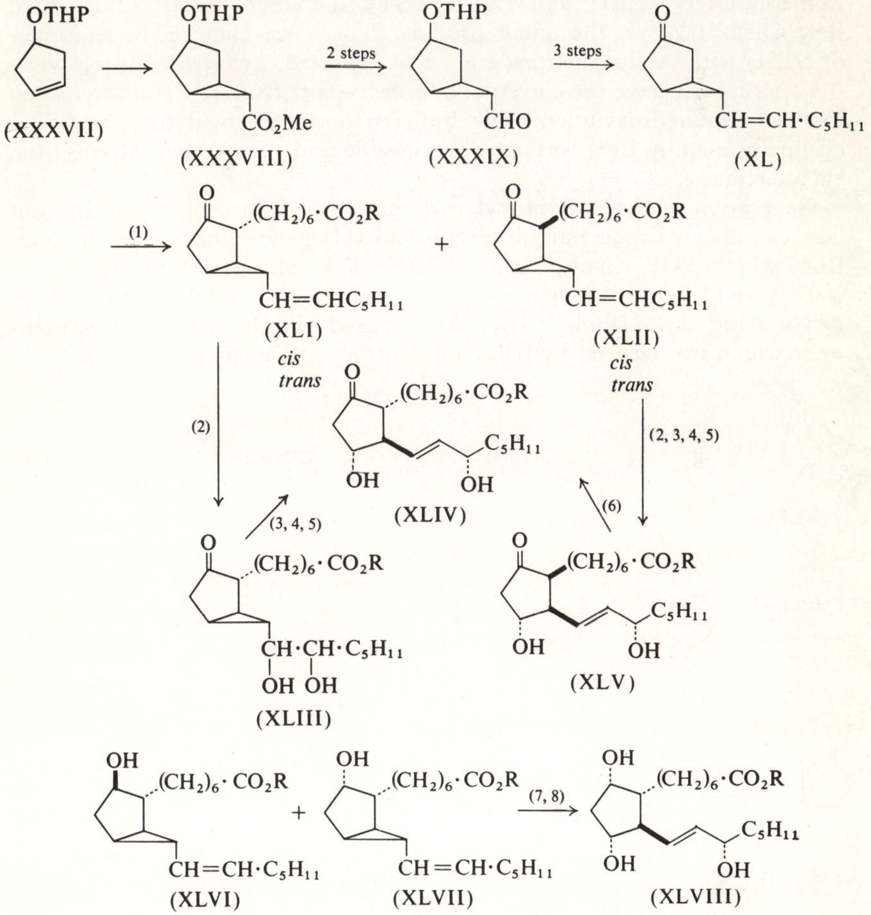

Figure 1.7. (1) $KOBu^t/I(CH_2)_6CO_2R$; (2) OsO_4; (3) $MeSO_2Cl/pyr$; (4) aq. acetone; (5) chromatography; (6) KOAc; (7) *m*-chloroperbenzoic acid; (8) HCO_3H

products did show biological activity. Nevertheless, the basic concept independently attracted two industrial research groups to explore its potential.

In a detailed account[32] the Smith, Kline and French group concluded that the published method was not a practical route but by suitable modification PGB_1 as well as biologically active PGF_1 isomers could be obtained.

The Upjohn group, in collaboration with Just, although confirming the impracticality of the first route, nevertheless developed a workable synthesis and have been able to prepare PGE_1, E_2 and E_3 as well as several analogues[33]. Limited space does not permit of a detailed account of the early work, and only the successful syntheses are discussed below (*see Figure 1.7*).

The aldehyde (XXXIX), as a mixture of isomers about the protected carbinol group, was obtained by a conventional sequence from the cyclopentene (XXXVII) via the ester (XXXVIII). It gave a mixture of *cis*- and *trans*-olefinic ketones (XL) which were separately alkylated to give the isomeric esters (XLI) and (XLII). The α-epimer (XLI), of required stereochemistry, was the minor product (35%). Reaction of the *cis*-isomer of (XLI) with osmium tetroxide gave, as expected, two *erythro*-glycols while the *trans*-isomer gave the corresponding *threo*-pairs (XLIII). It is interesting to note that hydroxylation using buffered performic acid (the conditions originally used by Just) was not stereospecific and gave all four glycols from each olefin.

Generation of the cyclopropyl carbonium ion by solvolysis, in aqueous acetone, of the *bis*-mesylate of glycols (XLIII) gave a mixture of products from which PGE_1 methyl ester (XLIV; R = Me) could be isolated in 5–10% yield. Since it is not possible to hydrolyse the methyl ester without dehydrating the β-ketol, PGE_1 was prepared via the 1,1,1-trichloroethyl ester which was cleaved by reduction with zinc in acetic acid.

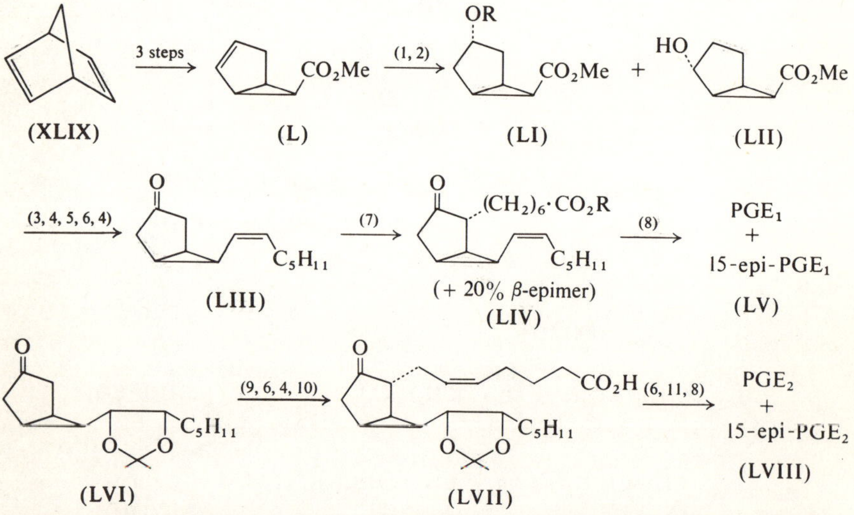

Figure 1.8. (1) $B_2H_6/H_2O_2/^-OH$; (2) dihydropyran/H+; (3) $LiAlH_4$; (4) Jones oxidation; (5) $Ph_3P=CHC_5H_{11}$; (6) H+; (7) $I(CH_2)_6CO_2R$/base; (8) solvolysis of glycol *bis*-mesylates; (9) $BrCH_2 \cdot C \equiv C \cdot (CH_2)_3 CH_2 \cdot O \cdot THP/KOBu^t$; (10) $H_2/Pd/BaSO_4$/pyr.; (11) esterification

8-Iso-PGE$_1$ was similarly prepared from the β-alkylated ketone (XLII). It is known that 8-iso-PGE$_1$ readily isomerises to PGE$_1$[8].

Reduction of the ketone (XLI) gave a mixture of the two epimeric alcohols (XLVI) and (XLVII) in the ratio of 9:1. A low yield (2–3%) of PGF$_{1\alpha}$ was obtained from the latter by solvolysis of the corresponding epoxide.

This route, although providing small amounts of two of the primary prostaglandins, is clearly not a practical one in its present form and attempts were made to improve on the low yields obtained in the solvolytic step.

It will be noted that the glycol mixture (XLIII) has the *exo*-configuration with respect to the bicyclic system. The observation that minor by-products, of presumed *endo*-configuration, gave better yields of prostaglandins has led to an improved synthesis via *endo*-derivatives[34] (*see Figure 1.8*).

The *endo*-ester (L), readily obtained from norbornadiene[35] (XLIX), was hydroborated to give a mixture of alcohols (LI; R = H) and (LII). The former, protected as its tetrahydropyranyl ether (LI; R = THP), was converted by the sequence shown into the olefinic ketone (LIII) which was then alkylated with ω-iodoheptanoate to give the bicyclic ester (LIV).

Two of the steps in this *endo*-series were found to be stereoselective in the required direction. Firstly, the Wittig reaction gave only the *cis*-olefin, and secondly, alkylation of the ketone (LIII) gave predominantly (4:1) the required α-epimer (LIV).

Solvolysis of the glycol *bis*-mesylates as described previously gave PGE$_1$ and 15-epi-PGE$_1$ esters in the remarkable yield of 35–40%.

A modification (LVI–LVIII) gave PGE$_2$ and its 15-epimer in a similar yield[36]. The 5,6-*cis*-double bond was introduced through the corresponding acetylene. As before the prostaglandin free acids were obtained by reductively

cleaving the 1,1,1-trichloroethyl esters. PGE$_3$ methyl ester has also been obtained by a further modification of this route[37].

The 3-oxa and 15-methyl analogues of PGE$_1$ (LIX and LX respectively) and of the other prostaglandin series were prepared by appropriate modifications to this route. The former cannot be degraded by metabolic β-oxidation of the acid side-chain and the latter, of course, cannot be a substrate for the 15-dehydrogenase enzyme. Both compounds are biologically active though they are less potent than the corresponding natural compounds[38]. It will be of particular interest to see if either of these structural changes significantly increases the molecule's half-life *in vivo*.

Due to the detailed, meticulous work of the Upjohn chemists, this has been the first route to be developed to the stage where a range of primary prostaglandins and analogues can be made. It remains to be seen whether it can be used to prepare these analogues in sufficient number (in terms of type and quantity) for meaningful structure–activity–selectivity relationships to be determined. It is clear, however, that this synthesis is not really suited to the preparation of prostaglandins on a large scale, and it has now been superseded by more efficient routes.

The Bicyclo[2,2,1]heptane Route—This synthesis, developed by the Harvard group[39], is perhaps the most important practical route to date. The aim was to design a synthesis such that the primary prostaglandins (E_1, E_2, $F_{1\alpha}$, $F_{2\alpha}$) could be obtained from a common intermediate (*see Figure 1.9*).

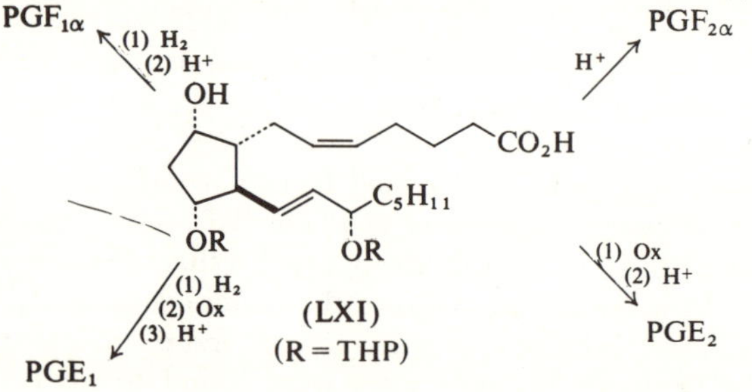

Figure 1.9

The choice of a suitably protected derivative of $PGF_{2\alpha}$ (LXI) as this key intermediate is particularly interesting. Provided that the protecting groups are labile to mild acid hydrolysis (e.g. tetrahydropyranyl ethers), this approach overcomes the difficulties associated with the β-ketol system of the PGE's. Moreover, for the synthesis of prostaglandins of the F_α series *per se*, it is clearly more economical to use a route in which the 9-hydroxy group is stereospecifically incorporated than to use procedures involving reduction of the corresponding PGE (which give a mixture of both 9α- and 9β-alcohols). Finally, it was anticipated that the less hindered *cis*-double bond could be selectively hydrogenated to give prostaglandins of the 1-series. The synthesis of the $PGF_{2\alpha}$ derivative (LXI) (*see Figure 1.10*) has so many interesting features that it will be discussed in some depth.

7-Methoxymethylbicycloheptenone (LXIV) is the key intermediate to this route. Its use builds in complete stereochemical control of the four contiguous asymmetric centres of $PGF_{2\alpha}$ (C-8, 9, 11 and 12) and also facilitates the selective protection required for the stepwise elaboration of the prostaglandin side-chains.

The bicycloheptene system was obtained by a cupric ion catalysed Diels–Alder reaction between 5-methoxymethylcyclopentadiene (LXII) and α-chloro-acrylonitrile (this reaction and the alkylation of cyclopentadiene are discussed below in more detail). Hydrolysis of the resulting chloronitrile (LXIII) under strongly basic conditions gave the required ketone (LXIV). Only the required C-7 isomer is obtained, presumably due to steric control during the Diels–Alder reaction.

A Baeyer–Villiger reaction on ketone (LXIV) in the presence of bicarbonate* gave the lactone (LXV). Hydrolysis of the lactone function gave the

---
* Probably necessary to prevent acid-catalysed rearrangement of the lactone (LXV) to cyclopentene (LXXII) via the carbonium ion shown on p. 16.

carboxylate (LXVI) which afforded the iodo-lactone (LXVII, R = Ac) on treatment with iodine. Closure of the new lactone produces a 5,5-bicyclic system and hence ensures the *cis* configuration as shown. De-iodination of (LXVII) with tributyltin hydride, followed by demethylation (boron tribromide) and Collins oxidation of the resulting primary alcohol (LXVIII; R = Ac, R¹ = CH₂OH), gave the labile aldehyde (LXVIII; R = Ac,

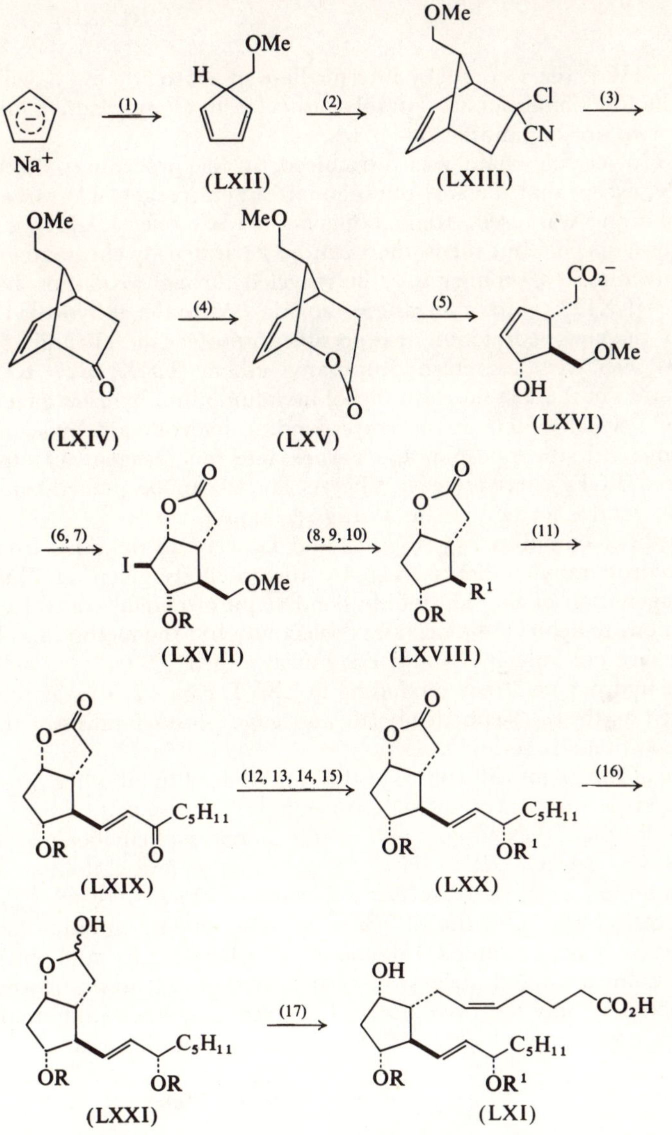

Figure 1.10. (1) ClCH$_2$OMe; (2) H$_2$C=CCl·CN/Cu^{2+}; (3) KOH/DMSO; (4) *m*-chloroperbenzoic acid/NaHCO$_3$; (5) KOH; (6) KI$_3$; (7) acylation; (8) Bu$_3$SnH; (9) BBr$_3$; (10) Collins oxidation; (11) (MeO$_2$) PO·CHCOC$_5$H$_{11}$; (12) Zn(BH$_4$)$_2$; (13) chromatography; (14) K$_2$CO$_3$; (15) dihydropyran/H$^+$; (16) Bu$^i_2$AlH; (17) Ph$_3$P=CH(CH$_2$)$_3$·CO$_2^-$

(LXV) ⟶ [structure with CO₂H] ⟶ (LXXII)

R^1 = CHO), the second key intermediate in the synthesis. It will be noted that the functionalities are suitably protected to allow selective introduction of the two prostaglandin side-chains.

The lower side-chain was introduced by the procedures described previously, except that the acyl-phosphonate anion (reagent 11) rather than the phosphorane was used. Reduction of the 15-ketone (LXIX) is of course non-stereospecific but the isomers can be separated by chromatography and the unwanted 15β-epimer may be recycled through oxidation back to the ketone (LXIX) with manganese dioxide. With the 9-hydroxyl group as part of the lactone function, it is possible to protect the 11- and 15-hydroxyl groups selectively as tetrahydropyranyl ethers (LXX; R = R^1 = THP). Reduction of the lactone with di-isobutylaluminium hydride gave the lactol (LXXI) which reacts as the corresponding hydroxy aldehyde in a Wittig reaction with the phosphorane carboxylate ion (reagent 17) to give the required $PGF_{2\alpha}$ derivative (LXI). As far as can be judged, only the *cis*-double bond is formed in this Wittig reaction.

The prostaglandins $F_{1\alpha}$, E_1, $F_{2\alpha}$ and E_2 were all obtained from the bis-tetrahydropyranyl ether (LXI) by the methods given in *Figure 1.9*.[39c] Hydrogenation of the 5,6-double bond requires careful control of reaction conditions to achieve the necessary selectivity but the method has been used to prepare not only PGE_1 and $F_{1\alpha}$ but also their 5,6-tritiated analogues.

The hydroxy acid (corresponding to LXVI, *Figure 1.10*) was conveniently resolved as the (+)-ephedrine salt and gave prostaglandins of the natural and enantiomeric series[39b].

One of the major difficulties in this, and indeed in all other prostaglandin syntheses, is the separation of C-15 epimers. A recent finding by Corey's group is that the 15-epimers of the mono-*p*-phenylbenzoates (LXX; R = *p*·Ph·C_6H_4CO, R^1 = H) are more easily separated than are either the corresponding acetates or several other esters which were tried. While this modification alleviates the difficulty to some extent, chromatography and recycling of the unwanted 15β-epimer is still necessary and this could be troublesome on an industrial scale. The ideal solution, however, would be the production of only the 15α-epimer. Two extremely interesting methods have now been reported which go far towards achieving this aim.

(LXXIII)

THE SYNTHESIS OF PROSTAGLANDINS

The first[41] stems from the discovery that stereospecific reduction of the 15-ketone (LXIX; R = $p\cdot$Ph$\cdot$C$_6$H$_4$CO) can be achieved using asymmetric borohydride ions. That derived by treating the borane (LXXIII, prepared from limonene and thexylborane) with t-butyllithium was particularly effective and gave a 4·5 : 1 ratio of 15α- to 15β-alcohols, when the reduction was carried out at −120°C.

The second method[42] derives from a modification of the original synthesis, aimed at the preparation of prostaglandins of the 3-series. It is shown in *Figure 1.11*. The allylic alcohol side-chain is introduced by reacting an oxido-ylide (LXXVI, for 2-series compounds; LXXV for 3-series compounds) of correct absolute stereochemistry at the potential prostaglandin C-15 with the

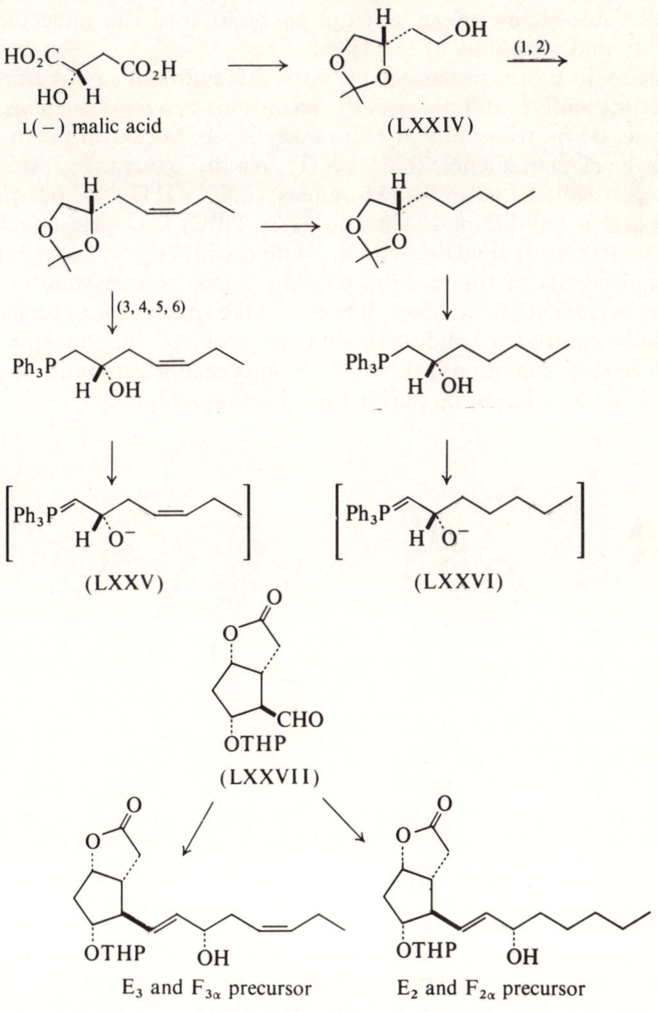

Figure 1.11. (1) Collins oxidation; (2) EtCH=PPh$_3$; (3) H$^+$; (4) TsCl/pyr; (5) NaI; (6) Ph$_3$P

optically active aldehyde (LXXVII). The two oxido-ylides were prepared by a multi-step process from the hydroxy-acetonide (LXXIV), which was itself derived from L(−)-malic acid. Previous work[43] had shown that oxido-ylides condense with carbonyl compounds to give only one of the two possible *trans*-olefins. In this case a 35–50% yield of each of the intermediates shown was obtained which were carried through to their respective prostaglandins by the usual reaction sequence.

This general route to the prostaglandins clearly meets the previously established criteria. It is suitable for large-scale production of most of the primary prostaglandins and is much more efficient than biosynthesis in this respect. Moreover, by changing the phosphonate (reagent 11, *Figure 1.10*) and/or the phosphorane (reagent 17) it is possible to modify the prostaglandin side-chains in an attempt to build into the molecule metabolic stability and selectivity of biological effect.

The main difficulties associated with this synthesis (apart from the use of rather expensive and potentially hazardous reagents such as tributyltin hydride, boron tribromide, Collins reagent, etc.) occur in the first two steps. 5-Alkylcyclopentadienes (e.g. LXII) readily rearrange through a 1,5-hydrogen shift to their 1-alkyl-isomers (LXXVIII). (5-Methylcyclopentadiene has a half-life of 20–25 min at 25–28°C)[44]. Cupric fluoroborate was used to accelerate the Diels–Alder reaction with respect to the isomerisation, and indeed makes the reaction possible. Some isomerisation of 5-methoxymethylcyclopentadiene does however take place even under the most carefully controlled conditions, and the required bicycloheptene (LXIII) produced is contaminated with an appreciable amount of the isomer (LXXIX) and has to be purified by chromatography.

It has been recently claimed, however[45], that if thallous cyclopentadiene is used in place of the sodium salt in the preparation of (LXII), the later isomerisation is markedly reduced and the product from the subsequent Diels–Alder reaction contains 75–85% of the required isomer (LXIII). Moreover, the thallium method permits preparation of the benzyloxymethyl analogue of (LXII) which can be transformed into the benzyl ether analogue of the methyl ether (LXVII). This means that, after deiodination, the benzyl ether may be cleaved by hydrogenolysis (this avoids the use of boron tribromide).

The Indane Route—This synthesis of E_1 by the Merck group[46] is the most recent and is clearly the handiwork of steroid chemists. It is, however,

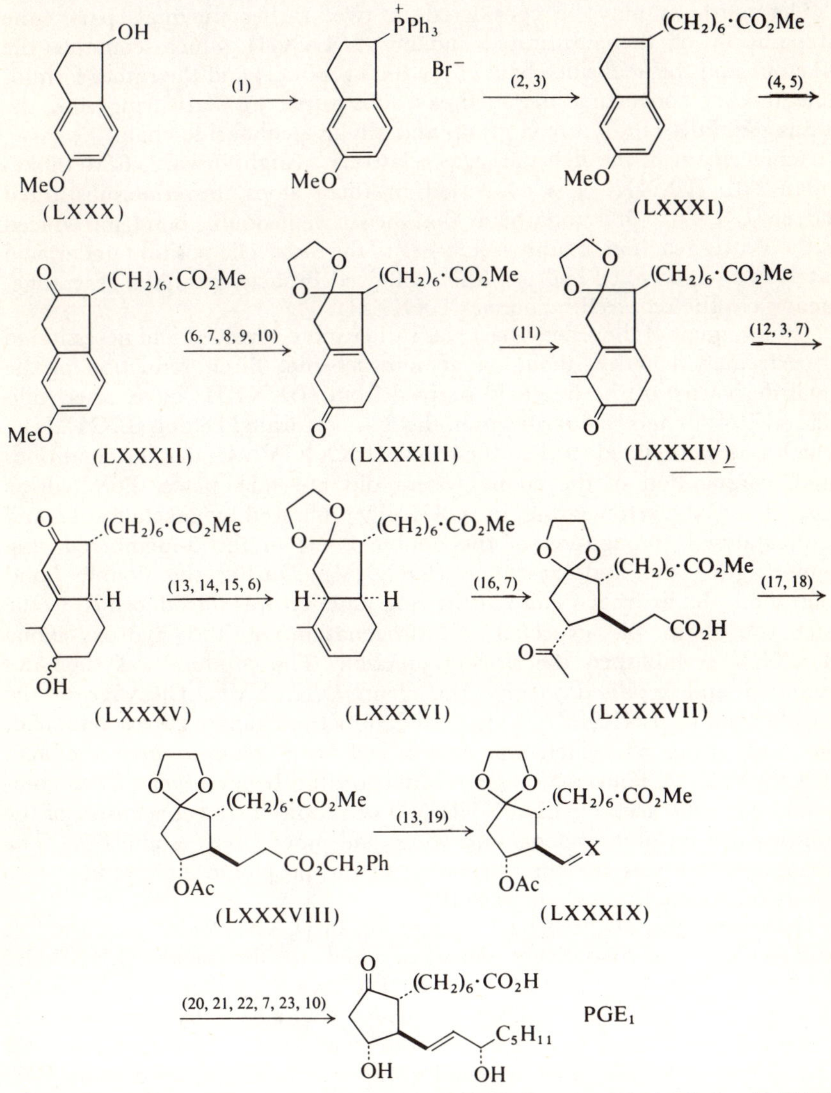

Figure 1.12. (1) $Ph_3P \cdot HBr$; (2) $MeO_2C(CH_2)_5CHO/KOBu^t$; (3) H^+; (4) OsO_4; (5) TsOH; (6) $(HOCH_2)_2/H^+$; (7) ^-OH; (8) Birch reduction; (9) CH_2N_2; (10) dil. AcOH; (11) $MeI/Ph_3C \cdot Li$; (12) $LiAlH(OBu^t)_3$; (13) H_2/Pd; (14) $MeSO_2Cl/pyr$; (15) Me_2SO/Δ; (16) $NaIO_4/KMnO_4$; (17) $PhCH_2N_2$; (18) CF_3CO_3H/Na_2HPO_4; (19) $Pb(OAc)_4/Cu^{2+}/h\nu$; (20) $OsO_4/NaIO_4$; (21) $(MeO)_2PO \cdot CH \cdot CO \cdot C_5H_{11}$; (22) $NaBH_4$; (23) chromatography

somewhat disappointing since it would appear to offer no improvement over the existing routes. The full scheme is shown in *Figure 1.12*.

In common with Corey's route, the second ring of the bicyclic system is cleaved to leave the 11-hydroxyl group and a short stem to which one of the prostaglandin alkyl side-chains is ultimately attached. Whereas in Corey's synthesis it was the acid side-chain, in the present sequence it is the allylic alcohol side-chain which is finally introduced. Again this approach gives complete stereochemical control (except at C-15).

The synthesis may be considered in two, rather unequal, parts: the preparation of the substituted indane (LXXXII) which contains the 9-ketone and the acid side-chain of the final product; and the rather formidable task of converting the methoxy substituted aromatic ring into the prostaglandin's 11α-hydroxyl group and allylic alcohol side-chain.

Achievement of the first part was relatively straightforward. 6-Methoxy-indan-3-ol (LXXX) was converted in three steps into the substituted indene (LXXXI) (it is noteworthy that the *exo*cyclic double bond, introduced in the Wittig reaction, readily isomerises to the *endo*cyclic position under acid catalysis), which after hydroxylation followed by acid-catalysed rearrangement gave the required indanone (LXXXII).

Achievement of the second part was rather more involved and necessitated an extremely lengthy, though ingenious scheme. Birch reduction of the ketal derivative of the free acid derived from (LXXXII) gave, after mild acid hydrolysis and re-esterification, the βγ-unsaturated ketone (LXXXIII), which was methylated to yield the enone (LXXXIV). Under the conditions used, conjugation of the enone system did not take place. Borohydride reduction of the ketone group in (LXXXIV) followed by deketalisation and acid-catalysed conjugation of the double bond to the 5-membered ring ketone, gave the hydroxy-enone (LXXXV). During the double bond migration, the hydrogen atom at the ring junction was introduced *cis* to the ester side-chain (as expected). Hydrogenation of the hydroxy-enone (LXXXV) established the *cis*-ring junction. The product was then dehydrated and ketalised to give the olefin (LXXXVI). Cleavage of the double bond in (LXXXVI) using periodate/permanganate gave a keto-acid, the acyl group of which was isomerised with base to give the acid (LXXXVII). A Baeyer–Villiger reaction on the benzyl ester of this compound gave the acetate (LXXXVIII). The required stereochemistry of the substituents on the cyclopentane ring had now been established. The remaining task was the transformation of the propionic ester residue into the allylic alcohol side-chain of PGE_1.

Hydrogenolysis of the benzyl ester group in (LXXXVIII) gave the free acid which was oxidatively decarboxylated to the olefin (LXXXIX; $X = CH_2$). The vinyl group was cleaved to give the aldehyde (LXXXIX; $X = O$) to which the required side-chain was added by the previously described procedures. C-15 epimers were separated, the methyl ester was hydrolysed with base, and the ketal group was finally cleaved using 50% aqueous acetic acid to give PGE_1. It is interesting that the 11-ketal function can be hydrolysed without dehydration of the resulting β-ketol.

This synthesis, although interesting because of the general approach, and the array of reagents and reactions used, is clearly not a practical proposition on account of its sheer length—29 steps from the methoxyindanol (LXXX).

Other Approaches—Two further syntheses have recently been described by Corey and are conveniently reviewed under this general heading. Unfortunately both have rather low yield key-steps and are therefore not competitive with his bicycloheptene route.

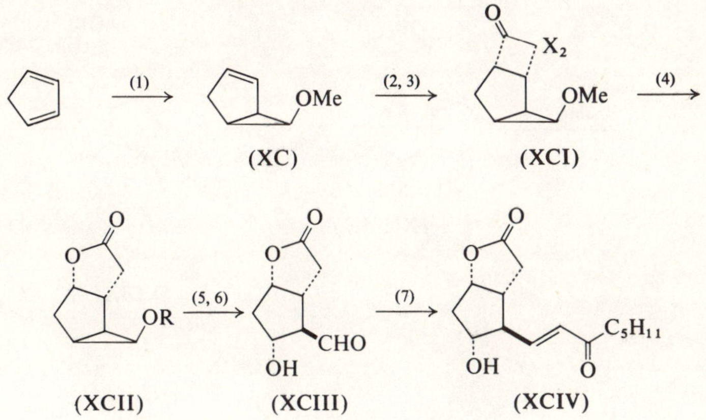

Figure 1.13. (1) MeOCH$_2$Cl/MeLi; (2) Cl$_2$C=C=O; (3) Zn/AcOH; (4) H$_2$O$_2$/AcOH; (5) BBr$_3$; (6) CrO$_3$/(NH$_4$)$_2$Ce(NO$_3$)$_6$; (7) (MeO$_2$)PO·CH·COC$_5$H$_{11}$

In the first approach[47] (*Figure 1.13*) cyclopentadiene is reacted with methoxycarbene to give a mixture of methoxycyclopropanes in which the *endo*-isomer (XC) predominates (4:1). Addition of dichloroketene to the remaining double bond takes place solely in the direction shown to give the dichlorocyclobutanone (XCI; X = Cl) which was reduced to (XCI; X = H) with zinc in acetic acid. (Note that only one of the four possible dichloroketene-addition products is formed.) The cyclobutanone (XCI; X = H) readily underwent a Baeyer–Villiger reaction to the lactone (XCII; R = Me) which was demethylated with boron tribromide to the cyclopropanol (XCII; R = H).

Thus, provided a suitable method for oxidatively hydrolysing the cyclopropane ring could be found, this route would be a short, efficient and highly stereospecific synthesis of the hydroxy-aldehyde (XCIII), a key intermediate in Corey's previous scheme (*see Figure 1.10*). However, although a vast array of reagents and conditions were tried, the best yield in the final steps was only 12% (because the aldehyde (XCIII) was too labile to be isolated the crude oxidation product was reacted with sodio-2-oxoheptylphosphonate and the yield was measured on the hydroxy-enone (XCIV)).

The second synthesis[48] (shown in *Figure 1.14*) is clearly related but depends upon the opening of an epoxide (rather than a cyclopropane ring) with concomitant attachment of the potential allylic alcohol side-chain.

The required epoxide (XCVII) was prepared in six steps from cyclopentadiene. The lactone (XCV), obtained via the cyclobutanone sequence used previously, gave predominantly (89%) the *cis-syn-cis* epoxy-lactone

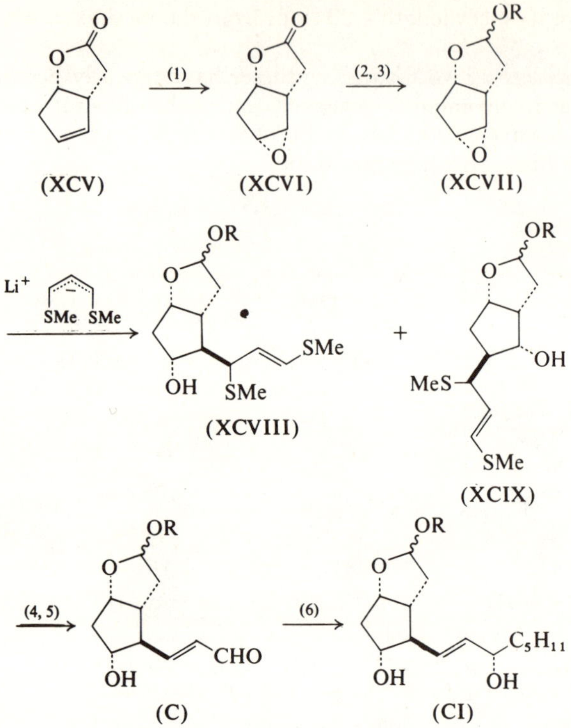

Figure 1.14. (1) AcO$_2$H/AcOH; (2) Bu$_2^i$AlH; (3) MeOH/BF$_3$·OEt$_2$; (4) HgCl$_2$/CaCO$_3$ in aq·CH$_3$CN; (5) chromatography; (6) C$_5$H$_{11}$·Li

(XCVI) on treatment with peracetic acid. The surprisingly high stereo-specific epoxidation from the apparently more hindered side of the olefin (XCV) occurred only in acetic acid as solvent and was confirmed by an unambiguous synthesis of the epoxide (XCVI).

The opening of the epoxide ring with 1,3-bis(methylthio)allyllithium (the nucleophilic equivalent of the unit —CH=CH·CHO) unfortunately took place in both possible directions and gave the mixture of thio-ethers (XCVIII and XCIX). Hydrolysis of the mixture in the presence of mercuric chloride gave two α,β-unsaturated aldehydes from which the required isomer (C) could be separated by chromatography. Reaction with n-amyllithium gave a mixture of secondary alcohols from which the α-isomer (CI; R = Me) could be separated, hydrolysed to the lactol (CI; R = H), and taken through to PGF$_{2α}$ by the sequence shown in Figure 1.10.

If it were possible to control the direction of the epoxide ring cleavage, this route could become very attractive because of its directness.

PROSTAGLANDINS FROM A NON-MAMMALIAN SOURCE

During a study of the metabolites of various marine organisms, Weinheimer and Spraggins[12] found that a Caribbean gorgonia, *plexaura homomalla*, contained surprisingly large amounts of 15-epi-PGA$_2$ together with its methyl ester acetate (CII; R = H or Me, R¹ = H or Ac). Several organisations became sufficiently interested to explore the commercial possibilities of this as a source of raw material for the preparation of PGE_2 and $PGF_{2\alpha}$. The transformation of the metabolite into PGE_2 (and $F_{2\alpha}$) was recently reported[38] and is shown in *Figure 1.15*.

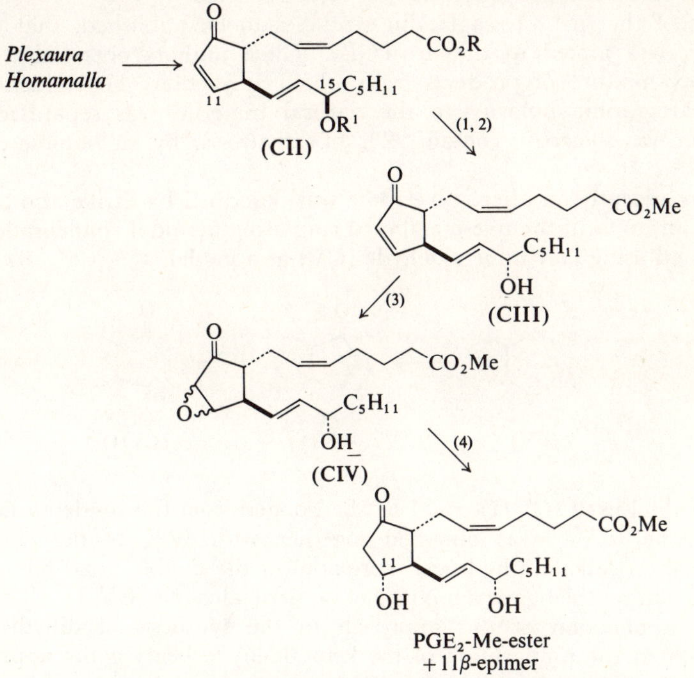

Figure 1.15. (1) $MeSO_2Cl$; (2) aq·acetone; (3) $H_2O_2/^-OH$; (4) $Cr(OAc)_2$

Epimerisation of the 15-hydroxyl group was achieved by solvolysis of the 15-mesylate (CII; R = Me; R¹ = $MeSO_2$) in aqueous acetone. The resulting hydroxy dienone (CIII) could be selectively epoxidised at the conjugated double bond to give a mixture of α and β-epoxides (CIV). Reduction of the epoxides with chromous acetate then gave a mixture of PGE_2 methyl ester and its 11-epimer. $PGF_{2\alpha}$ was obtained by a slightly modified process.

It will be interesting to see if the preparation of prostaglandins from this source effectively competes with total synthesis. Whatever the outcome, this finding has certainly stimulated searches for other non-mammalian sources of prostaglandins or potential intermediates.

SYNTHESIS OF 13,14-DIHYDROPROSTAGLANDINS

13,14-Dihydroprostaglandin E_1 is a metabolite of PGE_1 itself and retains a good deal of the primary compound's biological activity[49]. Thus, exploration of structure–activity–specificity relationships within the 13,14-dihydro series is of interest and depends upon the availability of an attractive synthesis.

In principle, omission of the double bond in PGE_1 should greatly simplify the synthetic problem. While this is true merely in terms of construction of the molecule, the difficulties of stereochemical control remain and may even be accentuated. It is certainly more difficult to separate the C-15 epimeric alcohols in the dihydro series, and for this reason the total synthesis of pure natural 13,14-dihydro-PGE_1 has yet to be reported.

One of the first prostaglandin syntheses to be published, that of Beal et al.[50], was aimed at dihydro-PGE_1. These authors obtained a rather complex mixture of products from which a fraction (11%) identical in chromatographic polarity to the natural material was separated. This fraction was shown to contain 22% of dihydro-E_1 by an isotopic dilution assay.

More recently an alternative route was described by Strike and Smith[51]. The plan, to form the five-membered ring using an aldol condensation, was evaluated using laevulinic aldehyde (CV) as a model.

(CV) → (CVI) → (CVII)

The aldol itself (CVII) could not be isolated from the condensation, and trans-enone (CVI) was obtained together with 10% of the cis-isomer. Epoxidation (alkaline hydrogen peroxide) of the double bond followed by hydrogenation then gave a mixture of isomeric alcohols (CVII).

The application of this approach to the synthesis of dihydro-PGE_1 necessitated the preparation of the keto-aldehyde bearing the appropriate side-chains. This was achieved by fairly conventional methods and is shown in *Figure 1.16*.

Carbonylation of the Grignard derivative of the protected ethynyl carbinol (CVIII) followed by hydrogenation and re-protection gave the aldehyde (CIX). Reaction of this aldehyde with propynylmagnesium bromide and condensation of the derived mesylate (CX) with the anion from the substituted t-butyl acetoacetate (CXI) gave the acetylenic ketone (CXII; $R = CO_2Bu^t$). This molecule now contained all the required carbon atoms.

One interesting point in the above sequence is the use of the propynyl group which serves both to activate the mesylate in (CX) and as a convenient masking group for required aldehyde in (CXIV).

The t-butyloxycarbonyl group in (CXII; $R = CO_2Bu^t$) was removed by heating with calcium iodide and the resulting compound (CXII; $R = H$) was hydrogenated to give the cis-olefin (CXIII). Ozonolysis of the double bond afforded the required keto-aldehyde (CXIV) which was cyclised as above to give what was assumed to be 13,14-dihydro PGE_1 as a mixture of the four possible 11- and 15-epimers (CXV). This mixture showed biological activity.

As the separation of some epimers (particularly C-15) of the 13,14-dihydro-

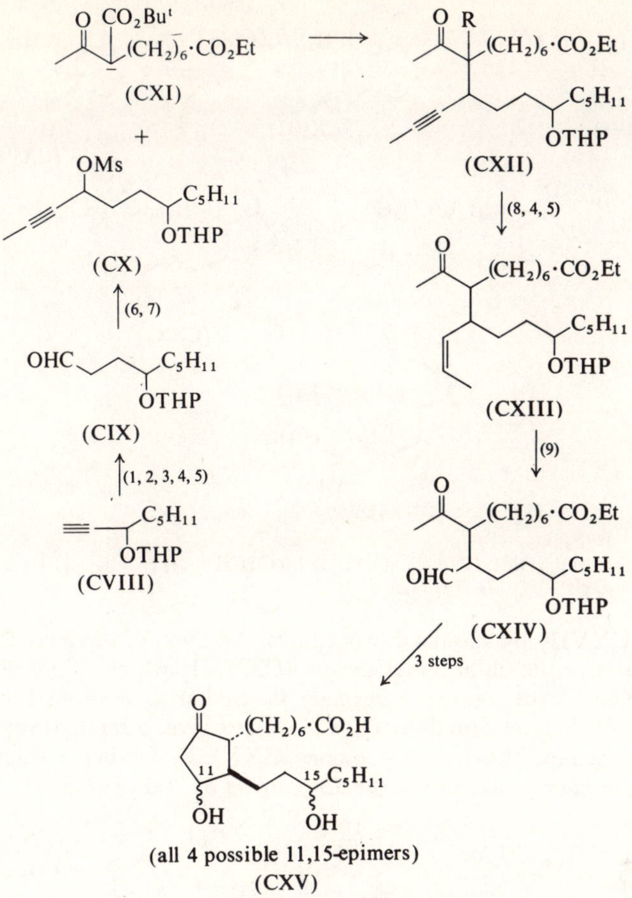

THE SYNTHESIS OF PROSTAGLANDINS

Figure 1.16. (1) EtMgBr; (2) HCO_2Et; (3) H^+; (4) dihydropyran/H^+; (5) H_2/Pd; (6) $CH_3C{\equiv}C{\cdot}MgBr$; (7) $MeSO_2Cl$/pyr; (8) $CaI/150°C$; (9) O_3/Zn

prostaglandins appears to be so difficult, the synthesis of pure compounds probably necessitates either hydrogenation of the appropriate pure prostaglandin (or intermediate), introduction of a side-chain of preformed correct absolute stereochemistry at C-15, or an enzymic reduction of the 15-ketone[52].

SYNTHESIS OF 11-DESOXY-PROSTAGLANDINS

This series has been explored by the Ayerst group in Canada who have obtained compounds exhibiting some of the pharmacological activities of the natural molecules. They have described two syntheses, one of which[53] (of 11-desoxy-$PGF_{1\beta}$), although conventional and relatively unimportant by present standards, is included here because it was the first total synthesis of a biologically active prostaglandin. Key steps of the route are given in *Figure 1.17*.

The cyclopentenone-acid (CXVI) was converted into the ester-acid

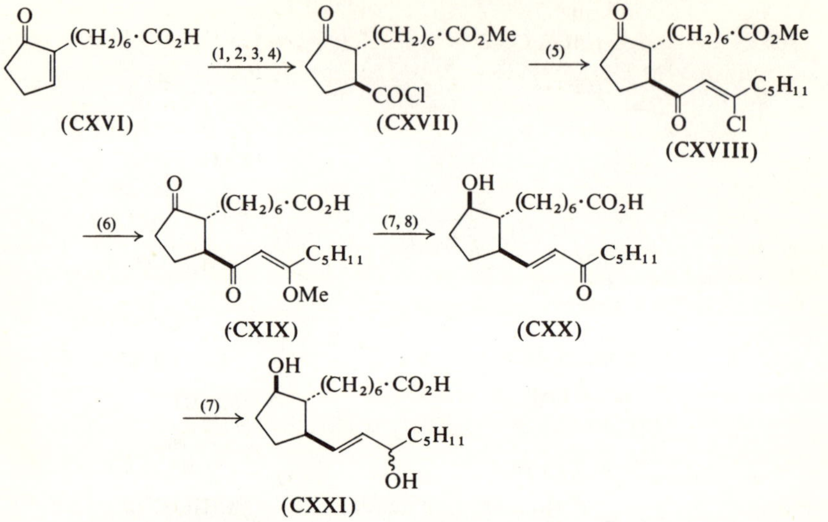

Figure 1.17. (1) Me$_2$C(OH)·CN; (2) $^-$OH; (3) MeOH/H$^+$; (4) SOCl$_2$; (5) HC≡C·C$_5$H$_{11}$/AlCl$_3$; (6) MeOH/$^-$OH; (7) NaBH$_4$; (8) H$^+$

chloride (CXVII) by standard procedures. Addition of the acid chloride to hept-1-yne gave the chlorovinyl ketone (CXVIII) which afforded the enol ether (CXIX) with sodium hydroxide in methanol. Sodium borohydride reduced both ketone functions in (CXIX) to give, after hydrolysis of the enol ether group, the hydroxy-enone (CXX). A further reduction with borohydride gave a mixture of the 15-epimers of PGF$_{1\beta}$ (CXXI).

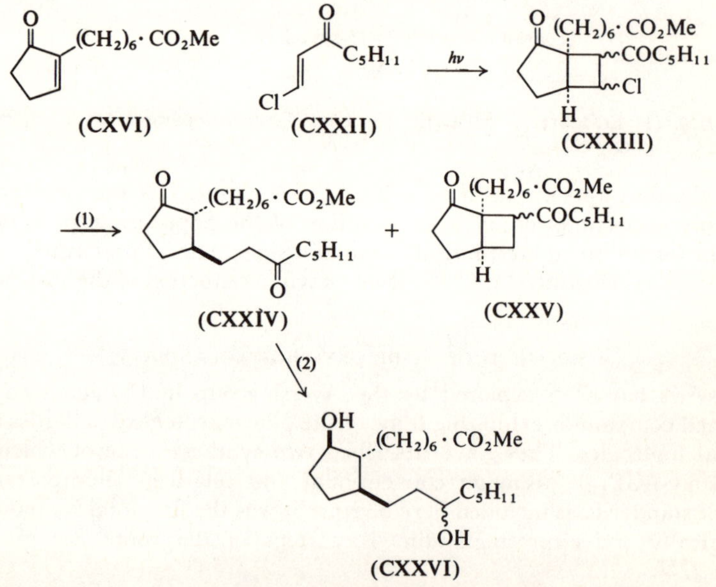

Figure 1.18. (1) Zn/AcOH; (2) NaBH$_4$

Two points are noteworthy: firstly, the novel method for introducing the allylic-alcohol side-chain and secondly, that reduction of the cyclopentane carbonyl gives mainly the 9β-alcohol in the 11-desoxy series.

The mixture of $PGF_{1\beta}$ epimers shows hypotensive activity in cats.

The second and more recent synthesis reported by the Canadian group[54] is aimed at 11-desoxy-13,14-dihydro-prostaglandins (*see Figure 1.18*). It is a completely novel approach in which the key step is the photo-addition of the chlorovinyl ketone (CXXII) to the substituted cyclopentenone (CXVI) to give the cyclobutane (CXXIII) of the stereochemistry indicated. Reduction with zinc in acetic acid gave, as minor product, the dechlorinated cyclobutane (CXXV) and, as major product, the ring-opened diketone (CXXIV). Thus, in two steps the prostanoic acid structure had been synthesised with the side-chains in the required *trans*-relationship. Borohydride reduction of the diketone (CXXIV) gave a mixture of four isomeric diols which could be separated into C-9 epimeric pairs. As before the 9β-pair (CXXVI) predominated. The diols also lowered blood pressure in experimental animals.

Manuscript received June 1971.

REFERENCES

1. VON EULER, U. S., *Klin. Wschr.*, **14**, 1182 (1935)
2. BERGSTRÖM, S., RYHAGE, R., SAMUELSSON, B. and SJÖVAL, J., *Acta. chem. scand.*, **17**, 2271 (1963); *J. biol. Chem.*, **238**, 3555 (1963)
3. PICKLES, V. R., *Nature*, **224**, 221 (1969); see Editorial Reviews in *Lancet*, 31 Jan., 223 (1970); *Br. Med. J.*, 31 Oct., 253 (1970)
4. BERGSTRÖM, S., *Science*, **157**, 382 (1967)
5. SAMUELSSON, B., *Angew. Chem. Int. Edn.*, **4**, 410 (1965)
6. BERGSTRÖM, S. and SAMUELSSON, B., eds, 'Prostaglandins', *Proc. 2nd Nobel Symp., Stockholm*, (1966)
7. HAMBERG, M. and SAMUELSSON, B., *J. biol. Chem.*, **241**, 257 (1966)
8. DANIELS, E. G., KRUEGER, W. C., KUPIECKI, F. P., PIKE, J. E. and SCHNEIDER, W. P., *J. Am. chem. Soc.*, **90**, 5894 (1968)
9. GRANSTRÖM, E. and SAMUELSSON, B., *Eur. J. Biochem.*, **10**, 411 (1969); *J. Am. chem. Soc.*, **91**, 3398 (1969); HAMBERG, M. and ISRAELSSON, U., *J. biol. Chem.*, **245**, 5107 (1970)
10. BEERTHIUS, R. K., NUGTEREN, D. H., PABON, H. J. J. and VAN DORP, D. A., *Recl. Trav. chim. Pays-Bas Belg.*, **87**, 461 (1968)
11. PACE-ASCIAK, C. and WOLFE, C. S., *Chem. Commun.*, 1234 (1970)
12. WEINHEIMER, A. J. and SPRAGGINS, R. L., *Tetrahedron Lett.*, 5185 (1969)
13. HORTON, E. W., *Physiol. Rev.*, **49**, 122 (1969)
14. SPEROFF, L. and RAMWELL, P. W., *Am. J. Obstet. Gynec.*, **107**, 111 (1970)
15. MOREAU, D., *New Scient.*, 3 Sept., 468 (1970); see Editorial reviews in ref. 3 and also *Chem. Engng News*, 16 Feb., 42 (1970); *Am. Drugg.*, 30 June, 39 (1969)
16. FERRIERA, S. H. and VANE, J. R., *Nature*, **216**, 873 (1967)
17. LAPIDUS, M., GRANT, N. H. and ALBURN, H. E., *J. Lipid Res.*, **9**, 371 (1968)
18. DANIELS, E. G. and PIKE, J. E., in *Abstracts of Symposium on Prostaglandins, Worcester Foundation*, eds., Ramwell, P. W. and Shaw, J. E., Interscience/Wiley, New York, 402 (1968)
19. PIKE, J. E., LINCOLN, F. H. and SCHNEIDER, W. P., *J. org. Chem.*, **34**, 3552 (1969)
20. HAMBERG, M. and SAMUELSSON, B., *J. Am. chem. Soc.*, **91**, 2177 (1969); DAWSON, W., JESSUP, S. J., GIBSON, W. M., RAMWELL, P. W. and SHAW, J. E., *Br. J. Pharmac.*, **39**, 585 (1970)
21. NAKANO, J., ÄNGGÅRD, E. and SAMUELSSON, B., *Eur. J. Biochem.*, **11**, 386 (1969) and refs. cited therein
22. RAMWELL, P. W., SHAW, J. E., CLARKE, G. B., GROSTIC, M. F., KAISER, D. G. and PIKE, J. E., *Progress in the Chemistry of Fats and Other Lipids*, Pergamon, Oxford, **9** (part II), 231 (1968); AXEN, U. *A. Rep. med. Chem.*, 290 (1967); BAGLI, J. F., *A. Rep. med. Chem.*, 170 (1969)

23. COREY, E. J., ANDERSON, N. H., CARLSON, R. M., PAUST, J., VEDEJS, E., VLATTAS, I. and WINTER, R. E. K., *J. Am. chem. Soc.*, **90**, 3245 (1968)
24. COREY, E. J. and ACHIWA, K., ibid., **91**, 1429 (1969)
25. COREY, E. J., VLATTAS, I., ANDERSON, N. H. and HARDING, K., ibid., **90**, 3247; 5947 (1968)
26. COREY, E. J., VLATTAS, I. and HARDING, K., ibid., **91**, 535 (1969)
27. RAMWELL, P. W., SHAW, J. E., COREY, E. J. and ANDERSON, N. H., *Nature*, **221**, 1251 (1969); SHIO, H., RAMWELL, P. W., ANDERSON, N. H. and COREY, E. J., *Experientia*, **26**, 355 (1970)
28. MIYANO, M. and DORN, C. R., *Tetrahedron Lett.*, 1615 (1969)
29. FINCH, N. and FITT, J. J., ibid., 4639 (1969)
30. WIBERG, K. B. and ASHE, A. J., ibid., 1553 (1965)
31. JUST, G. and SIMONOVITCH, C., ibid., 2093 (1967)
32. HOLDEN, K. G., HWANG, B., WILLIAMS, K. R., WEINSTOCK, J., HARMAN, M. and WEISBACH, J. A., ibid., 1569 (1968)
33. (a) SCHNEIDER, W. P., AXEN, U., LINCOLN, F. H., PIKE, J. E. and THOMPSON, J. L., *J. Am. chem. Soc.*, **90**, 5895 (1968); **91**, 5372 (1969); (b) JUST, G., SIMONOVITCH, C., LINCOLN, F. H., SCHNEIDER, W. P., AXEN, U., SPERO, G. B. and PIKE, J. E., ibid., **91**, 5364 (1969)
34. AXEN, U., LINCOLN, F. H. and THOMPSON, J. L., *Chem. Commun.*, 303 (1969)
35. MEINWALD, J., LABANA, S. S. and CHADHA, M. S., *J. Am. chem. Soc.*, **85**, 582 (1963)
36. SCHNEIDER, W. P., *Chem. Commun.*, 304 (1969)
37. AXEN, U., THOMPSON, J. L. and PIKE, J. E., ibid., 602 (1970)
38. BUNDY, G. L., *Symp. Prostaglandins, New York Acad. Sci.*, Abstr. 4, Sept. 1970
39. (a) COREY, E. J., WEINSHENKER, N. M., SCHAAF, T. K. and HUBER, W., *J. Am. chem. Soc.*, **91**, 5675 (1969); (b) COREY, E. J., SCHAAF, T. K., HUBER, W., KOELLIKER, U. and WEINSHENKER, N. M., ibid., **92**, 397 (1970); (c) COREY, E. J., NOYORI, R. and SCHAAF, T. K., ibid., **92**, 2586 (1970)
40. COREY, E. J. *Symp. Prostaglandins, New York Acad. Sci.*, Sept. 1970; *see also* footnote 12 of ref. 41
41. COREY, E. J., ALBONICO, S. M., KOELLIKER, U., SCHAAF, T. K. and VARMA, R. K., *J. Am. chem. Soc.*, **93**, 1491 (1971)
42. COREY, E. J., SHIRAHAMA, H., YAMAMOTO, H., TERASHIMA, S., VENKATESWARLU, A. and SCHAAF, T. K., ibid., 1490
43. COREY, E. J. and YAMAMOTO, H., ibid., **92**, 226, 3523 (1970)
44. MIRONOV, V. A., SOBOLEV, E. V. and ELIZAROVA, A. N., *Tetrahedron*, **19**, 1939 (1963)
45. COREY, E. J., KOELLIKER, U. and NEUFFER, J., *J. Am. chem. Soc.*, **93**, 1489 (1971)
46. TAUB, D., HOFFSOMMER, R. D., KUO, C. H., SLATES, H. L., ZELAWSKI, Z. S. and WENOLER, N. L., *Chem. Commun.*, 1258 (1970)
47. COREY, E. J., ARNOLD, Z. and HUTTON, J., *Tetrahedron Lett.*, 307 (1970)
48. COREY, E. J. and NOYORI, R., ibid., 311 (1970)
49. ÄNGGÅRD, E. and SAMUELSSON, B., *J. biol. Chem.*, **239**, 4097 (1964)
50. BEAL, P. F., BABCOCK, J. C. and LINCOLN, F. H., *J. Am. chem. Soc.*, **88**, 3131 (1966)
51. STRIKE, D. P. and SMITH, H., *Tetrahedron Lett.*, 4393 (1970)
52. ÄNGGÅRD, E. and LARSSON, C. *Eur. J. Biochem.*, **14**, 66 (1971)
53. BAGLI, J. F. and BOGRI, T., *Tetrahedron Lett.*, 5 (1966); BAGLI, J. F., BOGRI, T., DEGHENGHI, R. and WIESNER, K., ibid., 465 (1965)
54. BAGLI, J. F. and BOGRI, T., ibid., 1639 (1969)

2
BIOGENETIC-TYPE SYNTHESIS OF TERPENES

T. Money

MONOTERPENES	29
SESQUITERPENES	32
DITERPENES	46
TRITERPENES	55

THE VARIETY of complex structures exhibited by natural products has provided a constant challenge to the synthetic capabilities and ingenuity of organic chemists. Since the appearance of Robinson's 'biogenetic-type' synthesis of tropinone there has been a growing interest in meeting this challenge by using synthetic strategy based on the postulated or known biosynthesis of the particular natural product. The first, authoritative review of biogenetic-type syntheses appeared in 1961 (VAN TAMELEN, E. E., *Fortschr. Chem. org. NatStoffe*, **19**, 245 (1961)); the present review attempts to describe developments in the terpene area which have been reported during the period 1961 to 1970.

MONOTERPENES

The monocyclic and bicyclic monoterpenes are considered to be derived in nature by cyclisation of geranyl, neryl or linaloyl pyrophosphate[1,8]. Experimental support for some of these proposals has been provided. Acid hydrolysis of neryl phosphate (I; R = PO_3H_2), neryl pyrophosphate (I; R = $P_2O_6H_3$) or corresponding linalool derivatives (II; R = PO_3H_2 or $P_2O_6H_3$) yielded a mixture of acyclic alcohols and α-terpineol (III)[2,3]. Under similar conditions the corresponding geraniol derivatives (I, with *trans* double bond) yielded only acyclic products. A related study, however, has shown

that a low yield of α-terpineol can also be obtained from the geraniol derivative[4]. Similar results have been obtained by other workers who have shown that spontaneous decomposition of geranyl (IV) and neryl diphenylphosphate (V) in ether solution results in the formation of limonene (VI), terpinolene (VII), myrcene (VIII), *cis*-β-ocimene (IX) and *trans*-β-ocimene (X)[5]. The formation of monocyclic compounds from (IV) has been explained by invoking intermediate formation of linaloyl diphenylphosphate (II; R = PO(OPh)$_2$). The conversion of monocyclic into bicyclic monoterpenes has been proposed as a reasonable biosynthetic process[1], but various attempts[2-4,6], involving a variety of terpenyl derivatives (XI), have failed to provide laboratory support for this proposal. A recent synthesis of camphor (XIII) from dihydrocarvone enol acetate (XII) has prompted the suggestion

XI; Z = OH, Cl, phosphate, pyrophosphate or p-nitrobenzoate

that a similar reaction (e.g. via the enol phosphate derivative), could occur in the natural system[7]. Similar transformations can be considered for analogous compounds in the sesquiterpene series (page 42). The suggestion that bicyclic monoterpenes could arise directly from acyclic precursors[1,8] is supported by the photochemical conversion of myrcene (VIII) into β-pinene (XIV)[9]. Free radical cyclisation, reminiscent of that involved in the biosynthesis of higher terpenes (page 34), has been accomplished using

geranyl acetate (XV) as substrate[10]. Treatment of (XV) with benzoyl peroxide, cuprous chloride and cupric benzoate in benzene or acetonitrile yielded (XVI). Monoterpenes possessing carbon skeletons identical to (XVI) are known[11] and similar processes may be involved in their biosynthesis. Ionic cyclisation of epoxide (XVII) is reported[12] to yield the related cyclohexenol (XVIII) and analogous cyclisations will be described later in the sections dealing with sesquiterpenes, diterpenes and triterpenes.

The postulated biosynthetic relationship between chrysanthemic acid derivatives (XIX–XXII) and monoterpenes possessing santolinyl (XXIII), artemisyl (XXIV) and lavandulyl (XXV, XXVI) skeletons[13] has received some experimental support[14] and provided the basis of a convenient biogenetic-type synthesis of monoterpenes of each type. The conversion of a santolina-type compound (e.g. XXVII) to artemisia-type monoterpenes (e.g. XXVIII–XXX) by 1,2-shift of a vinyl group has been reported[15] and supports the view that a similar transformation could occur in nature.

SESQUITERPENES

The biosynthesis of the sesquiterpenes has been the subject of considerable speculation and an excellent review of the various proposals is available[16]. In general, the numerous structural classes of sesquiterpenes are believed to be derived from farnesyl (XXXI) or nerolidyl pyrophosphate (XXXII). Some support for this proposal has been obtained[17,17a] but, as yet, there is scant evidence for the chemical feasibility of the cyclisations which have been postulated to account for the structural diversity found in this group of natural products. Several attempts to produce appropriate chemical analogies for these processes are outlined below.

(XXXI) OP_2O_6H

(XXXII) $OP_2O_6H_3$

The acid-catalysed cyclisation of farnesol and nerolidol has been investigated and the results indicate that a complex mixture consisting of isomeric bisabolenes (XXXIII) and cadinenes (XXXIV) is obtained[18,19]. The biosynthesis of a number of sesquiterpenes is considered to involve γ-bisabolene (XXXV), or its biological equivalent, as an intermediate[16]. The formation of this structural type (e.g. bisabolol (XXXVII)), from nerolidyl diphenylphosphate (XXXVI) has been demonstrated[20]. The biogenetically patterned cyclisation of acyclic terpene epoxides has provided an elegant synthetic route to sesquiterpenes of the bicyclofarnesol class[21,22]. Treatment of *trans,trans*-farnesyl acetate monoepoxide (XXXVIII) with boron trifluoride etherate yielded a 6 : 1 ratio of bicyclic acetates (XXXIX) and (XL), which were easily converted into drimenol (XLI) and epidrimenol (XLII) respectively[22]. Methyl *trans,trans*-farnesate monoepoxide (XLIII) behaved similarly and yielded esters (XLIV) and (XLV)[21]. Later studies have shown

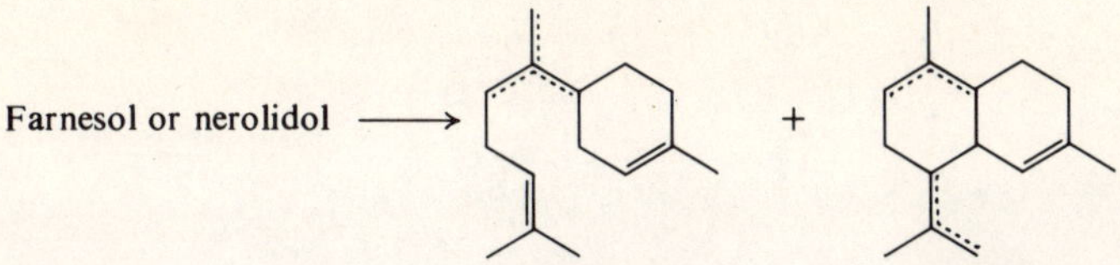

Farnesol or nerolidol ⟶ (XXXIII) + (XXXIV)

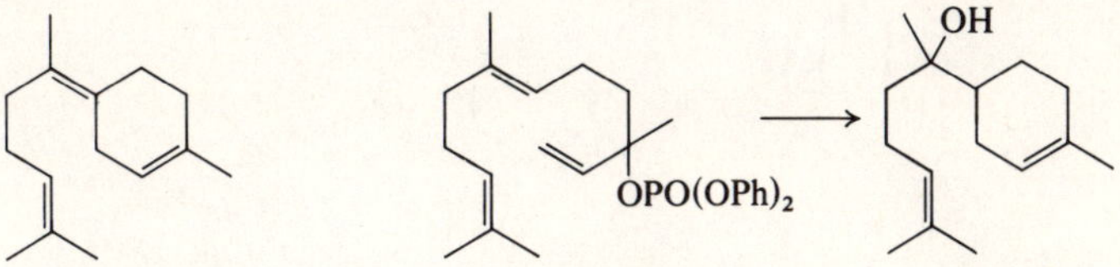

(XXXV)   (XXXVI) OPO(OPh)₂ ⟶ (XXXVII)

that the relative amounts of isomers (XLIV) and (XLV) depend on conditions: in cold phosphoric acid the relative abundance of (XLIV) and (XLV) is 38:62 while in boron trifluoride and benzene the mixture is rich (75–90%) in isomer (XLIV)[23]. Extension of this basic theme has afforded biogenetic-type routes from umbelliprenin (XLVI) to farnesiferol C (L), the isomer (XLIX) of farnesiferol B (LII), and a stereoisomer (XLVIII) of farnesiferol A (LI)[21]. Farnesiferol A (LI) and its isomer (LIV) are the products when *trans,cis*-umbelliprenin terminal epoxide (LIII) is treated in a similar way[21].

The successful use of ion-exchange resins to promote polyene cyclisation

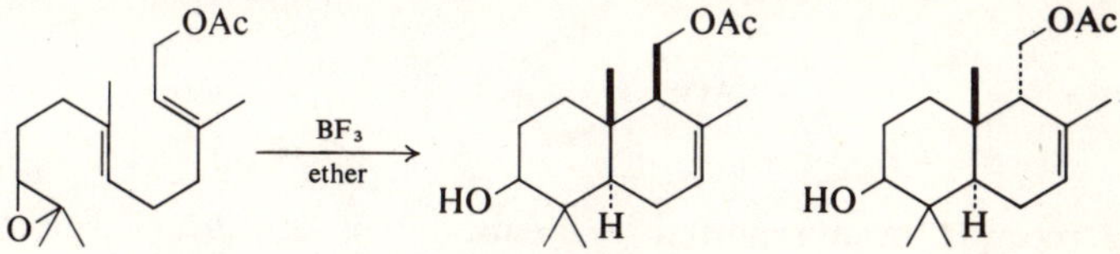

(XXXVIII) —BF₃/ether→ (XXXIX)   (XL)

↓   ↓

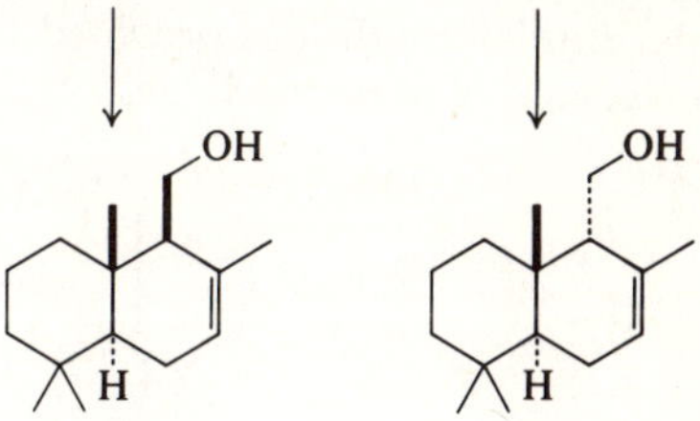

(XLI)   (XLII)

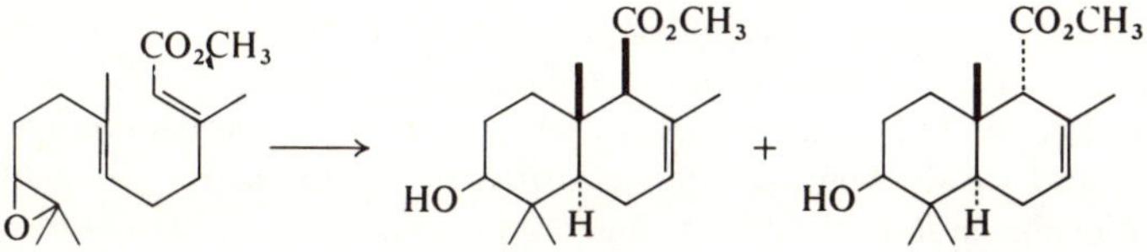

(XLIII) ⟶ (XLIV) + (XLV)

33

has recently been reported[24]. Treatment of *trans,trans*-farnesic acid (LV; R = H) with amberlite IR-120 or XE-100 in dioxane gave the monocyclic product (LVIII; R = H), while the corresponding ester (LV; R = CH$_3$) under similar conditions provided (LVIII; R = CH$_3$), (LVII) and (LVI). Conversion of monocyclic ester (LVIII; R = CH$_3$) into bicyclic esters (LVII) and (LVI) was also achieved by using amberlite XE-100 in acetic acid. A similar route to bicyclic sesquiterpenes of this type is provided by the free radical oxidative cyclisation of farnesyl acetate (LIX) to bicyclic product (LX)[25]. This reaction (cf. p. 30) is structurally and stereochemically selective and would seem to offer considerable promise for the synthesis of fused alicyclic systems. A biogenetic-type synthesis of drimenin (LXVI) from monocyclofarnesic acid (LVIII; R = H)[26] has been reported[27].

Acid-catalysed cyclisation of (LVIII; R = H) yielded drimenic acid (LXI; R = H) and the rearranged bicyclic acid (LXII)[28]. Photo-oxidation (Rose Bengal sensitiser) of the methyl ester (LXI; R = CH$_3$) yielded a mixture of esters (LXIII–LXV). Subsequent hydrolysis of (LXIII) resulted in concomitant lactonisation and the formation of a mixture of ($\pm$)-drimenin (LXVI) and ($\pm$)-isodrimenin (LXVII).

The postulated biosynthesis of α-chamigrene (LXX) and β-chamigrene (LXXI) involves intermediate formation of spiro-carbonium ion (LXIX) (or its biological equivalent) which may be derived by cyclisation of γ-bisabolene (XXXV) or cis-monocyclofarnesol (LXVIII)[16,29,30]. Indirect support for the latter pathway has been provided[31] by the reported biogenetic-type synthesis of α-chamigrene (LXX) from cis- or trans-monocyclofarnesol. The synthesis also provides a general route to spiro-carbocyclic systems of this type. The synthesis of the biologically active sesquiterpene abscissin II (LXXIV) from acid (LXXII) via the epidioxide (LXXIII) supports the suggestion that similar reactions may be involved in the natural route[32]. Acid-catalysed cyclisation of laurinterol (LXXV) yields the non-isoprenoid skeleton of aplysin (LXXVI) and supports the presumed biosynthetic relationship between these compounds[33].

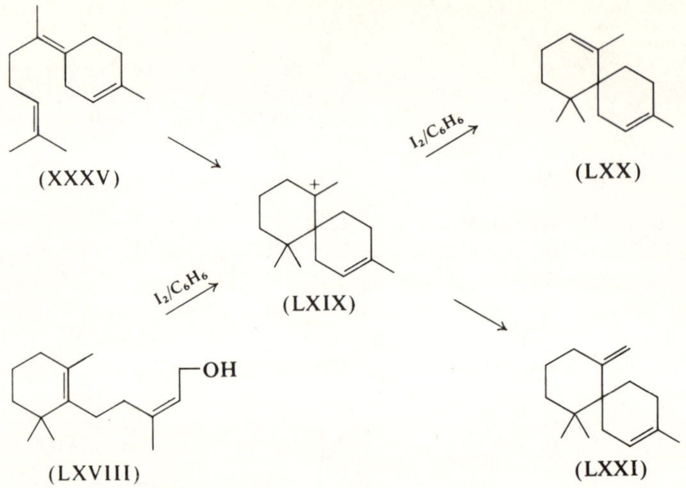

(XXXV) → (LXIX) —I_2/C_6H_6→ (LXX)

(LXIX) —I_2/C_6H_6→ (LXVIII)

(LXIX) → (LXXI)

It has been suggested that ten-membered ring intermediates (e.g. LXXVII; Z = biological leaving group) are involved in the biosynthesis of eudesmane, elemane and guaiane classes of sesquiterpenes[16,34-36]. Eudes-

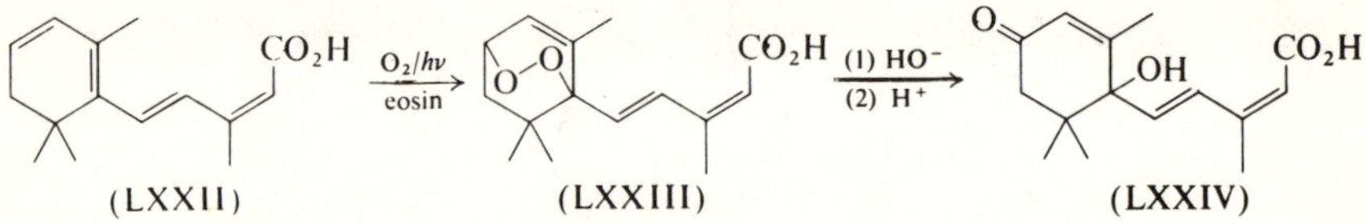

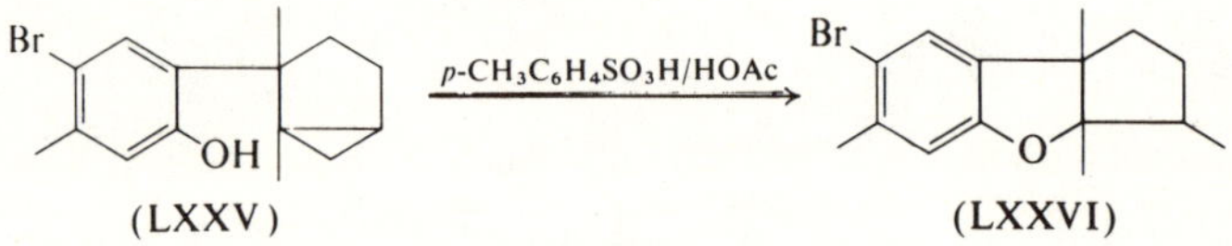

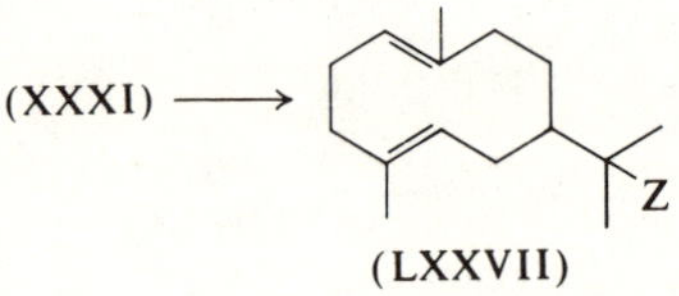

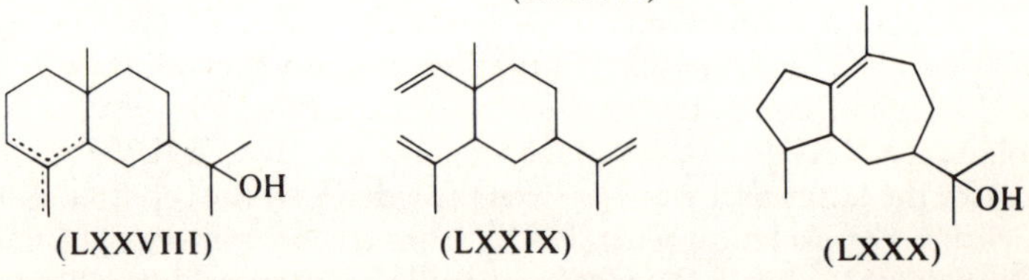

(LXXII) —$O_2/h\nu$, eosin→ (LXXIII) —(1) HO⁻ (2) H⁺→ (LXXIV)

(LXXV) —$p\text{-}CH_3C_6H_4SO_3H/HOAc$→ (LXXVI)

(XXXI) ⟶ (LXXVII)

(LXXVIII)   (LXXIX)   (LXXX)

mols (LXXVIII), β-elemene (LXXIX) and bulnesol (LXXX) are representative examples of each type. The quest for chemical support for this suggestion has been successful and has provided a biogenetic-type synthetic route to compounds belonging to each class. Thus treatment of triene (LXXXI) under a variety of conditions yielded a series of eudesmane-type compounds (LXXXII–LXXXVI)[37]. Conversion of the triene (LXXXI) into monoepoxide (LXXXVII) followed by treatment with aqueous acid produced similar results and yielded the diol (LXXXVIII) and alcohol (LXXXIX)[38]. A similar cyclisation is involved in the conversion of pyrethrosin (XC) to the bicyclic ketones (XCI) and (XCII)[35]. By contrast the cyclisation mode was directed towards the formation of the guaiane carbon skeleton when monoepoxide (XCIII) was treated with acid[38]. Compounds (XCIV) and (XCV) were obtained and a similar cyclisation was involved in the reported conversion of dihydroparthenolide (XCVI) into the guaianolide (XCVII)[39]. Further cyclisation of guaiane derivatives has been proposed for the biosynthesis of certain tricyclic sesquiterpenes, and this was

illustrated in the laboratory by the conversion of bulnesol (L) to β-patchoulene (XCVIII)[40]. Subsequent transformation of the latter to patchouli alcohol (XCIX) has been reported[41].

A synthetic route to elemane-type compounds (CI; R = H$_2$) and (CI; R = O), involving Cope rearrangement of ten-membered ring precursors (C; R = H$_2$) and (C; R = O), is based on the suggested biosynthetic route to this class of compound[37].

BIOGENETIC-TYPE SYNTHESIS OF TERPENES

The co-occurrence of the cyclodecadienone (CII) and the cadinane-type sesquiterpene preisocalamanediol (CIII) and the thermal conversion of (CII) into dehydroisocalamanediol (CIV) supports proposals made for the biosynthetic relationship between compounds of each type[42]. A new biosynthetic proposal for tricyclic vetiver (zizaane) sesquiterpenes (e.g. CVIII) involving cyclodecadienol (CV) and hinesol (CVI) has recently been described[43,43a]. Support for a crucial rearrangement [CVII → CVIII] on the proposed sequence has been provided by the laboratory transformation of mesylate (CIX) to tricyclic ketone (CX)[43,44] (the stereochemistry of the

BIOGENETIC-TYPE SYNTHESIS OF TERPENES

hydrogen atom at the ring junction differs in the two reports). Photosensitised oxidation of thujopsene (CXI) is reported to give thujopsadiene (CXIII) via hydroperoxide (CXII), and it has been suggested that a similar sequence could be involved in the biosynthesis of (CXIII)[45]. Analogous photo-oxidation of thujopsenol (CXIV) followed by reductive decomposition of the intermediate hydroperoxide (CXV) yields several oxidation products[46]. The sesquiterpene mayurone (CXVI) was a major component of the reaction mixture and a similar biosynthetic route to (CXVI) has been suggested[46].

The postulated biosynthesis of the cedrane group of sesquiterpenes (e.g. cedrene (CXVIII)) involves cyclisation of a spiro-intermediate (CXVII; Z = biological leaving group)[16]. Several successful syntheses of the cedrane skeleton have been accomplished using the biosynthetic postulate as a model. Synthetic intermediates (CXIX–CXXII) have been converted into cedrene (CXVIII) or cedrol (CXXV) and represent key steps in the reported synthesis[47,48] of these compounds. Acid-catalysed cyclisation of α-acoradiene (CXXVI) and α-acorenol (CXXVII)[49] yields (−)-α-cedrene (CXVIII), and in a similar fashion α-alaskene (CXXVIII) gives α-cedrene (CXVIII) in 70% yield[50]. The latter reaction contrasts with the behaviour of β-alaskene (CXXIX) which yields a mixture of isomeric olefins when treated with anhydrous formic acid[50]. Acid-catalysed cyclisation of β-acoradiene (CXXX) or β-acorenol (CXXXI) yields a new tricyclic sesquiterpene (CXXXII) which occurs in conjunction with β-acoradiene in *juniperus rigida*[51]. A similar cyclisation presumably occurs in nature.

An alternative hypothesis for the biosynthesis of a group of bicyclic, tricyclic and tetracyclic sesquiterpenes has recently been proposed[52] and forms the basis of a general synthetic route to members of the group. The basic feature of the postulate is that certain bicyclic and tricyclic sesquiterpenes could be constructed by cyclisation of an appropriate enol phosphate intermediate. Thus it is envisaged that cyclisation of dihydrocryptomerion enol phosphate (CXXXIV) could be involved in the biosynthesis of campherenone (CXXXV) and epicampherenone (CXXXIX)*. These compounds are sesquiterpene analogues of camphor (cf. p. 30), and the expectation that they could be biosynthetic precursors of β-santalene (CXXXVII), α-santalene (CXXXVIII), and epi-β-santalene (CXLI) is supported by recent synthetic work[54] and known transformations of analogous monoterpene series. A further extension of the theme to include a group of tricyclic and tetracyclic sesquiterpenes can be made. Thus, campherone enol phosphate (CXLII) is envisaged as a reasonable intermediate in the biosynthesis of copacamphor (CXLIII)*, epicopacamphor (ylangocamphor) (CXLVII)* and longicamphor (CLI)*. By processes identical to those indicated above the tricyclic ketones (CXLIII), (CXLVII) and (CLI) could be transformed further into a group of related sesquiterpenes†: copaborneol (CXLIV), copacamphene (CXLV), cyclocopacamphene (CXLVI), epicopaborneol (CXLVIII)*, sativene (CXLIX), cyclosativene (CL), longiborneol (CLII), longifolene (CLIII) and longicyclene (CLIV). It is also of interest to note that oxidative cleavage of copacamphor (CXLIII) provides an alternative biosynthetic route to the basic structure (CLV) of the picrotoxins[54a].

* Unknown in nature.
† Structures do not necessarily depict absolute configurations.

BIOGENETIC-TYPE SYNTHESIS OF TERPENES

Cyclisations and interconversions analogous to those postulated above have been investigated in the laboratory[7,53,54]. Dihydrocryptomerion enol acetate (cf. CXXXIV) has yet to be cyclised to campherenone but enol acetate (CLVI) cyclises to (CLVII) and (CLVIII) which can be converted

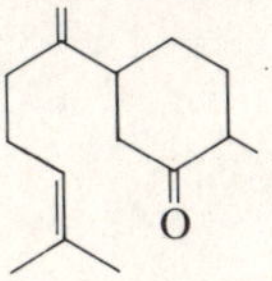

(CXXXIII)

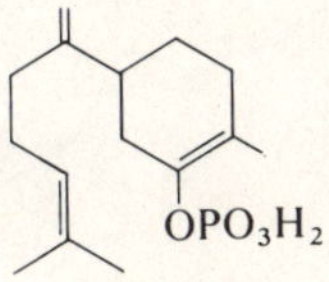

(CXXXIV)

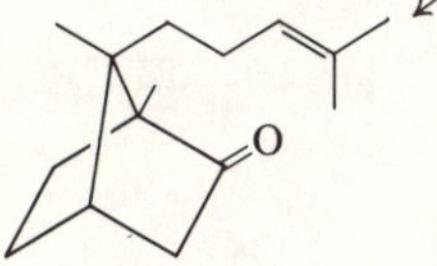

(CXXXV) 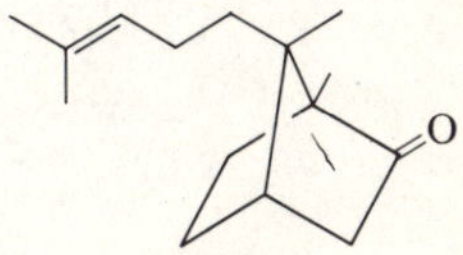 (CXXXIX)

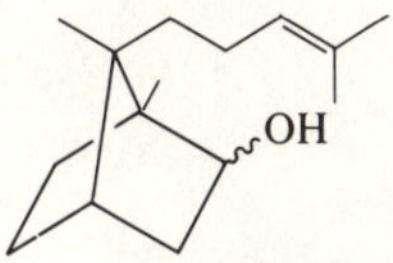

(CXXXVI) 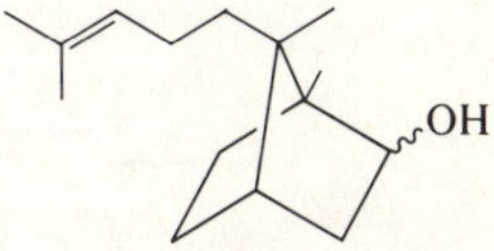 (CXL)

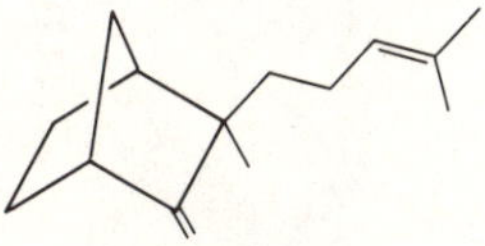

(CXXXVII) 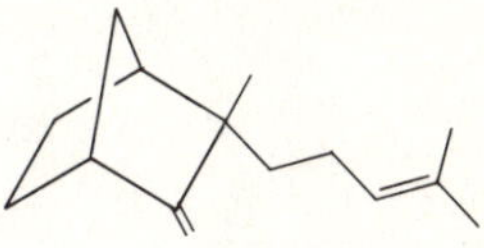 (CXLI)

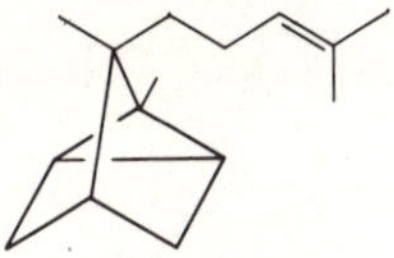

(CXXXVIII) 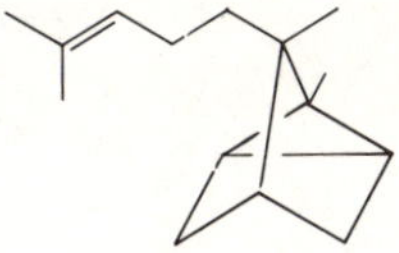 (CXXXVIII)

into (±)-campherenone (CXXXV) and (±)-epicampherenone (CXXXIX) respectively[53]. Reduction of (±)-campherenone (CXXXV) with lithium aluminum hydride yielded (±)-isocampherenol (CLIX) which was subsequently heated with pyridine and toluene-*p*-sulphonyl chloride to provide

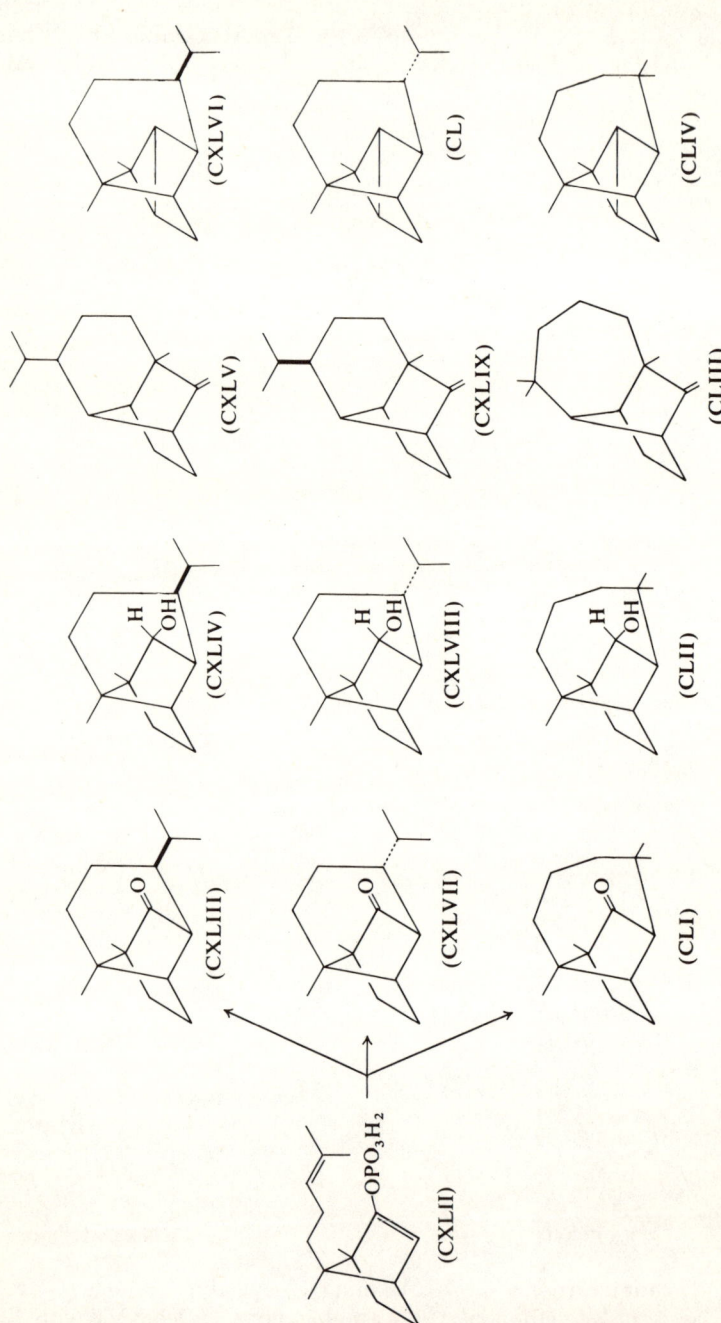

(±)-β-santalene (CXXXVII)[54]. The synthesis of (±)-α-santalene (CXXXVIII) was accomplished when a methanolic solution of hydrazone (CLX) was heated with mercuric oxide[54]. In the epicampherenone (CXXXIX) series analogous reactions yielded (±)-epi-β-santalene (CXLI) and (±)-α-santalene (CXXXVIII)[54]. Preliminary investigations have also shown that campherenone (CXXXV) can be converted in high yield to the corresponding enol acetate and that the latter, on treatment with boron trifluoride, provides a mixture of ketones whose constitution remains to be determined.

DITERPENES

The postulated biosynthesis[55-58] of the diterpenes involves the pyrophosphates of geranylgeraniol (CLXIII) and geranyllinalool (CLXIV) as basic isoprenoid precursors. Subsequent cyclisation and rearrangement processes could lead successively to bicyclic, tricyclic, tetracyclic and pentacyclic diterpenes. Structures (CLXV, CLXVI–CLXXIV) are representative examples of the labdane, pimarane, rosane, abietane, cassane, beyerane, kaurane, trachylobane and atisane classes of diterpene. Various features of this general scheme have received considerable support from *in vivo* studies using appropriate radioactive substrates[59-62]. In addition many attempts to provide a laboratory analogy for the postulated cyclisations and rearrangements have been successful and have provided, in several cases, attractive synthetic routes to diterpenes.

The formation of tricyclic and tetracyclic diterpenes from bicyclic labdane-type precursors is illustrated by formic acid-catalysed cyclisation of manool (CLXXV) and agathadiol (CLXXVI)[63]. Epimeric pimara-8,15-dienes (CLXXVII; R = CH$_3$ and CH$_2$OH) and 14α-hydroxyhibanes (CLXXVIII; R = CH$_3$ and CH$_2$OH) are produced. Two alternative explanations were considered for the formation of the tetracyclic compounds (CLXXVIII; R = CH$_3$ and CH$_2$OH) and route (a), i.e. (CLXXIX) → (CLXXX), was preferred on mechanistic grounds. However, route (b), involving cyclooctenyl intermediate (CLXXXI), has subsequently been supported by the results of these and other investigators (*see* below). A concurrent study[64] on the formic acid-catalysed cyclisation of manool (CLXXV) produced essentially identical results and provided additional laboratory support for the conversion of labdane-type compounds to diterpenes of the pimarane (CLXXVII) and beyerane (≡ hibane) (CLXXVIII) type. In addition, the preferred explanation for the formation of the tetracyclic alcohol (CLXXVIII) involved route (b), and this was elegantly supported by subsequent studies[65] involving radioactive starting materials (carbon-14 indicated by asterisk in (CLXXV) and (CLXXVIII)). Independent confirmation of this mechanism was provided by using deuterium-labelled manool (CLXXV) as starting material[66,67]. The formation of hibane derivatives, e.g. (CLXXVIII), via cyclooctenyl intermediate (CLXXXI), contrasts with the biosynthetic route which is believed to involve pimaradiene intermediates[62].

Other investigators[68] have also shown that manool and the related allylic primary alcohols can be cyclised to epimeric pimara-8,15-dienes

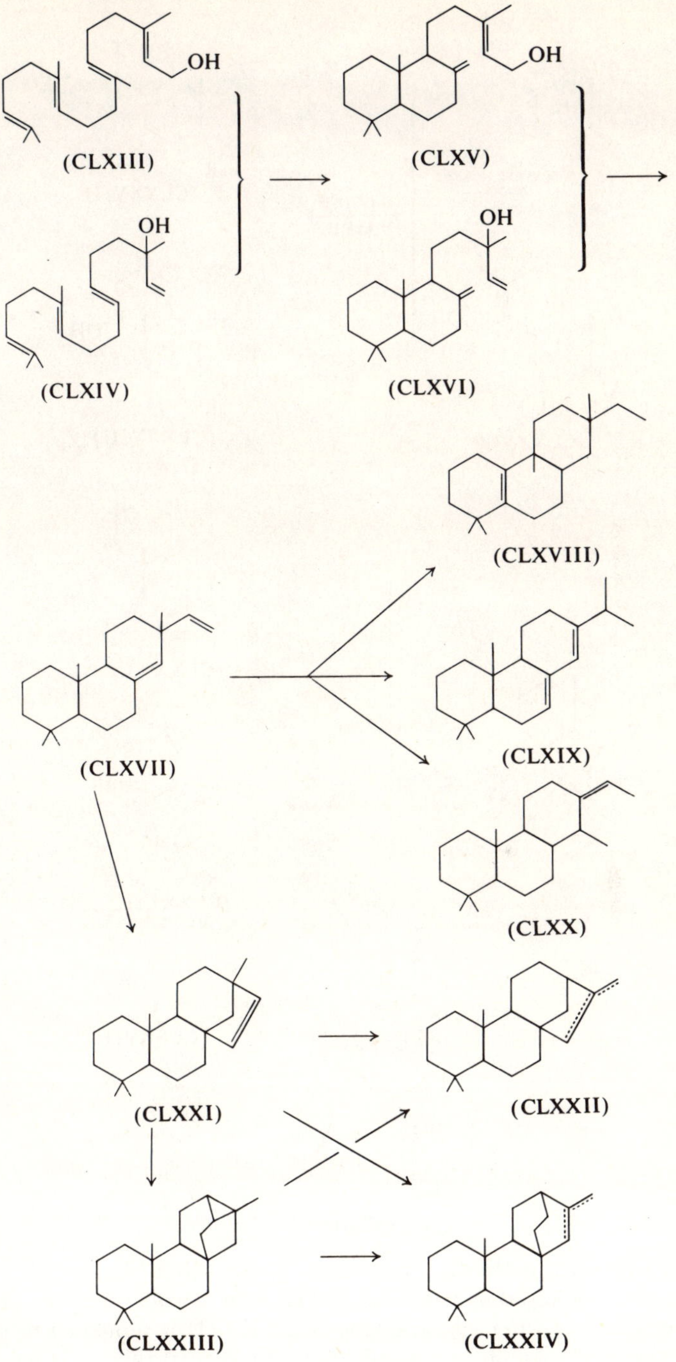
OH
(CLXIII)
OH
(CLXIV)
OH
(CLXV)
OH
(CLXVI)
(CLXVII)
(CLXVIII)
(CLXIX)
(CLXX)
(CLXXI)
(CLXXII)
(CLXXIII)
(CLXXIV)

(CLXXXII). In addition, they have shown that the rosane skeleton (exemplified by rosadiene (CLXXXIII)) can be obtained on prolonged acid treatment or when pimaradiene (CLXXXII) is subjected to the same acid conditions[68]. An elegant application of this rearrangement has resulted in the biogenetic-type synthesis of rosenonolactone (CLXXXV) and desoxy-

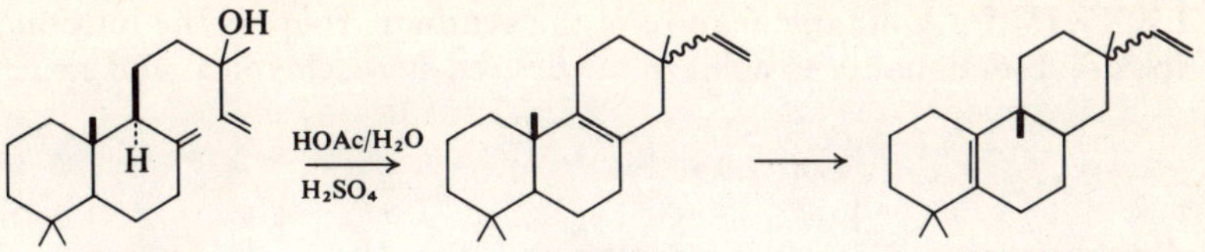

(CLXXV) → (CLXXXII) → (CLXXXIII)

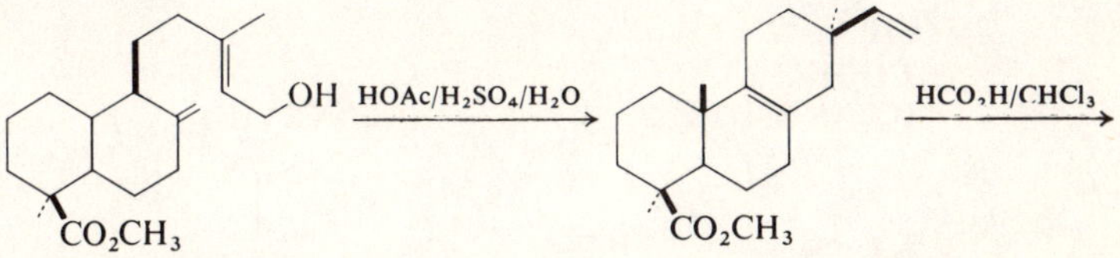

HOAc/H$_2$O, H$_2$SO$_4$

(CLXXXIV) → HOAc/H$_2$SO$_4$/H$_2$O → HCO$_3$H/CHCl$_3$ →

m-ClC$_6$H$_4$CO$_3$H/CHCl$_3$ → BF$_3$/C$_6$H$_6$, 0 °C →

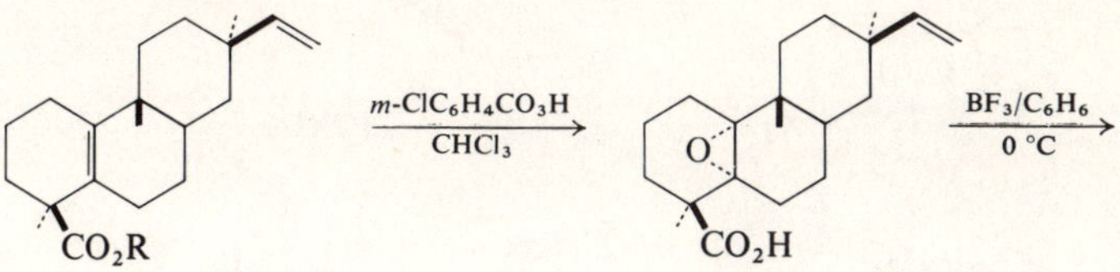

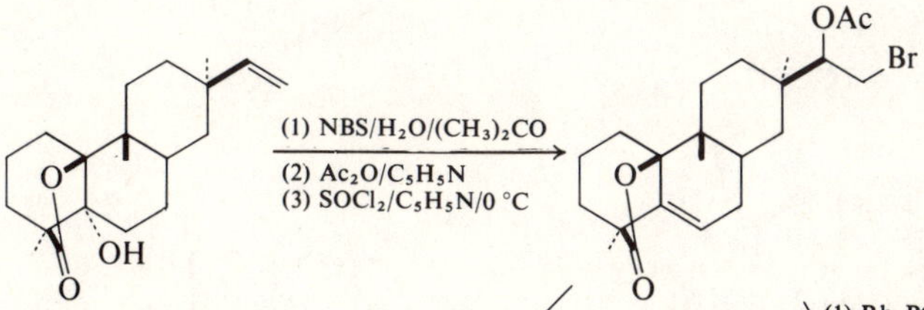

(CLXXXVIII)

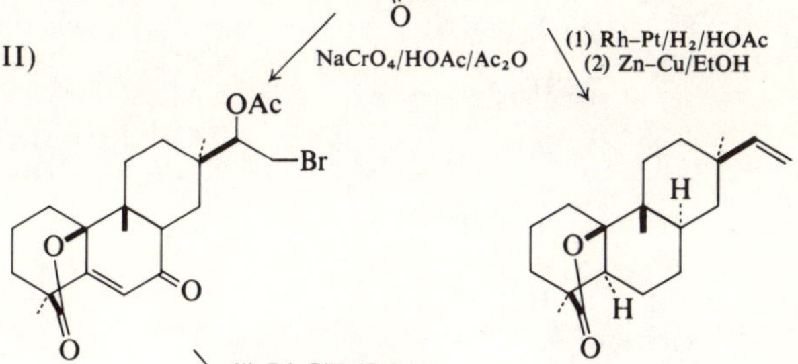

(1) NBS/H$_2$O/(CH$_3$)$_2$CO
(2) Ac$_2$O/C$_5$H$_5$N
(3) SOCl$_2$/C$_5$H$_5$N/0 °C

NaCrO$_4$/HOAc/Ac$_2$O

(1) Rh–Pt/H$_2$/HOAc
(2) Zn–Cu/EtOH

(CLXXXVI)

(1) Pd–C/H$_2$/EtOAc
(2) Zn–Cu/EtOH

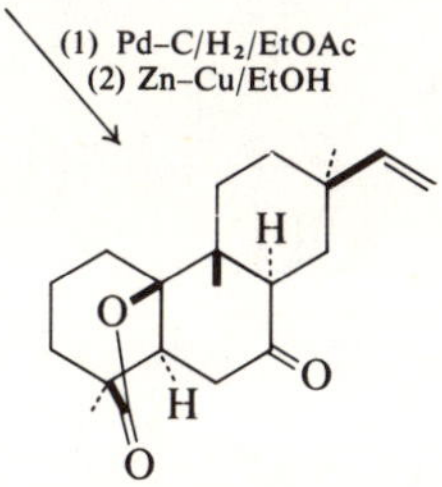

(CLXXXV)

rosenonolactone (CLXXXVI) from the methyl ester of isocupressic acid (CLXXXIV)[69]. A notable feature of the synthetic route is the introduction of epoxide functionality to achieve the desired stereochemical and structural

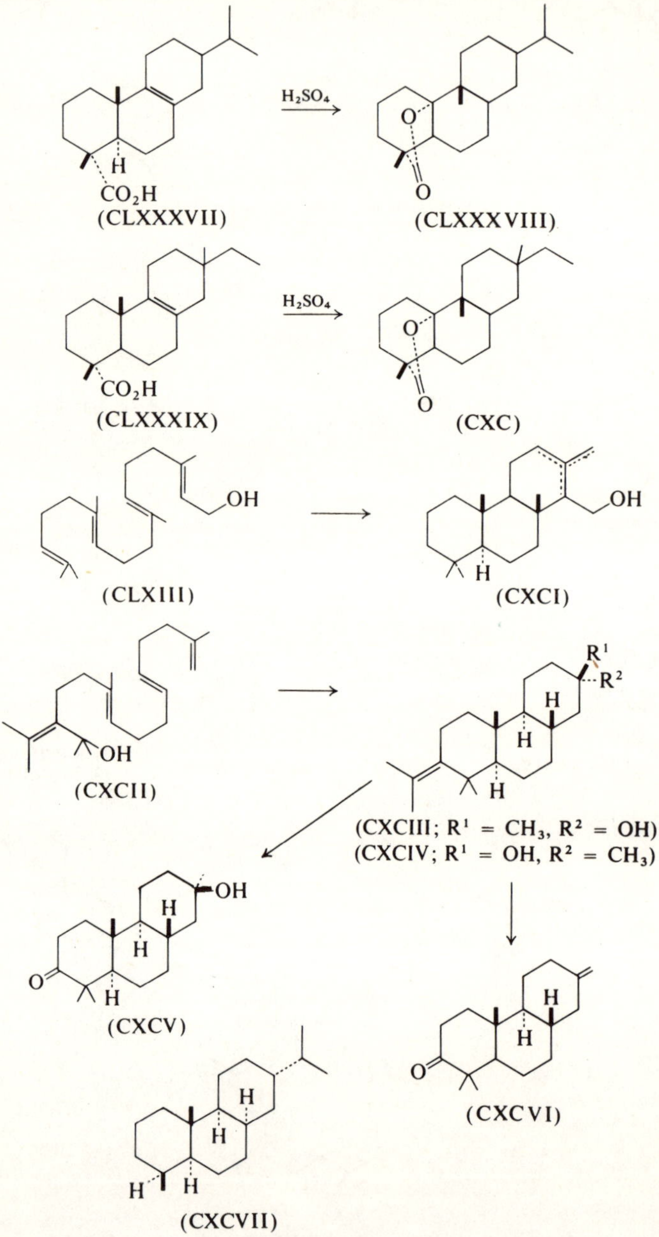

(CLXXXVII)   →[H₂SO₄]   (CLXXXVIII)

(CLXXXIX)   →[H₂SO₄]   (CXC)

(CLXIII)   →   (CXCI)

(CXCII)   →   (CXCIII; R¹ = CH₃, R² = OH)
(CXCIV; R¹ = OH, R² = CH₃)

(CXCV)

(CXCVI)

(CXCVII)

features required for eventual success. Lactone formation accompanied by 1,2-shift of the angular methyl group has previously been demonstrated in the conversion of dihydroabietic acid (CLXXXVII)[70] into the rearranged lactone (CLXXXVIII) and of dihydropimaric acid (CLXXXIX) into (CXC)[71]. The cyclisation of geranylgeraniol (CLXIII) can lead, in theory, to a tricyclic skeleton (CXCI) which is rarely found in diterpenes. A laboratory analogy for this process has been provided and is described in recent reviews[21,72]. More recently it has been shown that treatment of the tetraenol (CXCII) with trifluoracetic acid results in stereoselective cyclisation and formation of tricyclic alcohols (CXCIII) and (CXCIV)[73]. Ozonolysis of (CXCIV) yielded hydroxyketone (CXCV) while ozonolysis and dehydration of the mixture of alcohols gave enone (CXCVI). The method provides stereoselective entry into tricyclic systems and its application to the synthesis of (±)-fichtelite (CXCVII) has been described[72,74].

A postulated biosynthesis of α- (CXCIX) and β-levantenolide (CC) involves intermediate formation of the acyclic keto-acid (CXCVIII) from geranylgeraniol followed by stereospecific cyclisation[75]. Attempts to provide a laboratory analogy for this cyclisation have resulted in the biogenetic-type synthesis of (CXCIX) and (CC) from the acyclic butenolide (CCII)[75]. The possibility that a monocyclic intermediate is involved in the biosynthetic route is also supported by the cyclisation of (CCIII) to a mixture of α-(CXCIX) and β-levantenolide (CC)[76].

The postulated biosynthetic relationship between tricyclic and tetracyclic diterpenes has prompted synthetic studies in this area. Solvolytic cyclisation of the unsaturated, tricyclic toluene-*p*-sulphonate (CCIV) provides an atiseran-13-ol derivative (CCV) which can be dehydrated to alkenes,

(CCVII; R = CO$_2$Me) and (CCVIII; R = CO$_2$Me), closely related to atiserene (CCVII; R = CH$_3$) and isoatiserene (CCVIII; R = CH$_3$). Alternatively, more vigorous formolysis converts (CCV) into the hiban-12-ol isomer (CCVI) and thus provides entry into this series of tetracyclic diterpenes[77].

Biosynthetic theory implies that a synthetic relationship exists between tetracyclic and pentacyclic diterpenes (see p. 45) and many of these interconversions have been accomplished in the laboratory[78-88]. One of the earliest investigations demonstrated that treatment of trachylobane (CCIX; R = CH$_3$) or the ester (CCIX; R = CO$_2$CH$_3$) with acid results in cleavage of the cyclopropane ring in all possible modes and formation of compounds (CCX-CCXII) belonging to the hibane (syn. beyerane, stachane), atisane and kaurane series of tetracyclic diterpenes[78]. Acid-catalysed isomerisation

of (+)-stachene (CCXIII) (syn. (+)-hibaene) or trachylobane (CCXIV) produces an equilibrium mixture of (−)-isokaurene (CCXV), (−)-kaurene (CCXVI), (−)-isoatisirene (CCXVII) and (−)-atisirene (CCXVIII)[79]. Treatment of (−)-kaurene (CCXVI) under similar conditions (HCl—CHCl$_3$) produced (−)-isokaurene (CCXV) and a small amount of (−)-isoatisirene (CCXVII) and (−)-atisirene (CCXVIII)[79].

A similar study of diterpene interconversions has shown that (+)-hibaene (CCXIII) can be converted into 16-hydroxyisokaurene (CCXX) by treatment of the epoxide (CCXIX) with boron trifluoride etherate[80]. Subsequent oxidation of the alcohol (CCXX) and reduction of the corresponding ketone yielded (−)-kaurene (CCXVI). Conversion of (+)-hibaene to the tosylate (CCXXI) followed by solvolysis in buffered aqueous dioxane yields a mixture of (−)-kaurene (CCXVI), (−)-isokaurene (CCXV), and kauran-2-ol (CCXXII)[81]. It is interesting to note that atisirene or trachylobane could not be detected as products of this reaction[82]. A laboratory analogy for the conversion of hibane to trachylobane-type diterpenes has been provided[83] by the transformation of tosylhydrazone (CCXXIII) to methyltrachyloban-19-oate (CCXXIV), (CCXXV) and (CCXXVI). In the deamination of (CCXXVII) similar rearrangements occurred with the notable exception that trachylobane derivatives could not be detected[83]. The transformation of a phyllocladene derivative (CCXXX) into a compound possessing the neoatisirene skeleton (CCXXXI) has been accomplished[84]. In a similar vein the epoxide (CCXXXIII) from (−)-hibaene (CCXXXII) has been converted into the isokaurene derivative (CCXXXIV)[85] while the epoxide (CCXXXV) yielded (CCXXXVI)[86]. The conversion of kaurane to stachane (*syn.* beyerane, hibaene) derivatives has been reported by several groups of investigators. Treatment of (+)-kaurene (CCXXXVII) or (−)-hibaene (CCXXXII) with iodine in refluxing xylene produced an equilibrium mixture of (+)-kaurene (CCXXXVII), (+)-isokaurene (CCXXXVIII) and (−)-hibaene (CCXXXII)[85,87]. The conversion of kaurene derivative (CCXXXIX) to the corresponding hibaene derivative (CCXL)[88] and the synthesis of monogynol (CCXLV) from (CCXLI)[86] provide further examples of this rearrangement.

TRITERPENES

Since the previous review on biogenetic-type synthesis[89], a considerable amount of progress has been made in the biosynthesis of triterpenes and

(CCXXX) (CCXXXI) (CCXXXII) (CCXXXIII) (CCXXXIV) (CCXXXV) (CCXXXVI)

steroids[90-92]. In summary, it has been shown that squalene (CCXLVI) or squalene monoepoxide (CCXLVII) is the basic triterpene precursor which cyclises to produce all members of this group of natural products. Many investigations have been initiated to duplicate these cyclisation processes in the laboratory and thus provide a nonenzymic and stereoselective synthetic route to the tetra- and penta-cyclic triterpenes. The results obtained by two major groups of investigators have been reviewed[21,72], and only selected portions of these findings will be described again here. In particular it was found that squalene monoepoxide (CCXLVII), on short treatment with stannic chloride in benzene, yielded two tricyclic compounds (CCXLVIII) and (CCXLIX)[21,93]. Compounds of the type (CCL; Z = biological leaving group) could be important intermediates in the conversion of squalene to the tetracyclic triterpene systems since chemically precedented ring expansion followed by cyclisation and rearrangement (CCL → CCLII) provides an alternative biosynthetic sequence to lanosterol (CCLII)[92,94]. Another component of the reaction mixture is the bicyclic alcohol (CCLIV)[95] whose formation involves rearrangement of a type (cf. CCLIII) postulated in the biosynthesis of other terpene systems[95] e.g. rimuene, simiarenol, 3β-hydroxyglutinene-(5), alnusenone and the cucurbitacins.

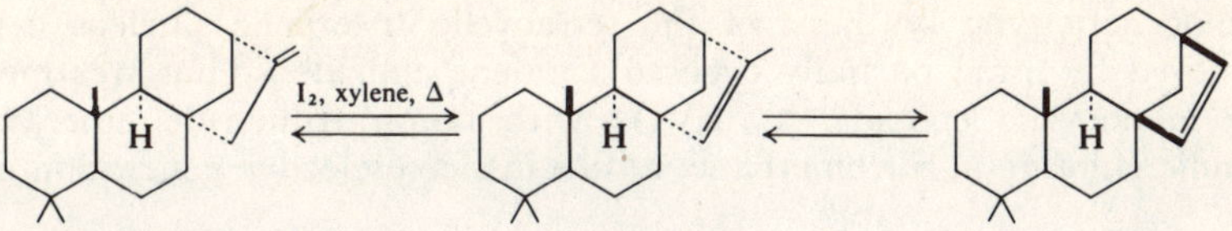

(CCXXXVII) (CCXXXVIII) (CCXXXII)

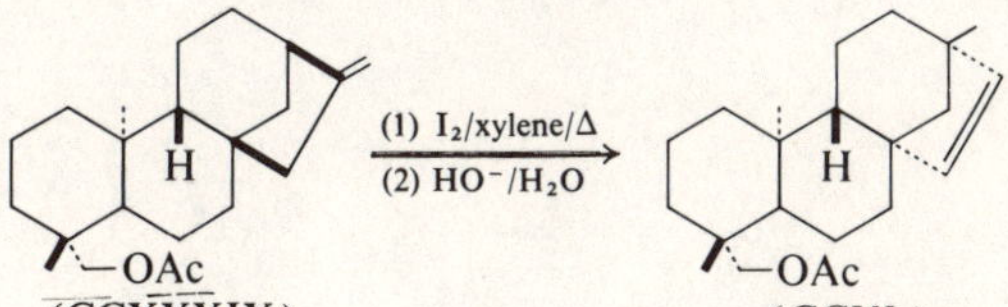

(CCXXXIX) (CCXL)

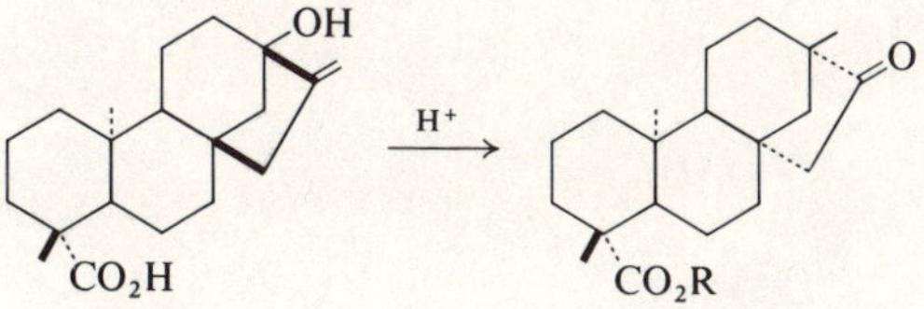

(CCXLI) (CCXLII)

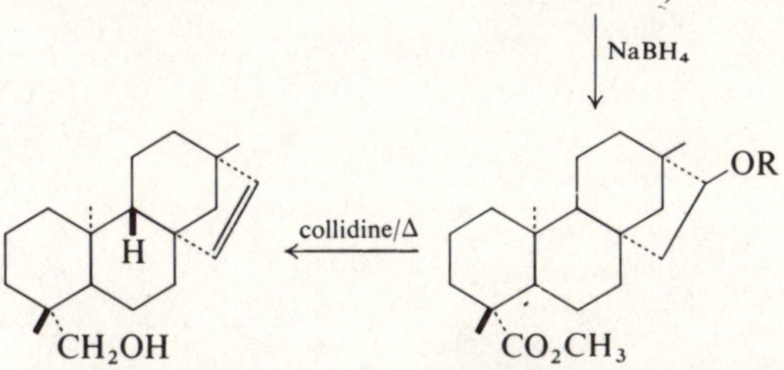

(CCXLV) (CCXLIII; R = H)
(CCXLIV; R = SO$_2$C$_7$H$_7$)

The tricyclic skeleton shown in (CCXLVIII) can also be produced *in vivo* by treatment of 18,19-dihydrosqualene-2,3-epoxide with a rat liver enzyme preparation[21]. Of considerable interest is the fact that malabaricol (CCLVa) and malabaricanediol (CCLVb) have recently been isolated from plants[96].

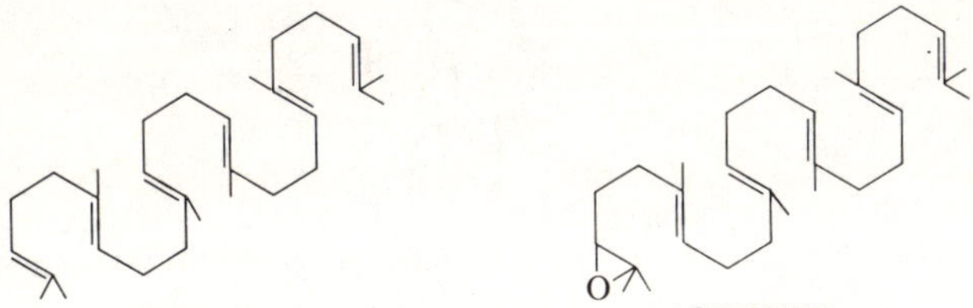

(CCXLVI) (CCXLVII)

In an elegant example of biogenetic-type synthesis, (±)-malabaricanediol (CCLVb) has been produced by acid-catalysed cyclisation of epoxydiol (CCLVI)[97].

Biogenetic-type synthesis of the tetracyclic triterpene nucleus can be achieved by using partially cyclised squalene epoxides. Thus treatment of the monocyclic epoxide (CCLVII) with boron trifluoride etherate, or stannic chloride in nitromethane, results in stereoselective generation of the

isoeuphenol system (CCLVIII)[98]. In a similar manner the bicyclic epoxide (CCLIX) can be converted into dihydro-9β-Δ^7-lanosterol (CCLXI)[99]. It is interesting to note that a cyclase preparation converts (CCLVII) and (CCLIX) to pentanorlanosterol (CCLX; R = H) and dihydrolanosterol

(CCLX; R = C_5H_{11}) respectively without formation of detectable amounts of (CCLVIII) and (CCLXI).

The pentacyclic triterpene hopenone-1 (CCLXIV) has been synthesised by a route which involves cyclisation of β-onocerin diacetate (CCLXII) to γ-onocerin diacetate (CCLXIII)[100,102], followed by ring contraction and conversion into hopenone-1 (CCLXIV)[101]. Prompted by the importance of squalene in triterpene biosynthesis, the cyclisation of suitably substituted polyolefins has been extensively studied[72,89,103] and has provided elegant, stereoselective biogenetic-type routes to diterpenes (p. 51), triterpenes and steroids. Using this technique the synthesis of 16,17-dehydroprogesterone (CCLXVII) has been accomplished[104]. The key step is the stereospecific, acid-catalysed cyclisation of the tetraeneol (CCLXV). Subsequent oxidative cleavage of rings A and D in the product (CCLXVI) and reclosure by aldol condensation provided the carbocyclic steroid framework (CCLXVII). Recent improvements[105] in the synthesis of starting material (CCLXV) have made this an efficient, biogenetic-type steroid synthesis. In a related study, Johnson and co-workers have shown that tetraene acetal (CCLXVIII), on treatment with stannic chloride in pentane, undergoes cyclisation to the tetracyclic compound (CCLXIX) in a highly stereoselective manner[106]. Further application of this process should provide an alternative synthetic entry into the triterpenes and steroids. A fascinating example of asymmetric induction has also been demonstrated using non-enzymic cyclisation of

BIOGENETIC-TYPE SYNTHESIS OF TERPENES

optically active acetal (CCLXX)[107]. The four isomers were separated into two pairs (CCLXXIa, b) and (CCLXXIIa, b) and were then oxidised to the corresponding octalones. From the axial octalol (CCLXXIIa, b; R = H) the octalone mixture was 92% (CCLXXIV) and 8% (CCLXXIII)

while the equatorial octalol (CCLXXIa, b; R = H) gave a mixture of 8% (CCLXXIV) and 92% (CCLXXIII). Thus the initial cyclisation produced (CCLXXIIb) and (CCLXXIa) in preference to (CCLXXIIa) and (CCLXXIb) by a factor of 9:1 in each case.

The ability of natural systems to introduce functionality into non-activated positions has stimulated efforts to emulate these reactions. The Barton reaction[108] has been used with some success but recently a new approach has made it possible to convert hexadecanol to 14-ketohexadecanol[109]. In the steroid area a more recent report[110] describes the conversion of 3α-cholestanol (CCLXXV) into 12-keto-3α-cholestanol (CCLXXVII). Photochemically induced interaction between the benzophenone carbonyl group and the C-12-methylene group in (CCLXXVI), followed by dehydration, oxidative cleavage and hydrolysis, yielded the ketone (CCLXXVII). Similarly, photolysis of (CCLXXVI) in acetonitrile, followed by lead tetraacetate cleavage and hydrolysis, provided a mixture of Δ^{14}-cholesten-3α-ol (CCLXXIX) and $\Delta^{8,14}$-cholesten-3α-ol (CCLXXX). The tertiary alcohol (CCLXXVIII) formed by specific photochemical reaction between the benzophenone carbonyl group and the C-14 methine group in

(CCLXXVI) has been suggested as an intermediate in this process[110]. Similar investigations[111] have also shown that the 3α-cholestanyl ester (CCLXXXI) undergoes direct photochemical hydrogen transfer to provide an efficient route to Δ^{14}- (e.g. CCLXXXIV) and Δ^{16}-steroids (e.g. CCLXXXV). The macrolide intermediates (CCLXXXII) and (CCLXXXIII) were also components of the reaction product mixture[111].

The enzymatic conversion of farnesyl pyrophosphate into squalene has been the subject of considerable study[112,135-138] and several mechanisms have been proposed[112-117,135-138] for the coupling process. One mechanism[113] involves a Stevens intramolecular rearrangement of an enzymically bound ylid (CCLXXXVI) to the thiosubstituted intermediate (CCLXXXVII), followed by reduction to squalene.

An alternative proposal[114-115] involving a five-centre intramolecular rearrangement of the unsymmetrical ylid (CCLXXXVIII) derived from a farnesylnerolidyl sulphonium salt could also explain the coupling process. Some laboratory analogies for these proposals have been provided[114,115,118-121] Careful treatment of the sulphonium salt (CCLXXXIX) with n-butyl lithium at −30°C gave the expected rearrangement product (CCXC) which could be reduced to the diene (CCXCI)[114].

The sulphide (CCXCII), on treatment with benzyne, has been converted into a mixture of sulphides which included (CCXCIII) and (CCXCIV)[115a]. These products can be explained by rearrangement of an intermediate ylid (CCXCIIa) or a sigmatropic reaction involving the covalent intermediate (CCXCIIb). Digeranyl sulphide (CCXCV) behaves in an analogous fashion under similar conditions[115b]. The extension of these studies to the synthesis of squalene from sulphonium ylid intermediates has been reported[118]. Reaction of difarnesyl disulphide (CCXCVI) with triphenylphosphine yielded farnesyl nerolidyl sulphide (CCXCVII) which was subsequently converted into 12-phenylthiosqualene (CCXCVIII) by treatment with benzyne. Reduction of (CCXCVIII) with lithium in liquid ammonia then gave natural squalene. Related studies[120,121] associated with the rearrangement of vinyl sulphur ylides have been reported, and the importance of these results in the monoterpene area has been emphasised.

An attractive synthesis[119] of squalene utilising phenylfarnesyl thioether (CCXCIX) is reminiscent of the suggested involvement of thiofarnesyl intermediates in the biological process[116]. Treatment of (CCXCIX) with butyl lithium, followed by alkylation of the intermediate anion (CCC) with farnesyl bromide, gave 12-phenylthiosqualene (CCCI) which was subsequently reduced to squalene.

Recent studies on the biosynthesis of squalene (CCXLVI) from farnesyl pyrophosphate (XXXI) have demonstrated that presqualene pyrophosphate

Flowsheet 1

(CCXLVIa) is an intermediate in this process (*Flowsheet 1*)[135-137]. Speculative proposals[137,138] have been made for the mechanism of biosynthesis of presqualene pyrophosphate and its conversion into squalene and recent synthetic studies have been influenced to some extent by these suggestions. In one route[139] decomposition of the allylic diazo compound (XXXIa) in

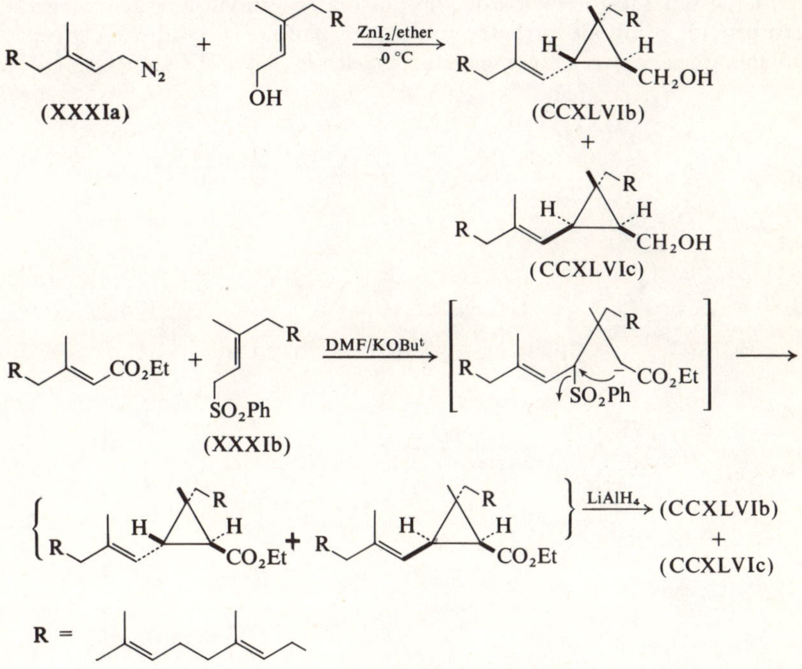

Flowsheet 2

the presence of farnesol provided presqualene alcohol (CCXLVIb) and the isomer (CCXLVIc). Another route[140] involving interaction between phenyl farnesyl sulphone (XXXIb) and ethyl farnesoate in the presence of base, yielded isomeric cyclopropyl esters which were reduced to the mixture of isomers, (CCXLVIb) and (CCXLVIc) (*Flowsheet 2*). Presqualene alcohol was separated from isomer (CCXLVIc) by preparative g.l.c. A more conventional synthesis[141] of presqualene alcohol is also shown below (*Flowsheet 3*) and involves intramolecular cyclisation of (XXXIc) to produce an appropriate cyclopropane derivative which could be converted into (CCXLVIb) and the isomer (CCXLVId). Pyrophosphorylation of the synthetic samples[139-141] of presqualene alcohol (CCXLVIb) yielded synthetic ester (CCXLVIa) which was converted enzymically into squalene. In contrast isomers (CCXLVIc) and (CCXLVId) yielded pyrophosphates which were not affected by the enzyme.

The biosynthesis of the complex molecular structures found in terpenes involves cyclisation of basic isoprenoid precursors (geranyl and farnesyl pyrophosphate and squalene) followed by a variable amount of secondary transformations of the initially formed product. The secondary transformations include oxidative and reductive processes, ring cleavage reactions and a great variety of rearrangements[122]. The synthesis of terpenes possessing complex structures far removed in type from the original cyclised intermediate is a considerable challenge to the organic chemist, and this is particularly evident in the triterpene area. It is possible, however, that synthetic schemes modelled on the postulated biosynthetic rearrangements can provide solutions to these problems. Examples of synthetic conversions, involving extensive rearrangement processes, have been described in a

Flowsheet 3

previous review[89] and these results have pointed the way to more recent developments. A recent example[123] in the triterpene field is the synthesis of fernadiene (CCCIV) and fern-8-ene (CCCV) by backbone rearrangement of a pentacyclic compound (CCCIII) whose structure is reminiscent of the postulated biosynthetic precursor (CCCII; Z = biological leaving group). The most frequently studied conversions have involved backbone rearrangements which result from treatment of triterpene or steroid epoxides with Lewis acids[124]. Much of the recent work has been concerned with the tetranortriterpenes which present an interesting problem in terpene biosynthesis and a challenge to the synthetic organic chemist. Representative members of this group of terpenes are azadirone (CCCVI), azadiradione (CCCVII), epoxyazadiradione (CCCVIII), gedunin (CCCIX), obacunone (CCCX), nomilin (CCCXI), ichangin (CCCXII), and limonin (CCCXIII). The classical investigations[125,126] are associated with limonin (CCCXIII) and it was suggested that the biosynthesis of this complex compound involved a euphol (CCCXIV)- or butyrospermol (CCCXV)-type precursor. Some of the reactions required to complete the total biosynthetic sequence may be outlined as follows: (a) backbone rearrangement of methyl groups promoted by cleavage of 7a,8a-epoxide and accompanied by formation of a double

bond between C-14 and C-15; (b) degradation of side-chain to yield a furan residue; (c) conversion of rings A and D to enones; (d) Baeyer–Villiger type cleavage of rings A and D (as enones); (e) intramolecular *trans*-lactonisation between ring A lactone and C-19-hydroxymethyl. Compounds (CCCVI–CCCXII) can be regarded as similar to the postulated intermediates involved in the biosynthesis of limonin (CCCXIII). Reactions of types (a)–(d) above have been accomplished in the laboratory[127–132] and a selection of relevant transformations is shown. Backbone rearrangements involving rings A, B, and C of the triterpene nucleus have been accomplished using 9a,11a-epoxides and are part of a general investigation[133] into a

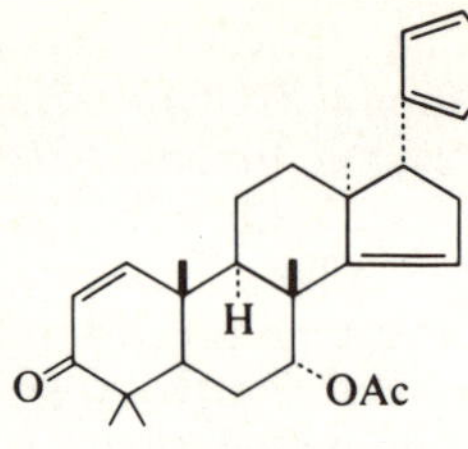

(CCCVI) 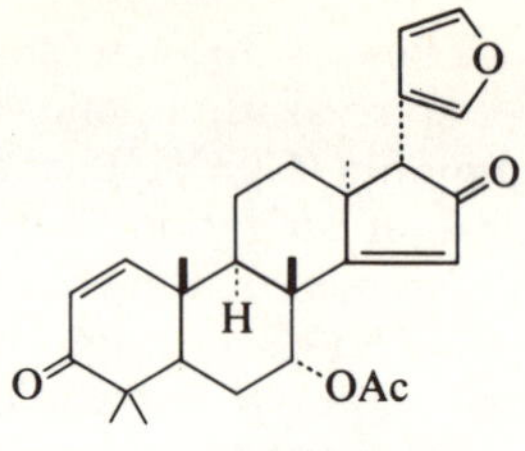 (CCCVII)

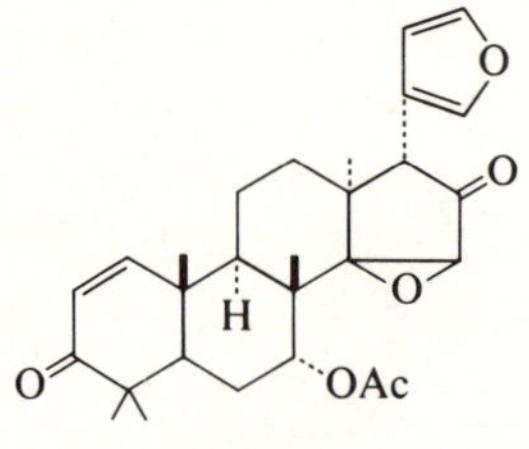

(CCCVIII) 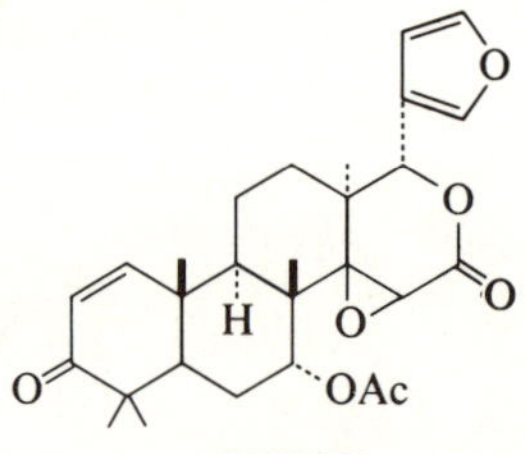 (CCCIX)

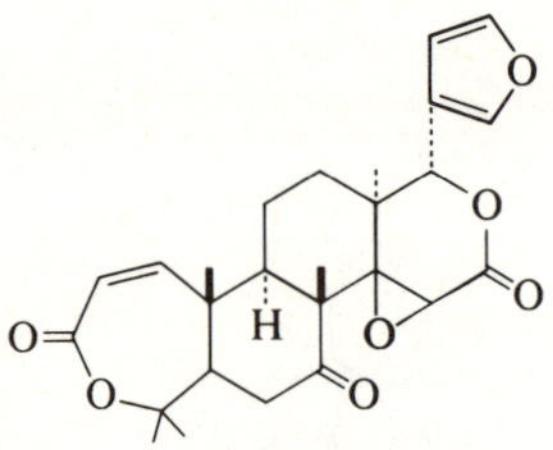

(CCCX) 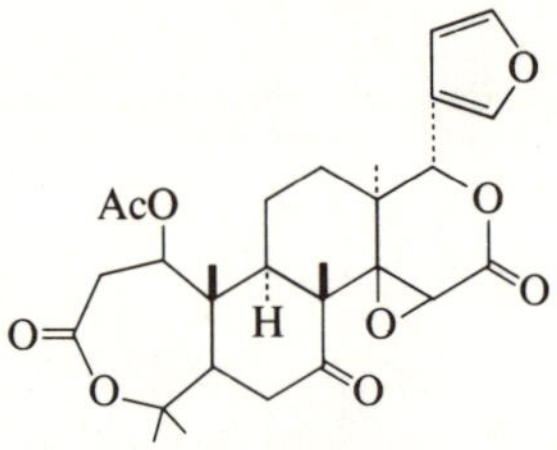 (CCCXI)

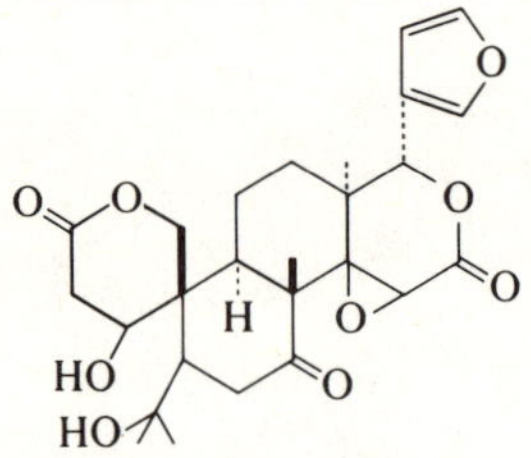

(CCCXII) 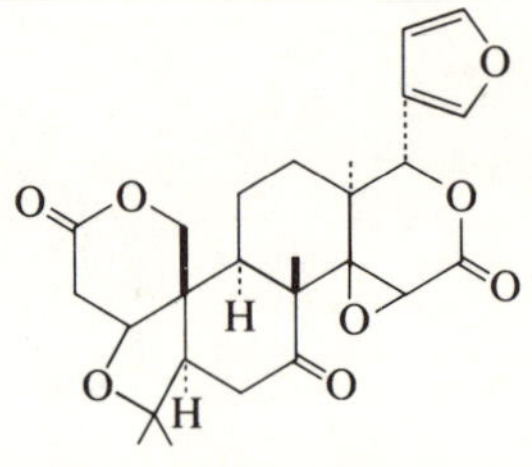 (CCCXIII)

biogenetic-type synthesis of the curcubitacins (e.g. CCCXXVI). A representative reaction in this study is the conversion of epoxide

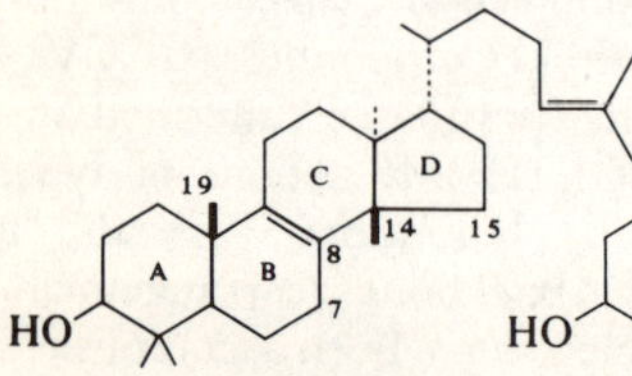

(CCCXIV) 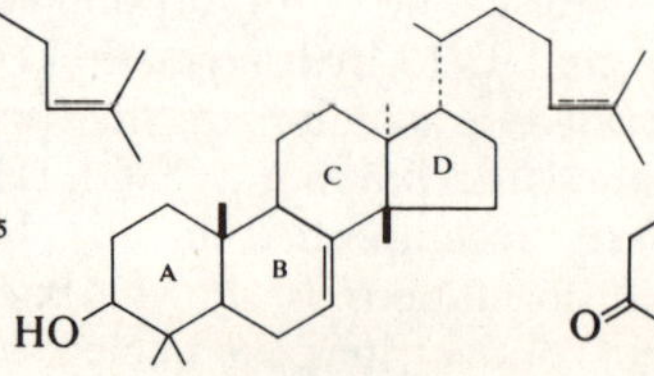 (CCCXV) 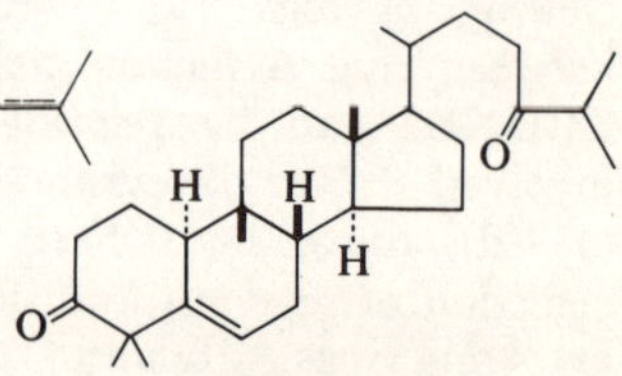 (CCCXXVI)

(CCCXXVII) into the rearranged compound (CCCXXVIII). Tetranortriterpenes in which ring B has been cleaved are represented by compounds such as andirobin (CCCXXIX) and methyl angolensate (CCCXXX). The suggested biosynthetic pathway involves a Baeyer–

(CCCXXII) →[H$_2$O$_2$/HO$^-$, Ref. 132]→ (CCCXXIII)

(CCCXXIV) →[C$_6$H$_5$CO$_3$H, Ref. 131]→ (CCCXXV)

(CCCVIII) →[C$_6$H$_5$CO$_3$H, Ref. 131]→ (CCCIX)

Villiger reaction on a 7-keto intermediate and a duplication of this proposal is included in the reported conversion of 7-oxo-7-deacetoxykhivorin (CCCXXXI) into methyl angelonsate (CCCXXX)[134]. Further studies by the

(CCCXXVII) →[BF$_3$/C$_6$H$_6$]→ (CCCXXVIII)

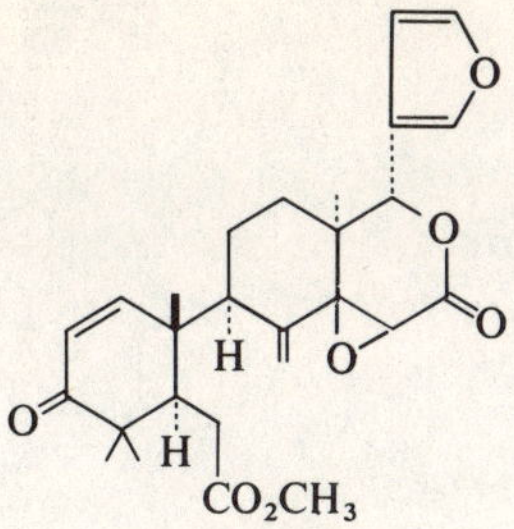

(CCCXXIX) (CCCXXX)

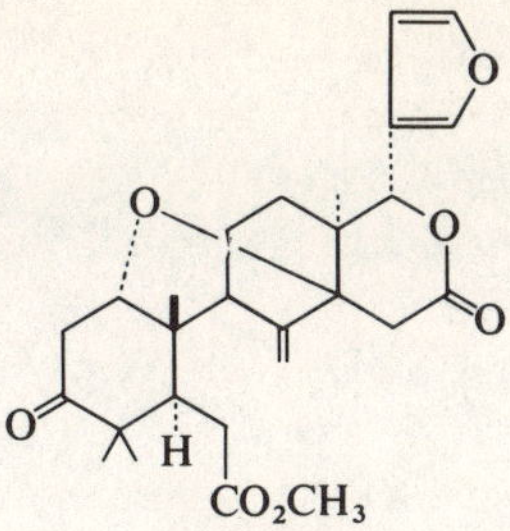

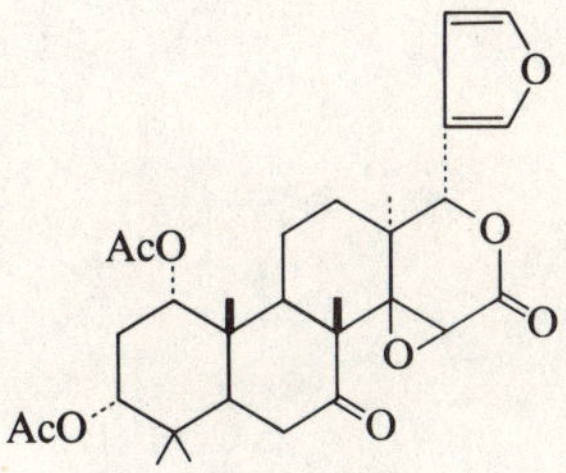

(CCCXXXI) →[CrCl₂] (CCCXXXII)

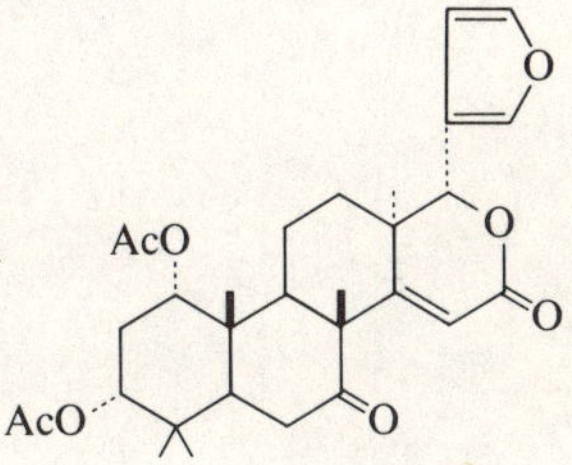

↓ CH₃CO₃H

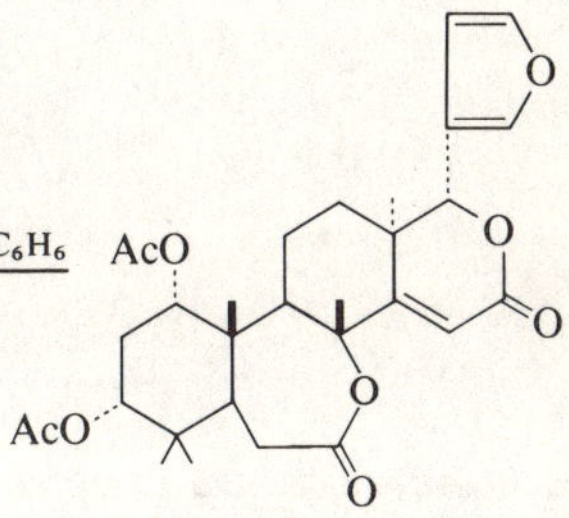

(CCCXXXIV) ←[p-CH₃C₆H₄SO₃H/C₆H₆] (CCCXXXIII)

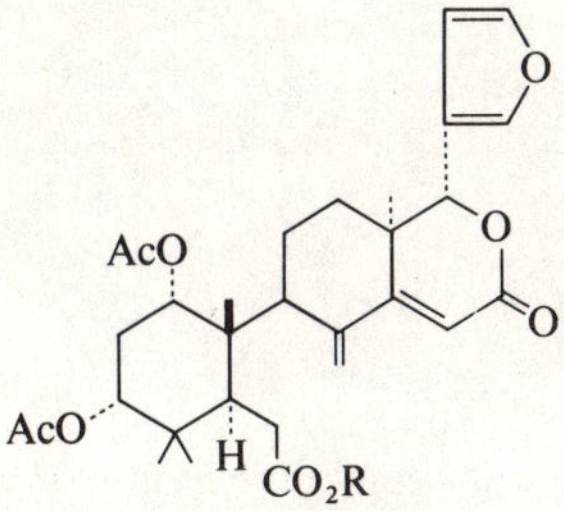

↓ mild hydrolysis

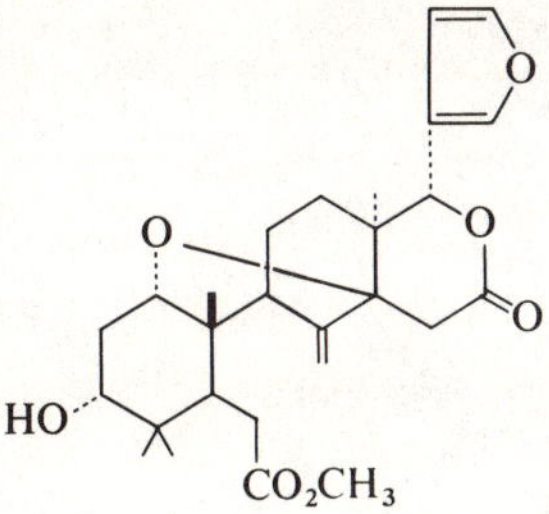

(CCCXXXV) →[oxidation] (CCCXXX)

same research group have resulted in an elegant biogenetic-type conversion of 7-oxo-7-deacetoxykhivorin (CCCXXXI) to mexicanolide (CCCXL)[142].

Manuscript received May 1971.

REFERENCES

1. RICHARDS, J. H. and HENDRICKSON, J. B., *The Biosynthesis of Steroids, Terpenes and Acetogenins*, W. A. Benjamin, New York, N.Y. (1964); BU'LOCK, J. D., *The Biosynthesis of Natural Products*, McGraw-Hill, London (1965); CLAYTON, R. B., *Q. Rev.*, **19**, 168, 201 (1965)
2. RITTERSDORF, W., *Angew. Chem. Int. Edn.*, **4**, 444 (1965)
3. CRAMER, F. and RITTERSDORF, W., *Tetrahedron*, **23**, 3015 (1967)
4. VALENZUELA, P. and CORI, O., *Tetrahedron Lett.*, 3089 (1967)
5. MILLER, J. A. and WOOD, H. C. S., *Angew. Chem. Int. Edn.*, **3**, 310 (1964); HALEY, R. C., MILLER, J. A. and WOOD, H. C. S., *J. chem. Soc. C*, 264 (1969)
6. WILCOX, C. F. and CHIBBER, S. S., *J. org. Chem.*, **27**, 2332 (1962)
7. FAIRLIE, J. C., HODGSON, G. L. and MONEY, T., *Chem. Commun.*, 1196 (1969)
8. RUZICKA, L., *Experientia*, **9**, 357 (1953); *Proc. chem. Soc.*, 341 (1959)
9. CROWLEY, K. J., *Proc. chem. Soc.*, 245 (1962)

10. BRESLOW, R., GROVES, J. T. and OLIN, S. S., *Tetrahedron Lett.*, 4717 (1966)
11. DE MAYO, P., *Mono- and Sesquiterpenoids,* Interscience, 86 (1959)
12. VAN TAMELEN, E. E., *Accts Chem. Res.*, **1**, 1 (1968); GOLDSMITH, D. J., *J. Am. chem. Soc.*, **84**, 3913 (1962); GOLDSMITH, D. J. and PHILLIPS, C. F., ibid., **91**, 5862 (1969)
13. BATES, R. B. and PAKNIKAR, S. K., *Tetrahedron Lett.*, 1453 (1965)
14. CROMBIE, L., HOUGHTON, R. P. and WOODS, D. K., *Tetrahedron Lett.*, 4553 (1967)
15. THOMAS, A. F., *Chem. Comm.*, 1054 (1970) and references cited
16. PARKER, W., ROBERTS, J. S. and RAMAGE, R., *Q. Rev. chem. Soc.*, **21**, 331 (1967)
17. HANSON, J. R. and ACHILLADELIS, B., *Chemy. Ind.*, 1643 (1967); (a) ACHILLADELIS, B., ADAMS, P. M. and HANSON, J. R., *Chem. Commun.*, 511 (1970)
18. NAVES, Y-R., *Helv. chim. Acta*, **49**, 1029 (1966)
19. GUTSCHE, C. D., MYCOCK, J. R. and CHANG, C. T., *Tetrahedron*, **24**, 859 (1968); cf. BRIEGER, G., NESTRICK, T. J. and MCKENNA, C., *J. org. Chem.*, **34**, 3789 (1969)
20. RITTERSDORF, W., *Angew. Chem. Int. Edn.*, **44**, 444 (1965)
21. VAN TAMELEN, E. E., *Accts Chem. Res.*, **1**, 111 (1968) and references cited
22. VAN TAMELEN, E. E. and SHARPLESS, K. B., *Tetrahedron Lett.*, 2655 (1967)
23. VAN TAMELEN, E. E. and MCCORMICK, J. P., *J. Am. chem. Soc.*, **91**, 1847 (1969)
24. MORIYAMA, H., SUGIHARA, Y. and NAKANISHI, K., *Tetrahedron Lett.*, 2851 (1968)
25. BRESLOW, R., OLIN, S. S. and GROVES, J. T., *Tetrahedron Lett.*, 1837 (1968)
26. STORK, G. and BURGSTAHLER, A. W., *J. Am. chem. Soc.*, **77**, 5068 (1955)
27. KITAHARA, Y., KATO, T., SUZUKI, T., KANNO, S. and TANEMURA, M., *Chem. Commun.*, 342 (1969)
28. KITAHARA, Y., KATO, T. and KANNO, S., *J. chem. Soc. C*, 2397 (1968)
29. ITO, S., ENDO, K., YOSHIDA, T., YATAGAI, M. and KODAMA, M., *Chem. Commun.*, 186 (1967)
30. OHTA, Y. and HIROSE, Y., *Tetrahedron Lett.*, 2483 (1968)
31. KANNO, S., KATO, T. and KITAHARA, Y., *Chem. Commun.*, 1237 (1967); *Tetrahedron*, **26**, 4287 (1970)
32. CORNFORTH, J. W., MILBORROW, B. V. and RYBACK, G., *Nature*, **206**, 715 (1956); MOUSSERON-CANET, M., MANI, J-C., DALLE, J-P. and OLIVE, J-L., *Bull. Soc. chim. Fr.*, 3874 (1966)
33. IRIE, T., SUZUKI, M., KUROSAWA, E. and MASAMUNE, T., *Tetrahedron Lett.*, 1837 (1966)
34. RUZICKA, L., *Experientia*, **9**, 357 (1953)
35. BARTON, D. H. R. and DE MAYO, P., *J. chem. Soc.*, 150 (1957); BARTON, D. H. R., BROCKMANN, O. C. and DE MAYO, P., *J. chem. Soc.*, 2263 (1960)
36. HENDRICKSON, J. B., *Tetrahedron*, **7**, 82 (1959)
37. BROWN, E. D., SOLOMAN, M. D., SUTHERLAND, J. K. and TORRE, A., *Chem. Commun.*, 111 (1967)
38. BROWN, E. D. and SUTHERLAND, J. K., *Chem. Commun.*, 1060 (1968)
39. GOVINDACHARI, T. R., GOSHI, B. S. and KANRAT, V. N., *Tetrahedron*, **21**, 1509 (1965)
40. BATES, R. B. and SLAGEL, R. C., *J. Am. chem. Soc.*, **84**, 1308 (1962) and references cited
41. BUCHI, G., MACLEOD, W. and PADILLA, J. O., *J. Am. chem. Soc.*, **86**, 4438 (1964); BUCHI, G. and MACLEOD, W., *J. Am. chem. Soc.*, **84**, 3205 (1962)
42. IGUCHI, M. and NISHIYAMA, A., *Tetrahedron Lett.*, 855 (1970)
43. MACSWEENEY, D. F., RAMAGE, R. and SATTAR, A., *Tetrahedron Lett.*, 557 (1970); (a) An alternative proposal for the biosynthesis of zizaane-type sesquiterpenes has recently been described: ANDERSON, N. H. and (in part) FALCONE, M. S., *Chemy Ind.*, 62 (1971) and references cited
44. KIDO, F., UDA, H. and YOSHIKOSHI, A., *Chem. Commun.*, 1335 (1969)
45. ITO, S., TAKESHITA, H., MUROI, T., ITO, M. and ABE, K., *Tetrahedron Lett.*, 3091 (1969)
46. TAKESHITA, H., SATO, T., MURIO, T. and ITO, S., *Tetrahedron Lett.*, 3095 (1969)
47. COREY, E. J., GIROTRA, N. N. and MATHEW, C. T., *J. Am. chem. Soc.*, **91**, 1557 (1969)
48. CRANDALL, T. G. and LAWTON, R. G., *J. Am. chem. Soc.*, **91**, 2127 (1969)
49. TOMITA, B. and HIROSE, Y., *Tetrahedron Lett.*, 143 (1970)
50. ANDERSON, N. H. and SYRDAL, D. D., *Tetrahedron Lett.*, 2277 (1970)
51. TOMITA, B., ISONO, T. and HIROSE, Y., *Tetrahedron Lett.*, 1371 (1970)
52. Presented by the author of the C.I.C.–A.C.S. Conference, Toronto, May, 1970
53. HODGSON, G. L., MACSWEENEY, D. F. and MONEY, T., *Chem. Commun.*, 766 (1971)
54. HODGSON, G. L., MACSWEENEY, D. F. and MONEY, T., *Tetrahedron Lett.*, 1972, in press; (a) Cf. EDWARDS, O. E., DOUGLAS, J. L. and MOOTOO, B., *Can. J. Chem.*, **48**, 2517 (1970)
55. RUZICKA, L., *Experientia*, **9**, 357 (1953); *Proc. chem. Soc.*, 341 (1959); *Pure appl. Chem.*, **6**, 493 (1963)

56. WENKERT, E., *Chemy. Ind.*, 282 (1955)
57. WHALLEY, W. B., *Tetrahedron*, **18**, 43 (1962)
58. MCCRINDLE, R. and OVERTON, K. H., *Adv. org. Chem.*, **5**, 47 (1965) and references cited
59. BIRCH, A. J., RICKARDS, R. W., SMITH, H., HARRIS, A. and WHALLEY, W. B., *Tetrahedron*, **7**, 241 (1959)
60. CROSS, B. E., GALT, R. H. B. and HANSON, J. R., *J. chem. Soc.*, 295 (1964)
61. ACHILLADELIS, B. and HANSON, J. R., *Chem. Commun.*, 488 (1969) and references cited
62. HANSON, J. R. and WHITE, A. F., *Chem. Commun.*, 103 (1969) and references cited
63. EDWARDS, O. E. and ROSICH, R. S., *Can. J. Chem.*, **46**, 1113 (1968)
64. WENKERT, E. and KUMAZAWA, Z., *Chem. Commun.*, 140 (1968)
65. FOURREY, J.-L., POLANSKY, J. and WENKERT, E., *Chem. Commun.*, 714 (1969)
66. EDWARDS, O. E. and MOOTOO, B. S., *Can. J. Chem.*, **47**, 1189 (1969)
67. Cf. HALL, S. F. and OEHLSCHLAGER, A. C., *Chem. Commun.*, 1157 (1969)
68. MCCREADIE, T. and OVERTON, K. H., *Chem. Commun.*, 288 (1968)
69. MCCREADIE, T., OVERTON, K. H. and ALLISON, A. J., *Chem. Commun.*, 959 (1969)
70. OURISSON, G., *Proc. chem. Soc.*, 274 (1964) (review) and references cited
71. WENKERT, E. and CHAMBERLIN, J. W., *J. Am. chem. Soc.*, **81**, 688 (1959)
72. JOHNSON, W. S., *Accts. Chem. Res.*, **1**, 1 (1968) and references cited
73. JOHNSON, W. S. and SCHAAF, T. K., *Chem. Commun.*, 611 (1969)
74. JOHNSON, W. S., JENSEN, N. P., HOOZ, J. and LEOPOLD, E. J., *J. Am. chem. Soc.*, **90**, 5872 (1968)
75. TANEMARA, M., SUZUKI, T., KATO, T. and KITAHARA, Y., *Tetrahedron Lett.*, 1463 (1970)
76. KATO, T., TANAMURA, M., SUZUKI, T. and KITAHARA, Y., *Chem. Commun.*, 28 (1970)
77. COATES, R. M. and BERTRAM, E. F., *Chem. Commun.*, 797 (1969)
78. HUGEL, C., LODS, L., MELLOR, J. M. and OURISSON, G., *Bull. Soc. chim. Fr.*, 2894 (1965)
79. APPLETON, R. A., MCALEES, A. J., MCCORMICK, A., MCCRINDLE, R. and MURRAY, R. D. H., *J. chem. Soc. C*, 2319 (1966)
80. KAPADI, A. H. and DEV., S., *Tetrahedron Lett.*, 1255 (1965); ibid., 1171 (1964)
81. SOBTI, R. R. and DEV., S., *Tetrahedron Lett.*, 3939 (1966)
82. Cf., APPLETON, R. A., GUNN, P. A. and MCCRINDLE, R., *Chem. Commun.*, 1131 (1968)
83. COATES, R. M. and BERTRAM, E. F., *Tetrahedron Lett.*, 5145 (1968)
84. BUCHANAN, J. G. ST. C. and DAVIS, B. R., *Chem. Commun.*, 1142 (1967)
85. YOSHIKOSHI, A., KITADANI, M. and KITAHARA, Y., *Tetrahedron*, **23**, 1175 (1967)
86. HANSON, J. R., *Tetrahedron*, **23**, 793 (1967)
87. Cf., BRIGGS, L. H., CAIN, B. F., DAVIS, B. R. and WILMSHURST, J. K., *Tetrahedron Lett.*, 17 (1959); BRIGGS, L. H., CAIN, B. F., DAVIS, B. R. and RUTLEDGE, P. S., *J. chem. Soc.*, 1850 (1962)
88. MORI, K. and MATSUI, M., *Tetrahedron*, **24**, 3095 (1968)
89. VAN TAMELEN, E. E., *Fortschr. Chem. org. NatStoffe*, **19**, 245 (1961)
90. CLAYTON, R. B., *Q. Rev. chem. Soc.*, **19**, 168, 201 (1965)
91. CORNFORTH, J. W., *Angew. Chem. Int. Edn.*, **7**, 903 (1968)
92. JAYME, M., SCHAEFER, P. C. and RICHARDS, J. H., *J. Am. chem. Soc.*, **92**, 2059 (1970) and references cited
93. Cf. KISHI, M., KATO, T. and KITAHARA, Y., *Chem. pharm. Bull., Tokyo*, **15**, 107 (1967)
94. Cf. VAN TAMELEN, E. E. and FREED, J. H., *J. Am. chem. Soc.*, **92**, 7206 (1970)
95. SHARPLESS, K. B. and VAN TAMELEN, E. E., *J. Am. chem. Soc.*, **91**, 1848 (1969) and references cited
96. CHAWLA, A. and DEV, S., *Tetrahedron Lett.*, 4837 (1967)
97. SHARPLESS, K. B., *J. Am. chem. Soc.*, **92**, 6999 (1970)
98. VAN TAMELEN, E. E., MILNE, G. M., SUFFNESS, M. I., RUDLER CHAUVIN, M. C., ANDERSON, R. J. and ACHINI, R. S., *J. Am. chem. Soc.*, **92**, 7202 (1970)
99. VAN TAMELEN, E. E. and MURPHY, J. W., ibid., **92**, 7204 (1970)
100. BARTON, D. H. R. and OVERTON, K. H., *J. chem. Soc.*, 2639 (1955)
101. SCHAFFNER, K., CAGLIOTTI, L., ARIGONI, D. and JEGER, O., *Helv. chim. Acta*, **41**, 152 (1958)
102. VAN TAMELEN, E. E., SCHWARTZ, M. A., HESSLER, E. J. and STORMI, A., *Chem. Commun.*, 409 (1966)
103. STORK, G. and BURGSTAHLER, A. W., *J. Am. chem. Soc.*, **77**, 5068 (1955)
104. JOHNSON, W. S., SENMELHACK, M. F., SULTANBAWA, M. V. S. and DOLAK, L. A., *J. Am. chem. Soc.*, **90**, 2994 (1968); further developments leading to the synthesis of ($\pm$)-progesterone

have recently been reported: JOHNSON, W. S., GRAVESTOCK, M. B., and MCCARRY, B. E., *J. Am. chem. Soc., 93*, 4332 (1971)

105. Cf. JOHNSON, W. S., LI, T., HARBERT, C. A., BARTLETT, W. R., HERRIN, T. R., STASKUN, B. and RICH, D. H., *J. Am. chem. Soc.,* 4461 (1970)
106. JOHNSON, W. S., WIEDHOOP, K., BRADY, S. F. AND OLSON, G. L., *J. Am. chem. Soc.,* **90**, 5277 (1968)
107. JOHNSON, W. S., HARBERT, C. A. and STIPANOVIC, R. D., *J. Am. chem. Soc.,* **90**, 5279 (1968)
108. Reviewed by AKHTAR, M. in *Advances in Photochemistry,* ed., Noyes, W. A., Jr., Hammond, G. S. and Pitts, J. N., Interscience, New York, Vol. 2, 263 (1964)
109. BRESLOW, R. and WINNIK, M., *J. Am. chem. Soc.,* **91**, 3083 (1969)
110. BRESLOW, R. and BALDWIN, S. W., *J. Am. chem. Soc.,* **92**, 732 (1970)
111. BALDWIN, J. E., BHATNAGAR, A. K. and HARPER, R. W., *Chem. Commun.,* 659 (1970)
112. POPJAK, G. and CORNFORTH, J. W., *Biochem. J.,* **101**, 553 (1966); CLAYTON, R. B., *Q. Rev. chem. Soc.,* **19**, 168 (1965); RISUGER, G. E. and DURST, H. D., *Tetrahedron Lett.,* 3133 (1968); FRANTZ, I. D. and SCHROEPFER, G. J., *A. Rev. Biochem.,* **36**, 691 (1967)
113. CORNFORTH, J. E., CORNFORTH, R. H., DONNENGER, C., POPJAK, G., RYBACK, G. and SCHROEPFER, G. J., *Proc. R. Soc. B,* **163**, 436 (1966)
114. BALDWIN, J. E., HACKLER, R. E. and KELLY, D. P., *J. Am. chem. Soc.,* **90**, 4758 (1968); *Chem. Commun.,* 537, 538 (1968); BALDWIN, J. E. and KELLY, D. P., *Chem. Commun.,* 899 (1968)
115. (a) BLACKBURN, G. M., OLLIS, W. D., PLACKETT, J. D., SMITH, C. and SUTHERLAND, I. O., *Chem. Commun.,* 186 (1968); (b) BLACKBURN, G. M. and OLLIS, W. D., ibid., 1261
116. KRISHNA, G., WHITLOCK, H. W., FELDBRUEGGE, D. H. and PORTER, J. W., *Archs. Biochem. Biophys.,* **114**, 200 (1966)
117. RISINGER, G. E. and DURST, H. D., *Tetrahedron Lett.,* 3133 (1968)
118. BLACKBURN, G. M., OLLIS, W. D., SMITH, C. and SUTHERLAND, I. O., *Chem. Commun.,* 99 (1969)
119. BIELLMANN, J. F. and DUCEP, J. B., *Tetrahedron Lett.,* 3707 (1969)
120. TROST, B. M. and LAROCHELLE, R., *Tetrahedron Lett.,* 3327 (1968) and references cited
121. BATES, R. B. and FELD, D., *Tetrahedron Lett.,* 417 (1968)
122. Cf. KING, J. F. and DE MAYO, P., *Molecular Rearrangements,* Part 2, ed., De Mayo, P., Interscience, New York (1964)
123. IGUCHI, K. and KAKISHAWA, H., *Chem. Commun.,* 1486 (1970)
124. BLUNT, J. W., HARTSHORN, M. P. and KIRK, D. N., *Tetrahedron,* **22**, 3195 (1966) and subsequent papers
125. ARIGONI, D., BARTON, D. H. R., COREY, E. J., JEGER, O., CAGLIOTTI, L., DEV, S., FERRINI, P. G., GLAZIER, E. R., MELERA, A., PRADHAN, S. K., SCHAFFNER, K., STERNHELL, S., TEMPLETON, J. F. and TOBINGA, S., *Experientia,* **16**, 41 (1960)
126. BARTON, D. H. R., PRADHAN, S. K., STERNHELL, S. and TEMPLETON, J. F., *J. chem. Soc.,* 255 (1961)
127. COTTERRELL, G. P., HALSALL, T. G. and WRIGLESWORTH, M. J., *Chem. Commun.,* 1121 (1967)
128. BUCHANAN, J. G. ST. C. and HALSALL, T. G., *Chem. Commun.,* 242 (1969)
129. BUCHANAN, J. G. ST. C. and HALSALL, T. G., *Chem. Commun.,* 48 (1969)
130. LAVIE, D. and JAIN, M. K., *Chem. Commun.,* 278 (1967)
131. LAVIE, D. and LEVY, E. C., *Tetrahedron Lett.,* 1315 (1970)
132. BUCHANAN, J. G. ST. C. and HALSALL, T. G., *Chem. Commun.,* 1493 (1969)
133. APSIMON, J. W. and ROSENFELD, J. M., *Chem. Commun.,* 1271 (1970) and references cited
134. CONNOLLY, J. D., THORNTON, I. M. S. and TAYLOR, D. A. H., *Chem. Commun.,* 1205 (1970)
135. POPJAK, G. in *Natural Substances Formed Biologically from Mevalonic Acid,* ed., Goodwin, T. W., Adademic Press, London, 20–31 (1970) and references cited
136. RILLING, H. C. and EPSTEIN, W. W., *J. biol. Chem.,* **245**, 4597 (1970)
137. RILLING, H. C., DALE PORTER, C., EPSTEIN, W. W. and LARSEN, B., *J. Am. chem. Soc.,* **93**, 1783 (1971)
138. VAN TAMELEN, E. E. and SCHWARTZ, M. A., *J. Am. chem. Soc.,* **93**, 1780 (1971)
139. ALTMAN, L. J., KOWERSKI, R. C. and RILLING, H. C., *J. Am. chem. Soc.,* **93**, 1782 (1971)
140. CAMPBELL, R. V. M., CROMBIE, L. and PATTENDEN, G., *Chem. Commun.,* 218 (1971)
141. COATES, R. M. and ROBINSON, W. H., *J. Am. chem. Soc.,* **93**, 1785 (1971)
142. CONNOLLY, J. D., THORNTON, I. M. S. and TAYLOR, D. A. H., *Chem. Commun.,* 17 (1971)

3
RECENT ADVANCES IN THE CHEMISTRY OF CANNABINOIDS

R. K. Razdan

INTRODUCTION	78
NUMBERING SYSTEMS	79
ISOLATION	79
SYNTHESIS AND STRUCTURE ELUCIDATION	80
Petrzilkás Synthesis	82
Mechoulam's Synthesis	83
Razdan's Synthesis	83
Tetrahydrocannabinolcarboxylic Acids	84
Cannabicyclol	86
Cannabichromene	87
CHEMICAL REACTIONS	88
ISOMERISATIONS AND CYCLISATIONS	91
BIOGENESIS	96
METABOLISM	96

INTRODUCTION

THE RESINOUS exudate of the female flowers of *Cannabis Sativa* L. (family Moraceae) forms the essential constituent of the psychotomimetically active drug variously known as hashish, charas, ganja, bhang, marihuana, etc. Depending on the country of origin and the mode of preparation, it is given different names; for example the unadulterated resin, which is smoked, is called hashish in the Middle East and charas in India. Bhang is a concoction made in water or milk from the flowering tops of the plant and is generally ingested by mouth, whereas marihuana refers to the dried flowering tops of the plant which are smoked in a pipe or cigarette. *Cannabis* preparations[1a,2] have been known to the Chinese, Indians, Persians, Assyrians and Scythians from time immemorial. Their recorded use in medicine goes back as far as the Indian medical writings, especially the Susrita, compiled 1000 B.C. and even today their use is popular both in the Ayurvedic and Tibbi systems of Indian medicine. The tincture of *Cannabis* was included in the U.S. pharmacopoeia until 1937 when it was removed and classified as a narcotic drug. The advent of the present day 'drug culture' has focused much attention on the use and misuse of this drug and has caused social, legal and medical problems all over the world. In spite of this background, the chemistry of this plant remained relatively obscure until 1963 when the structure of cannabidiol was established. Since then rapid progress has been made in the isolation, structure elucidation and synthesis of various constituents.

Mechoulam and Gaoni[2] have reviewed the chemistry of hashish until 1967 and proposed the term 'cannabinoids' for the typical C_{21}-group of compounds present in *C. Sativa*. It also encompasses their analogues and transformation products. More recently another review[3] and proceedings of a conference on the botany and chemistry of *Cannabis*[1] have been published.*, The present article deals mainly with the advances made in the chemistry of cannabinoids since 1967 although brief mention is made of earlier work for purposes of clarity and historical background.

NUMBERING SYSTEMS

The following numbering systems have been used. The Chemical Society numbering was used in older papers. The dibenzopyran numbering is used by *Chemical Abstracts* and is sometimes employed in conjunction with biphenyl numbering for cannabidiol derivatives. In the present review the monoterpene numbering system will be used as it can be more easily adapted for the cannabidiol types and iso-tetrahydrocannabinols (iso-THCs).

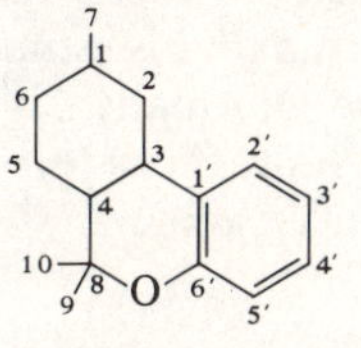

Monoterpenoid

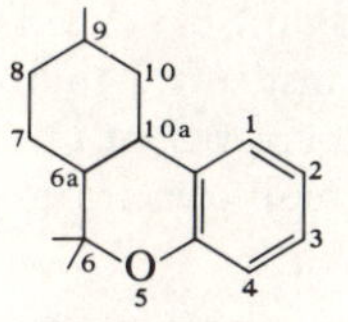

Dibenzopyran

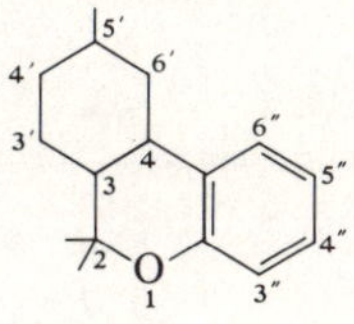

Earlier Chemical Society (London)

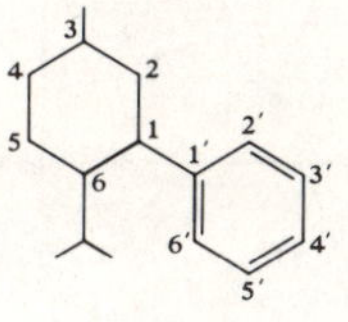

Biphenyl

ISOLATION[2]

Δ^1-Tetrahydrocannabinol (THC) is the main active constituent of hashish. For the isolation of this and other constituents, generally the alcoholic or the petroleum ether–benzene extract of the plant is separated into neutral and acidic fractions. These are then further purified by repeated column chromatography, preparative thin-layer chromatography and countercurrent distribution or a combination of these methods. Various supports have been used in column chromatography especially Florisil, acid-washed alumina–silver nitrate, silicic acid, silicic acid–silver nitrate and silica gel[5-11]. Japanese workers have also used cellulose powder impregnated with

* Another review has now appeared[79].

dimethylformamide[5] as column support. A novel solvent system[6], benzene–n-hexane–diethylamine (25:10.1), has also been claimed to separate cannabidiol, THC and cannabinol on a silica gel column. The following cannabinoids of known structure have been isolated from the plant so far (*Figure 3.1*).

In addition, Jacob and Todd[12] have described the isolation of cannabol, $C_{21}H_{30}O_2$ (*p*-phenylazobenzoate, m.p. 117–8°C), and Covello[13] has reported a crystalline material, m.p. 129–133°C, possessing a highly inebriating effect in the dog. Other materials including non-cannabinoid terpenes[14], nitrogen bases[15], phenolic compounds[16,17], and sugars[16] have been either isolated from or detected in *C. Sativa*. The proportion of different cannabinoids present varies considerably from sample to sample and is dependent upon the quality of soil, botanical variation, climatic conditions, geographic location[1b,4], mode of preparation and storage conditions, etc.

According to Gaoni and Mechoulam[4], a typical hashish sample contains the following (*Table 3.1*) neutral components. The content of active Δ^1-THC was found to vary from 1 to 5%. The acid fraction of the extracts contains the cannabinoid acids which are biologically inactive. However, on heating they decarboxylate to furnish the corresponding neutral cannabinoids. Hence the amount of Δ^1-THC acids should be taken into consideration when the amount of Δ^1-THC available on smoking is estimated. In hashish these acids constitute 1 to 3% of the material. Normally Δ^1-THC is the most common isomer present in various samples; however, a few varieties contain in addition small quantities of the other active isomer $\Delta^{1(6)}$-THC.

Table 3.1[4] CONTENT IN HASHISH[a], R_f VALUES (T.L.C.)[b], AND RETENTION TIMES (V.P.C.)[c] OF SOME NATURAL NEUTRAL CANNABINOIDS

	Yields[a]	R_f[b]	Retention time[c]
Cannabicyclol (IX)	0·11	0·62	4′33″
Cannabidiol (IVa)	3·74 (1·4) (2·5)	0·58	5′40″
$\Delta^{1(6)}$-THC (II)	Not detected	0·57	7′10″
Δ^1-THC (I)	3·30 (1·4) (3·4)	0·51	7′52″
Cannabinol (Va)	1·30 (0·3) (1·2)	0·47	10′12″
Cannabichromene (VIIIa)	0·19	0·43	5′35″
Cannabigerol (VIa)	0·30	0·42	9′20″

[a] As percentage of hashish, determined by v.p.c. (the numbers in parentheses are from two partial analyses of different batches); [b] Chromatoplates of silica gel, elution with petroleum ether (b.p. 40–60°C) and ether in a ratio of 4:1; [c] column 2% OV-17 on Gas-Chromosorb Q, N_2 flow 30 cc/min, column temperature 235°C.

SYNTHESIS AND STRUCTURE ELUCIDATION

The chemical structures and absolute configurations of Δ^1- and $\Delta^{1(6)}$-THCs (I and II), cannabidiol (IVa) cannabigerol (VIa), cannabinol (Va) and their corresponding acids have been firmly established and their total syntheses achieved. This work is reviewed in great detail by Mechoulam and Gaoni[2]. Currently emphasis has been on the stereospecific synthesis of the isomers (I) and (II) although a synthesis of (II) involving optical resolution has been described[18].

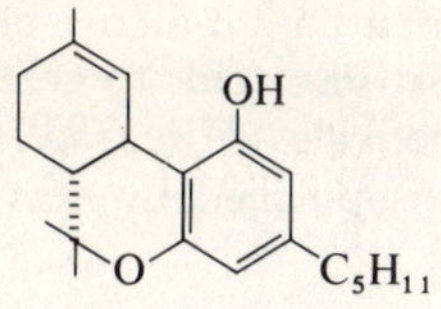

(I, Δ^1-THC)

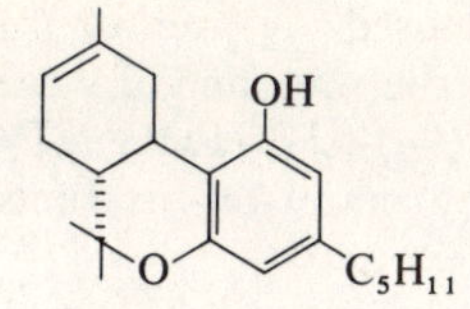

(II, $\Delta^{1(6)}$-THC)

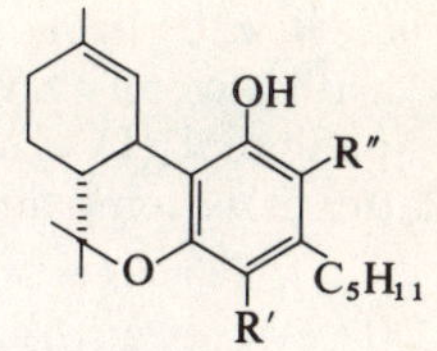

(IIIa, R' = H, R" = COOH, Δ^1-THC acid A
IIIb, R" = COOH, R' = H, Δ^1-THC acid B)

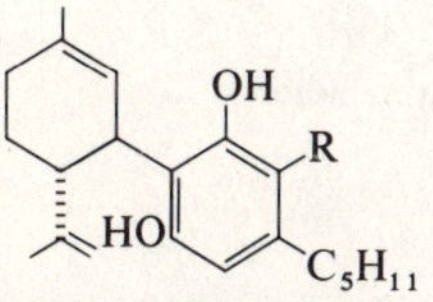

(IVa, R = H, cannabidiol
IVb, R = COOH, cannabidiolic acid)

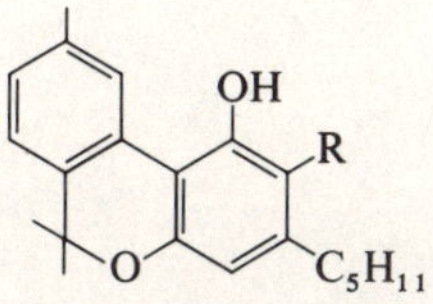

(Va, R = H, cannabinol
Vb, R = COOH, cannabinolic acid)

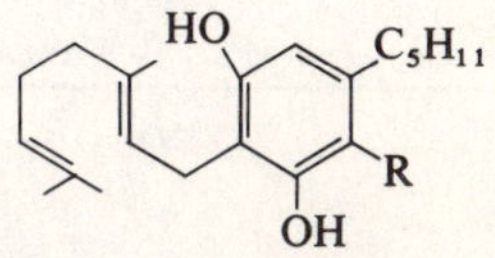

(VIa, R = H, cannabigerol
VIb, R = COOH, cannabigerolic acid)

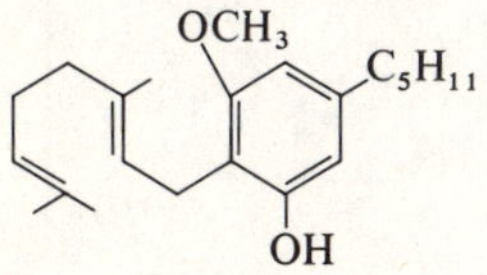

(VII, cannabigerol monomethyl ether)

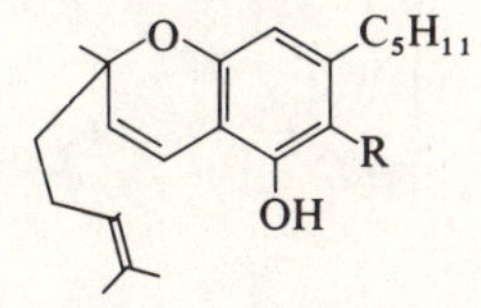

(VIIIa, R = H, cannabichromene
VIIIb, R = COOH, cannabichromenic acid)

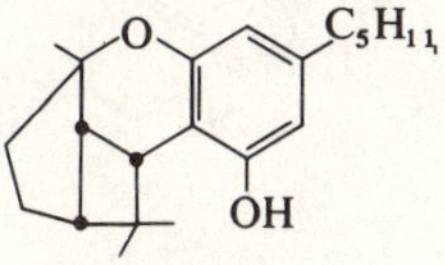

(IX, cannabicyclol)

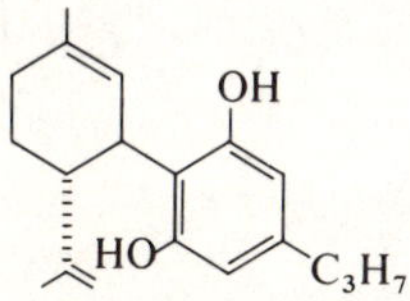

(X, cannabidivarin)

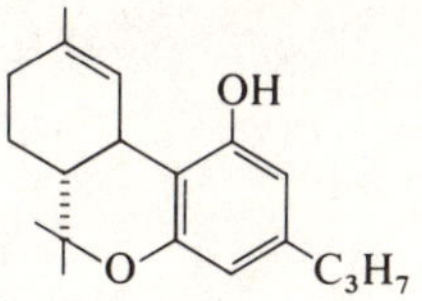

(XI)

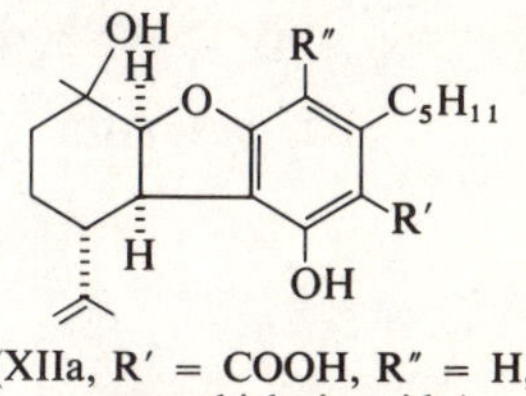

(XIIa, R' = COOH, R" = H, cannabielsoic acid A
XIIb, R' = H, R" = COOH, cannabielsoic acid B)

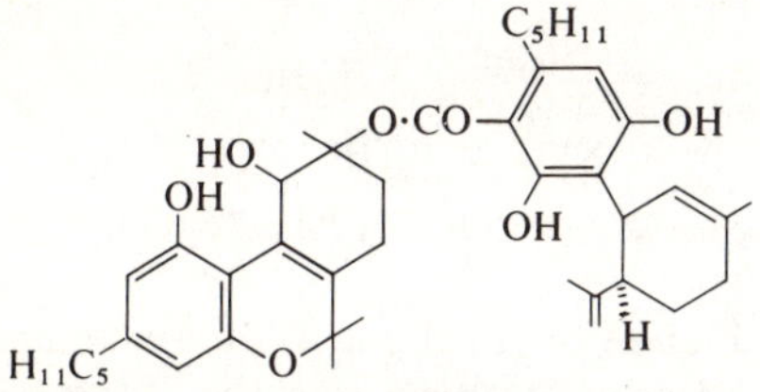

(XIII, cannabidiolic acid tetrahydrocannabitriol ester)

Figure 3.1

Petrzilka's Synthesis

Petrzilka et al.[19] have condensed (+)-cis or-trans-p-mentha-2,8-dien-1-ol (XIV) with olivetol (XV) in the presence of weak and strong acids to give (−)-cannabidiol (IVa) and (−)-$\Delta^{1(6)}$-THC (II) respectively (*Figure 3.2*). The latter is accompanied by many by-products and is purified by very

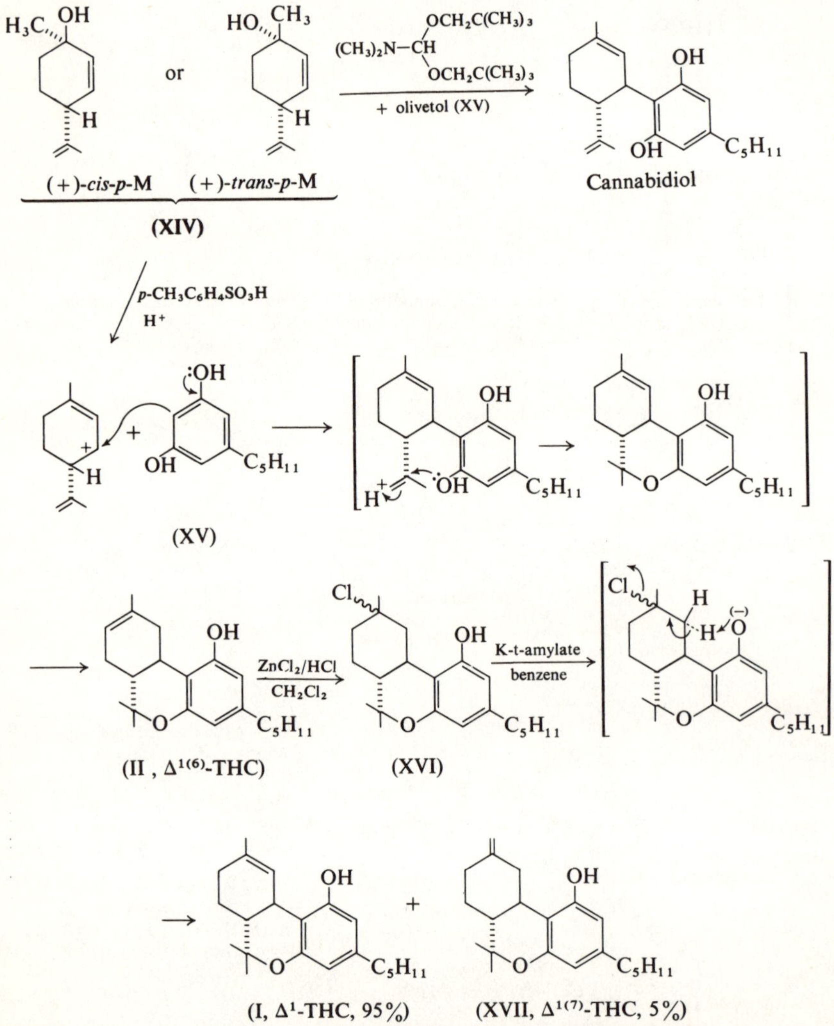

Figure 3.2. Synthesis of Δ^1-tetrahydrocannabinol (Δ^1-THC) by Petrzilka et al.

careful column chromatography on Florisil or silica gel. With zinc chloride and hydrogen chloride it is converted quantitatively into the chloro compound (XVI) which is said[19] to be transformed almost quantitatively into Δ^1-THC (I) by treatment with potassium-t-amylate in benzene. In our experience however this reaction must be carried out under very carefully controlled conditions and even under the best conditions a mixture of 95% (I) and 5% (−)-$\Delta^{1(7)}$-THC (XVII) is obtained as shown by gas chromatography. Furthermore (XVII) has been isolated from this mixture by chromatography on silver nitrate–silica gel. This constitutes the first total synthesis* of (−)-$\Delta^{1(7)}$-THC[20] although the synthesis of the racemic compound[21] has been reported. (−)-$\Delta^{1(7)}$-THC has not yet been found in C. Sativa.

Mechoulam's Synthesis

Mechoulam and his co-workers[22] used verbenol (XVIII) as their starting material, visualising stereochemical control of the reaction by the bulky dimethylmethylene bridge. $\Delta^{1(6)}$-THC thus obtained was converted into (XVI) which on treatment with sodium hydride in tetrahydrofuran gave a mixture of (I) and (II) (*Figure 3.3*) which was separated by careful chromatography.

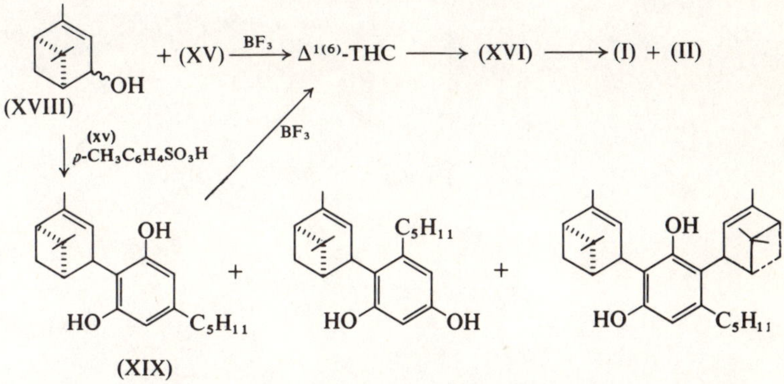

Figure 3.3. Synthesis of Δ^1-tetrahydrocannabinol (Δ^1-THC) by Mechoulam's group

Similarly, starting with (+)-verbenol, the unnatural (+)-$\Delta^{1(6)}$-THC was prepared[18,22] and found to be biologically inactive.

Razdan's Synthesis

Razdan and Handrick[23] have utilised (+)-*trans*-2-carene epoxide (XX) to provide the first one-step synthesis of (−)-Δ^1-THC. Under different conditions (II) can also be obtained, thus allowing an entry into cannabinoids via carane derivatives. The *cis* isomer (XXI) was separated from (I) by preparative gas chromatography. It is interesting to note that no cannabidiol (IVa) is formed in this reaction in contrast to Petrzilka's synthesis, suggesting a different mechanism (*Figure 3.4*).

Of the three syntheses described above, Petrzilka's process is at present being used for large-scale production of (I) because of the commercial availability of *p*-menthadienol (XIV). All three processes are obviously applicable to the preparation of various analogues of THC by using different resorcinols and such syntheses have actually been carried out by Petrzilka and his co-workers[19].

* Another total synthesis by a different route has just been reported[71].

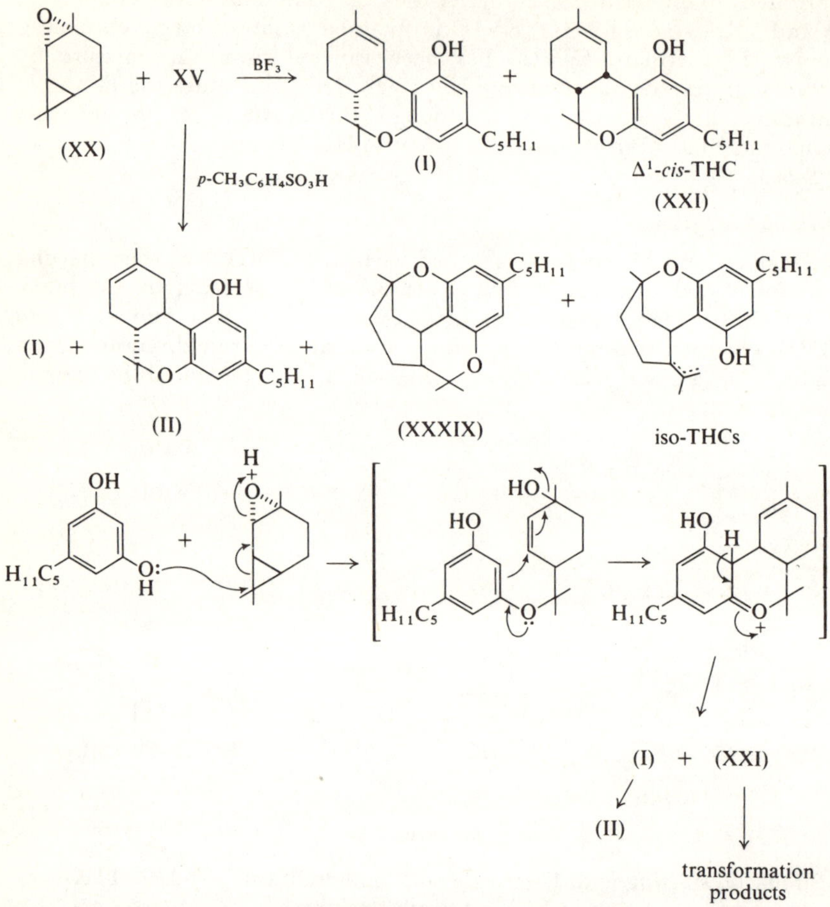

Figure 3.4. Synthesis of Δ^1-tetrahydrocannabinol (Δ^1-THC) by Razdan's group

Tetrahydrocannabinolcarboxylic acids
These acids, particularly those derived from Δ^1-tetrahydrocannabinol (IIIa and b), form an important group of compounds present in *C. Sativa*. Some of the discrepancies observed in the biological data from various sources are due to the presence of Δ^1-THC acids in the samples. This is because the inactive acids (IIIa) and (IIIb) are decarboxylated on heating (smoking) or on gas–liquid chromatography and are converted into the biologically active Δ^1-THC. The acid (IIIa) was isolated by two groups of workers[24,5]. It is now referred to as Δ^1-THC acid A (*Figure 3.1*) by Mechoulam *et al.*[25] as they have isolated the isomeric acid (IIIb) (Δ^1-THC acid B). Unlike (IIIa), the isomeric (IIIb) is crystalline, much more polar on thin-layer chromatography and shows a strong band at 1735 cm^{-1} (i.r.). The various interconversions of these acids are shown in *Figure 3.5*[26]. The $\Delta^{1(6)}$-THC acids A and B formed from (IIIb) are not known to occur naturally.

Taking advantage of the observation that the hydrogen atoms *ortho-* and

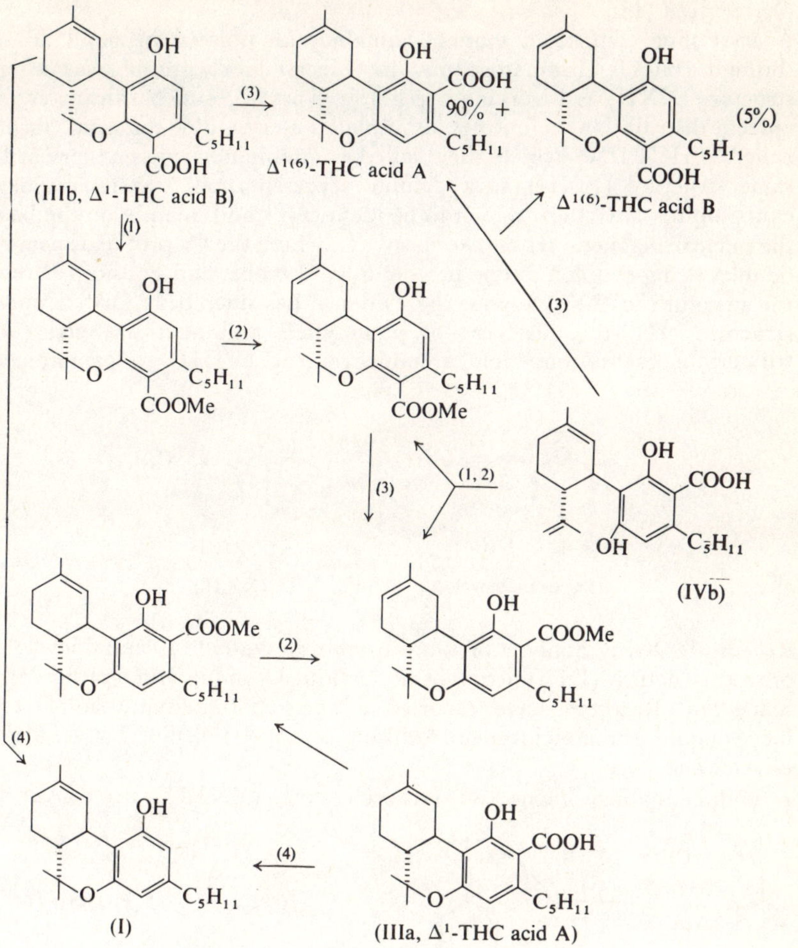

Figure 3.5. Interrelations of the THC acids

Reagents: (1) CH_2N_2; (2) toluene-p-sulphonic acid in benzene, 80°C; (3) BF_3 etherate, r.t. in CH_2Cl_2; (4) heat. (*From Mechoulam, R. et al., 'Some Aspects of Cannabinoid Chemistry'*[26], *by courtesy of J. & A. Churchill*)

para- to the hydroxyl groups in the resorcinol series are acidic[27], Mechoulam and Ben-Zvi[28] synthesised (IIIa) from (I) by treatment with methylmagnesium carbonate. Cannabidiolic acid (IVb) and cannabigerolic acid (VIb) were similarly synthesised from cannabidiol and cannabigerol respectively.

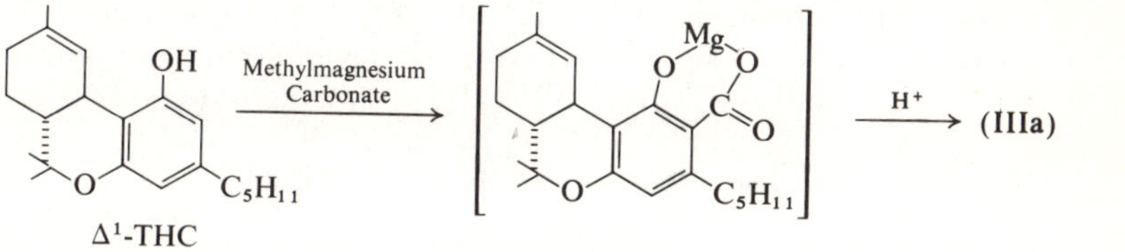

Cannabicyclol (IX)

A crystalline compound named cannabicyclol isolated in small amounts through column chromatography by Mechoulam's group[2] was assigned structure (XXII) as a working hypothesis. Korte's group[29], utilising counter-current distribution techniques also isolated a crystalline material (formerly called 'THC III')[30] which they called cannabipinol and put forward the same structure (XXII) as a definite representation. Cannabicyclol and cannabipinol have been shown to be identical[31a] and, mainly on the basis of the nuclear magnetic resonance spectrum, where the C_3 proton appears as a doublet being coupled to one proton only, Crombie and Ponsford[32a] revised the structure to (IX). Supporting evidence has since been forthcoming for structure (IX) on the basis of photolytic[33] and acid-catalysed (boron trifluoride) conversions[34] of cannabichromene (VIIIa) to cannabicyclol.

(IX, cannabicyclol)   (XXII)

Recently[35], X-ray analysis of the dibromo derivative of cannabicyclol has proved structure (IX) unequivocally*. Both Crombie and Ponsford[32] and Kane and Razdan[31] have reported a one-step total synthesis of cannabicyclol and cannabichromene utilising a novel reaction between citral, olivetol and pyridine.

Both cannabichromene and tetracyclic ether (XXXIX) are converted to

Citral $\xrightarrow[\text{pyridine}]{(XV)}$ (XXXIX) + Cannabichromene + (IX) + other products

iso-THCs (XL) and (XLI) on treatment with toluene-*p*-sulphonic acid. The reaction between substituted resorcinols or phloroglucinol and citral in pyridine to form tetracyclic ethers of type (XXXIX) is a general one which leads to substituted iso-THC derivatives[31b,32b]. The following appears to be the most likely mechanism for the pyridine reaction[32b]. The initial condensation product (XXIII) cyclises as shown to give cannabichromene which either forms cannabicyclol or proceeds via ion (XXIV) to the tetracyclic ether. The conversion of (VIIIa) to (IX) may be catalysed by metal ion and would involve changing (VIIIa) to the β-conformation as proposed[34] for the conversion of (VIIIa) to (IX) with boron trifluoride.

---

* The authors make certain comments regarding Kane and Razdan's paper (1968)[31a]; however, their comments placed the work out of context and ignored the historical background since most of the evidence appeared only after the publication of the paper[31a].

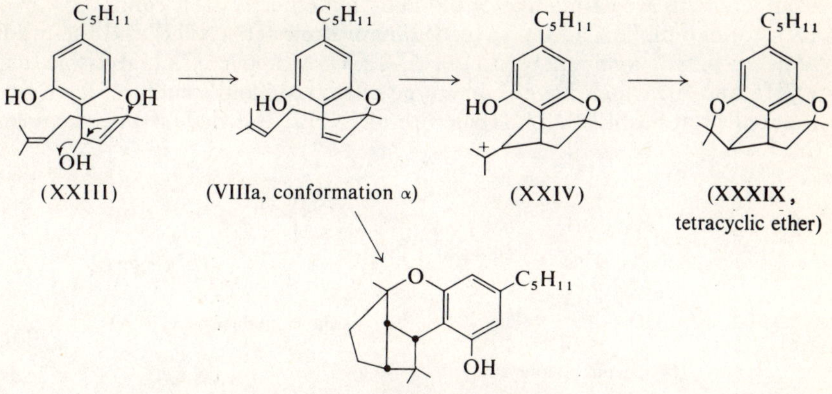

(XXIII) (VIIIa, conformation α) (XXIV) (XXXIX, tetracyclic ether)

(IX, cannabicyclol)

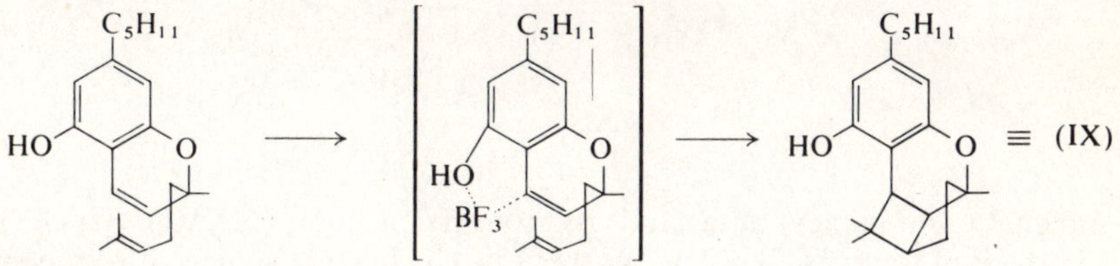

(VIIIa, conformation β)

Cannabichromene (VIIIa)

This is another minor constituent the structure of which has been firmly established[2,4]. It has been correlated with cannabigerol[36] (VIa) and cannabidiol[37] (IVa). In addition to its synthesis from citral, it has been synthesised by dehydrogenation of cannabigerol with chloranil[38] or dichlorodicyanobenzoquinone[39]. Both syntheses probably proceed via hydride abstraction, but the *o*-quinone methide intermediate (XXV) has also been proposed[39], which has interesting biogenetic implications. Contrary to earlier reports[29,40,41], cannabicyclol and cannabichromene do not show apparent optical rotation[4,37] when isolated as pure materials from the plant. The related cannabichromenic acid (VIIIb), originally reported[7] to be optically active, is probably also racemic[4]. This lack of optical activity in cannabichromene indicates that it is either an artifact or is being formed via the symmetrical intermediate (XXV). Since cannabicyclol has been formed from cannabichromene[33,34] its zero rotation is explicable, yet it remains to be established if it occurs naturally or is formed either on storage or during isolation.

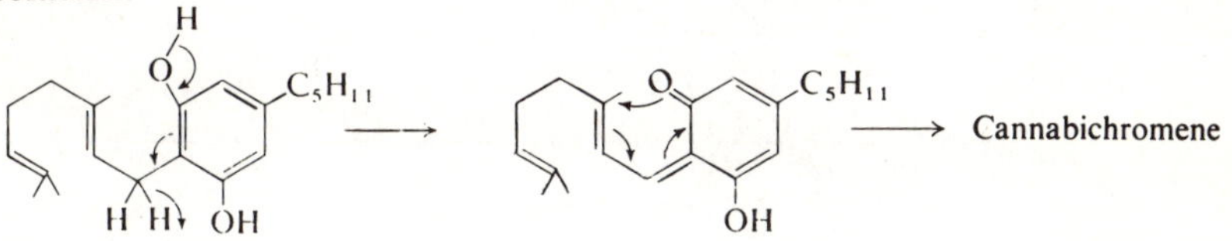

Cannabigerol (XXV) (VIIIa)

Other constituents which have been recently isolated and characterised from *C. Sativa* are cannabigerol monomethyl ether[8] (VII), cannabidivarin[10] (X), cannabidiolic acid tetrahydrocannabitriol ester[9] (XIII) and the mildly active n-propyl analogue[42] (XI) of Δ¹-THC. Moreover, cannabielsoic acids (XIIa and b), which have a new type of cannabinoid structure, have been isolated from hashish[43]. The structure of (XIIa) was deduced from nuclear

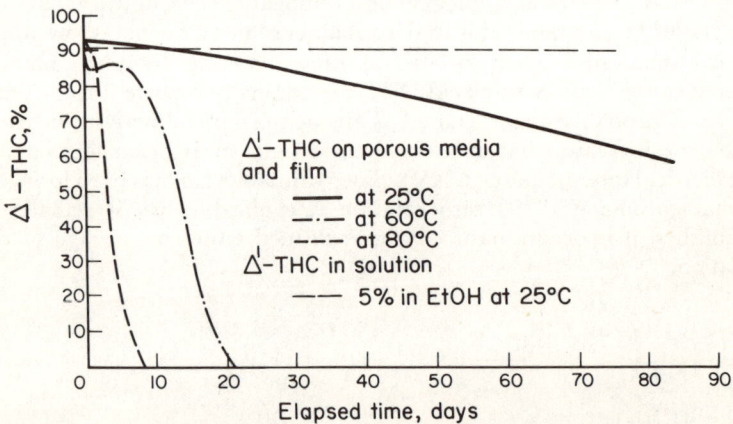

magnetic resonance data and its conversion into (XXVI) which was synthesised from (XIX). The synthesis of (XIIa) was achieved from cannabidiolic acid (IVb) by a novel photo-oxidative cyclisation process.

CHEMICAL REACTIONS

The cannabinoids being phenolic are susceptible to oxidation, for example, Δ¹-THC is a labile resinous substance which acquires a violet tinge after chromatography and goes dark on exposure to air. Similarly the active oily extract of the plant is reddish brown and darkens on prolonged storage. Stability studies of THCs (I and II)[44] under accelerated aging conditions

Figure 3.6. Effect of temperature on stability of Δ¹-tetrahydrocannabinol

have shown that Δ^1-THC (I) is mainly converted into cannabinol (Va) and is relatively much less stable than (II) particularly at elevated temperatures. A 1–5% solution of (I) in ethanol at 0°C was found to be most suitable for storage purposes (*Figure 3.6*).

The conversion of (I) to cannabinol was shown to be an oxidative process which could be envisioned as proceeding through the intermediate dienes (XXVIIa and b). This postulated mechanism has been confirmed by trapping the dienes with *N*-phenylmaleimide to give the adducts (XXVIIIa and b). The difference in the stability of (I) and (II) has been ascribed to

the benzylic-allylic proton in (I) (absent in II) which facilitates dehydrogenation to the dienes and hence to cannabinol[44].

The well-known Beam test for identification of cannabinoids is based on the oxidation of cannabidiol, cannabigerol and their acids[45]. Under the conditions of the test (O_2/alkali), cannabidiol (IVa) is oxidised to a mixture of the quinone (XXIX) and its dimers. Similarly (II) is oxidised to the quinone (XXX) with *m*-chloroperbenzoic acid. Recently Razdan and Kane[46] have reported the isolation of a peroxide from the pyridine-catalysed reaction of citral with olivetol discussed on page 86 and have proposed the novel dioxetane structure (XXXII), mainly on the basis of nuclear magnetic resonance data. The chemical shifts of the C-1 methyl singlet at 1·52 ppm and the C-6 proton at 4·95 ppm are in excellent agreement with those of the corresponding groups in the dioxetane[47] (XXXI). It shows a weak band at 850 cm^{-1} (i.r.) ascribed to O—O stretching which has also been reported in diethoxydioxetane[48]. On treatment of (XXXII) in the presence of hydrogen–10% palladium on charcoal in methanol, a rearranged product (XXXIII) was obtained probably by the mechanism shown. Further work is necessary to confirm structure (XXXII). Should this prove to be correct, it would suggest that alternative mechanistic pathways may be involved in the pyridine-catalysed reaction discussed earlier.

Various cannabinoids react differently to dehydrogenation with chloranil[38]. Thus, in contrast to Δ^1-THC which is converted into cannabinol (Va), Δ^1-cis-THC (XXI) and cannabidiol (IVa) remain unchanged under identical conditions. This difference is attributed[38] to the fact that in (I), the breaking pseudo-axial C_3—H bond will remain in constant overlap with the π-electrons of the aromatic ring and the double bond during abstraction of hydride thus lowering the energy of the transition state. On the other hand, in cannabidiol such overlap is possible only with the π-electrons of the double bond as the C_3—H bond is nearly parallel to the plane of the benzene ring. Similar factors are probably involved in the non-reactivity of (XXI).

Photochemical reactions in cannabinoids have had very limited value. The conversion of cannabichromene to cannabicyclol[33] and the synthesis of cannabielsoic acid[43] (XIIa) have already been mentioned. Other reactions[26] include a synthesis of iso-cannabichromene (XXXV) from (XXXIV) and addition of methanol across the double bond in cannabidiol (IVa) to give (XXXVI).

ISOMERISATIONS AND CYCLISATIONS

The presence of double bonds and free phenolic groups in cannabinoids makes them very labile. In recent years many interesting isomerisations and transformations have been observed (*Figure 3.7*). It is well known that Δ^1-*trans*-THC (I) is completely isomerised to $\Delta^{1(6)}$-*trans*-THC (II) when refluxed with toluene-*p*-sulphonic acid in benzene[2]. In contrast Δ^1-*cis*-THC (XXI) is converted into $\Delta^{4(8)}$-iso-THC (XLI) which was previously erroneously regarded as $\Delta^{1(6)}$-*cis*-THC (XXXVIII)[37,49]. Compound (XXXVIII) is as yet unknown and none of the *cis*- (except cannabicyclol IX) and iso-cannabinoids have been isolated from *C. Sativa*. Recently Razdan and Zitko[50] have shown that (XXI), (XXXIX) and iso-THCs (XL and XLI) are interconvertible. The equilibrium is however heavily in favour of iso-THCs, particularly (XLI). The key compound in these transformations seems to be the tetracyclic ether (XXXIX). Ring B of (XXXIX) cleaves preferentially[32b] to give (XL) which is mostly converted into (XLI), the more thermodynamically stable isomer. This is in complete agreement with Gaoni and Mechoulam's[37] observation that the acetate of (XXI) is unchanged by reaction with toluene-*p*-sulphonic acid, since the formation of (XXXIX) is blocked. This mechanism also explains why a transformation parallel to that observed in the *cis* series (to iso-THCs) does not take place in the *trans* series since the tetracyclic ether (XXXIX) with a *trans*-3a,4 ring junction cannot be formed.

An interesting interconversion of *cis*- into *trans*-THC (XXI → II) has been reported[50] in the presence of boron tribromide/methylene chloride. This must involve cleavage of the ether bond in (XXI) followed by inversion to

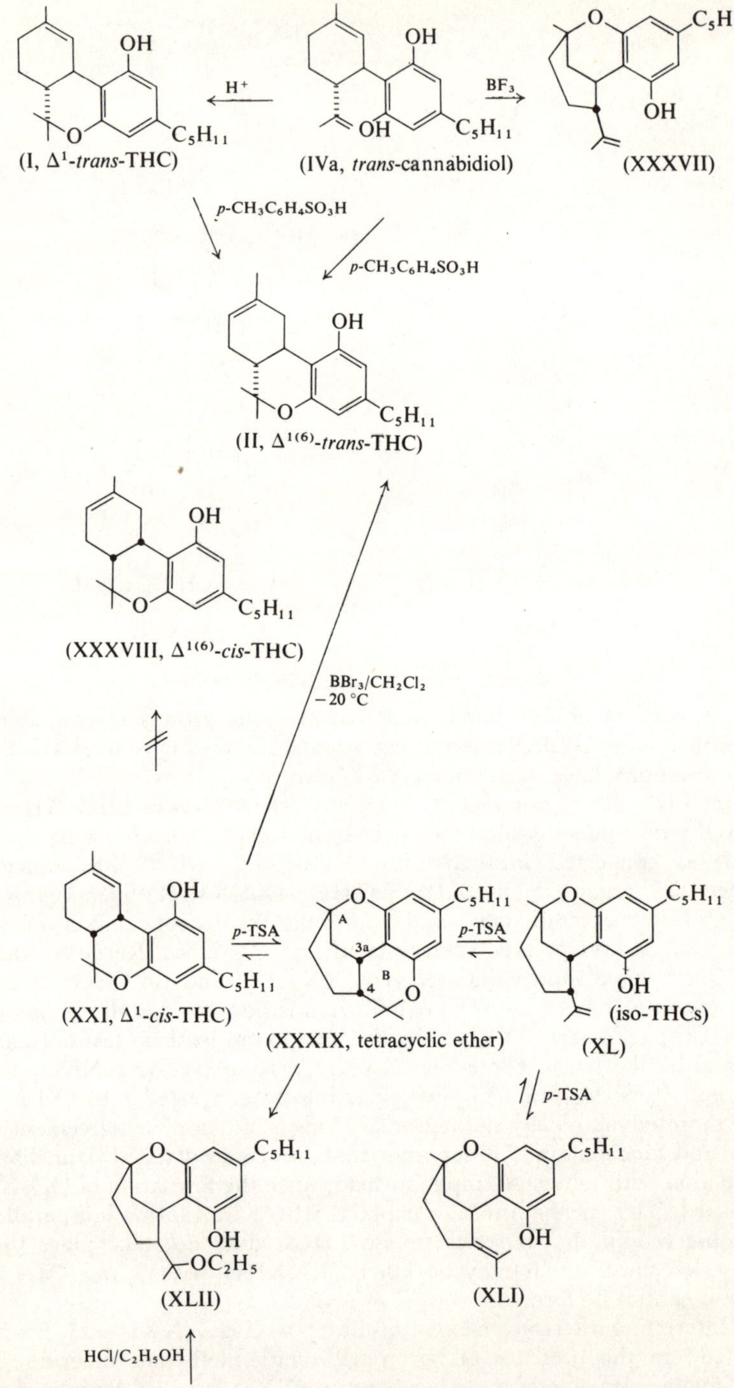

OH
C_5H_{11}
(I, Δ^1-*trans*-THC)
H⁺
OH
OH
C_5H_{11}
(IVa, *trans*-cannabidiol)
BF_3
O
C_5H_{11}
OH
(XXXVII)
p-$CH_3C_6H_4SO_3H$
p-$CH_3C_6H_4SO_3H$
OH
O
C_5H_{11}
(II, $\Delta^{1(6)}$-*trans*-THC)
OH
O
C_5H_{11}
(XXXVIII, $\Delta^{1(6)}$-*cis*-THC)
BBr_3/CH_2Cl_2
−20 °C
OH
O
C_5H_{11}
(XXI, Δ^1-*cis*-THC)
p-TSA
O
C_5H_{11}
A
3a
B
4
O
(XXXIX, tetracyclic ether)
p-TSA
O
C_5H_{11}
OH
(iso-THCs)
(XL)
p-TSA
O
C_5H_{11}
OH
OC_2H_5
(XLII)
HCl/C_2H_5OH
O
C_5H_{11}
OH
(XLI)

Figure 3.7

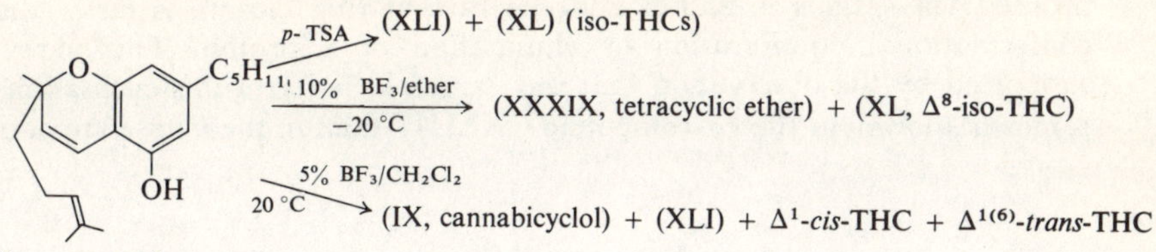

$$\text{(VIIIa, cannabichromene)} \xrightarrow{p\text{-TSA}} \text{(XLI)} + \text{(XL) (iso-THCs)}$$

$$\xrightarrow[-20\,°C]{10\%\ BF_3/\text{ether}} \text{(XXXIX, tetracyclic ether)} + \text{(XL, }\Delta^8\text{-iso-THC)}$$

$$\xrightarrow[20\,°C]{5\%\ BF_3/CH_2Cl_2} \text{(IX, cannabicyclol)} + \text{(XLI)} + \Delta^1\text{-}cis\text{-THC} + \Delta^{1(6)}\text{-}trans\text{-THC}$$

Figure 3.7 (continued). Some transformations in cannabinoids

the more thermodynamically stable *trans* ring junction and ring closure. Cannabidiol* (IVa) gives a number of products[2] on treatment with dilute HCl/C_2H_5OH including Δ^1-THC (I). However, on treatment with boron trifluoride the Δ^8-*trans*-iso-THC (XXXVII) is formed[37].

The tetracyclic ether (XXXIX) gives the cleavage product[32b] (XLII) with ethanolic hydrogen chloride. The same compound is obtained from cannabichromene as well. However, cannabichromene gives different products with toluene-*p*-sulphonic acid[31a,38] and boron trifluoride[34,32b] (*Figure 3.7*). As a general guide it has been proposed[50] that in cannabinoids where one of the oxygen atoms forms part of a pyran ring, acid catalysts like toluene-*p*-sulphonic acid, which can protonate a double bond, effect transformations with retention of stereochemistry at the C_3—C_4 ring junction. However, non-protonic acid catalysts like boron tribromide or boron trifluoride, effect interconversions with inversion at the C_3—C_4 ring junction to more thermodynamically stable forms.

Since a *trans* C_3—C_4 ring junction is thermodynamically more stable, the conversion of *trans*-cannabidiol (IVa) to (II) and Δ^8-*trans*-iso-THC (XXXVII) becomes understandable. Similarly, the products obtained from cannabichromene can be easily explained. The geometry is fixed[32b] because of the chromene ring in (VIIIa) and only iso-THCs (XLI) and (XL) are formed with toluene-*p*-sulphonic acid. With boron trifluoride cleavage of the chromene is a competitive reaction and thus some *cis*- and *trans*-THC are formed. Under the conditions where cleavage of the chromene ring is suppressed (low temperatures) only transformation products with unchanged stereochemistry at the C_3—C_4 ring junction (i.e. XXXIX and XL) were observed.

In addition to the transformations which have been reported, the cannabinoid 1,5-dienes undergo highly stereoselective cyclisations[36]. Reaction of geraniol with olivetol gives cannabigerol (VIa) whereas nerol gives *cis*-cannabigerol (XLIII). Cyclisation of (VIa) with 100% sulphuric acid gives 88% of the *trans* tricyclic compound (XLIV) and 3% of the *cis* tricyclic compound (XLV). Under the same conditions (XLIII) gave 71% (XLV) and 8·8% (XLIV). These results are in accordance with the Stork–Eschenmoser hypothesis that the stereochemical course of cyclisations of acyclic polyenes induced by acid is dictated by the configuration of the central double bond[51]. According to Mechoulam and Yagen[36] these cyclisations

---

* As far as the writer is aware *cis*-cannabidiol is not known; earlier reports are erroneous. *See* footnote (Ref. 4), p. 1104, Ref. 19. Similarly discussions regarding *cis*-cannabidiol in Ref. 50 should be ignored.

proceed via cations A and B and the rate of ring closure is faster than conformational equilibration or elimination of a proton. This view is supported by the observation that the stereoselectivity of this cyclisation is somewhat lower in the *cis*-compound (XLIII) than in the *trans*-compound.

In the case of (XLIII) rearrangement of B (containing a bulky pseudo-axial group) to A becomes competitive with ring closure to (XLV). In view of the high stereoselectivity of these cyclisations these authors[36] have also suggested that these reactions may represent a chemical model for related biochemical cyclisations.

Other reactions in cannabinoids, e.g. cleavage reactions (reverse Friedel–Crafts), aromatisation and aromatic substitution have been discussed by Mechoulam and Gaoni[2] in their earlier review and no new examples of

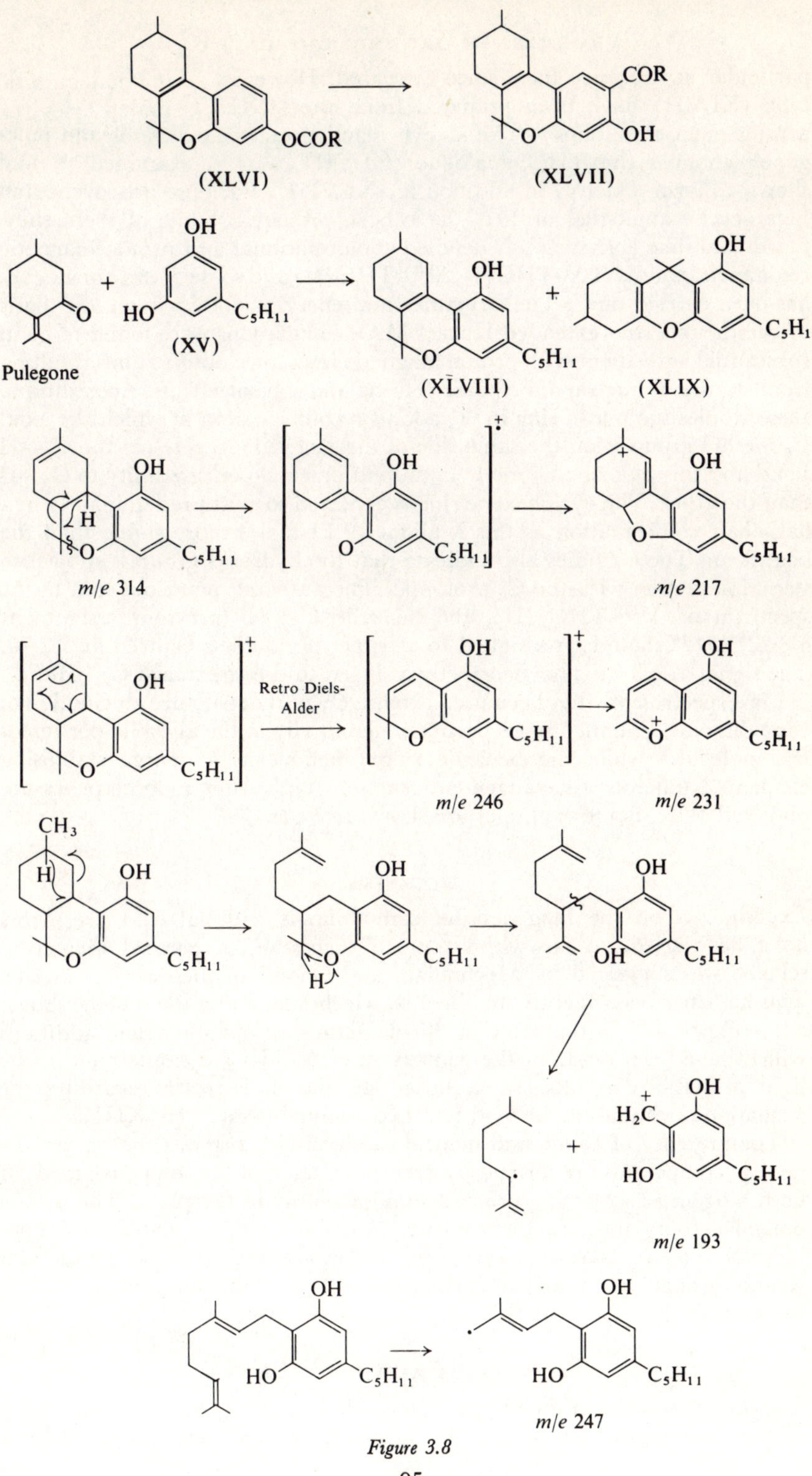

Figure 3.8

particular significance have since appeared. However, THC analogues of type (XLVII) have been prepared from esters (XLVI) under Fries rearrangement conditions[52]. The known reaction between olivetol and pulegone[53] to give the THC analogue (XLVIII) was re-examined[54,55] and shown to form (XLIX) in addition to (XLVIII). Various carbocyclic and heterocyclic analogues of THC have been prepared. Some of them show marihuana-like activity[3,56]. A detailed conformational and proton magnetic resonance analysis of Δ^1-THC (I), $\Delta^{1(6)}$-THC (II) and some of their analogues has been carried out[57]. The structures and energies obtained on the basis of Westheimer and extended Huckel M.O. calculations were found to be in substantial agreement with proton magnetic resonance observations resulting from studies of nuclear Overhauser effects and solvent effects. According to these studies the pyran ring in (I) adopts a conformation in which the axial C_8 methyl group is on the same side of the molecule (α side) as the C_3—H bond and this puts one C_8 methyl group in much closer proximity to C_3—H than the other. The cyclohexene ring is expected to exist predominantly in a half-chair conformation as this is about 19 kJ mol^{-1} more stable than the boat form. These studies also indicate that the hydroxyl group is subject to steric interaction with the C_2 proton(s). The authors[57] point out that in the spectrum of $\Delta^{1(6)}$-THC (II), the benzylic C_3—H (previous assignment δ 3·22)[58,18,19] should be assigned to a triplet of doublets centred at δ 2·70. The signal at δ 3·22 is now correctly assigned to the equatorial C_2—Hα.

Mass spectroscopy has been used extensively in the structure elucidation of cannabinoids. Fragmentation occurs preferentially in the alicyclic portion of the molecules while the benzene ring functions as a charge stabilising element. Numerous rearrangement reactions typical for monoterpenes are observed[59,60,61]. Some examples are shown in *Figure 3.8*.

BIOGENESIS

Experiments on the biogenesis of cannabinoids with labelled precursors have been tried, so far without much success[4,62]. A general biogenetic scheme was suggested by Mechoulam and Gaoni[2] in their review article. This has since been slightly modified by Mechoulam[3] and the scheme shown here (*Figure 3.9*) is the same as Mechoulam's except for a few additions which have been made in the conversion of Δ^1-THC to cannabinol in the light of stability results. In addition speculation is made regarding the formation of cannabidiolic acid tetrahydrocannabitriol ester (XIII).

The presence of acidic and neutral cannabinoids suggests that either the neutral compounds are formed by decarboxylation of the respective acids or both biogenetic sequences proceed simultaneously in the plant. The lack of optical activity in cannabichromene, cannabichromenic acid and cannabicyclol suggests that either the cyclisation leading to these compounds passes through a symmetric intermediate A or a nonenzymic process is involved in their formation[3].

METABOLISM

The intense interest in the pharmacology of (I) and (II) has led to the study of their metabolites. Tritiated (I) (aromatic protons) was prepared by a

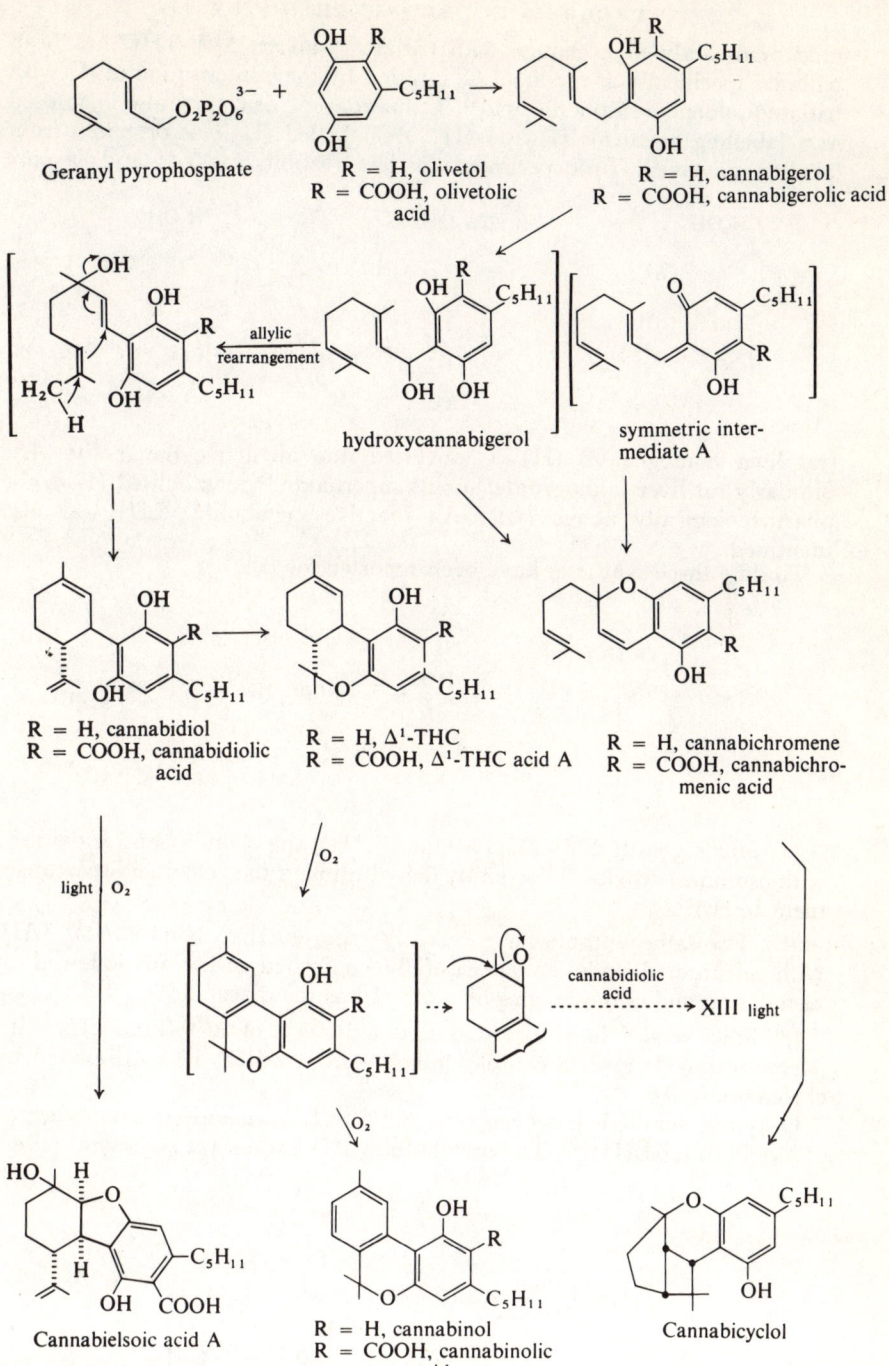

Figure 3.9. Proposed biogenesis of cannabinoids

mild acid-catalysed exchange with tritiated water[63]. $\Delta^{1(6)}$-THC has been tritiated specifically at the axial C_2 position by the isomerisation of (I) with tritiated toluene-p-sulphonic acid[64]. Trifluoroacetic acid[65] has also been used as a labelling agent for (I) and (II). ^{14}C-labelled (II) was prepared from labelled olivetol[63]. In experiments *in vivo* (rabbit, rat)[66,67] and *in vitro*

(L)   (LI)   (LII)

(rat liver homogenate) (II) is converted into an active metabolite (L). Similarly rat liver homogenate[68] or its supernatant[69] metabolised (I) to the pharmacologically active (LI). An inactive metabolite (LII) was also identified.

To date three syntheses have been reported for (L).

(LIII) → → → (L)

(a) Starting with $\Delta^{1(6)}$-THC acetate (LIII), the olefinic bond is oxidised with osmium tetroxide followed by dehydration of the glycol and rearrangement to (L)[70].

(b) The same authors have recently reported that oxidation of (LIII) with selenium dioxide under carefully controlled conditions followed by acetylation and chromatography gives (L) as the diacetate[66b].

(c) Foltz et al.[67] have reported that oxidation of $\Delta^{1(6)}$-THC (II) with selenium dioxide gives a complex mixture from which (L) was isolated by chromatography.

Only one synthesis has been reported for (LI) whereby it is obtained in 1% yield from (LIII)[66b]. The metabolite (LII) has not yet been synthesised.

(LIII) $\xrightarrow{(1) SeO_2}{(2) LiAlH_4}$ (LI) +

The pharmacology, structure–activity relationships and analytical aspects of cannabinoids have not been discussed being outside the scope of the present article. The pharmacology is unique and rapid progress is being made in studying these compounds in both animals and humans.

Manuscript received April 1971.

ADDENDUM

Significant progress has been made in the study of the metabolism of THCs. Five new metabolites of Δ^1- and $\Delta^{1(6)}$-THCs have been identified[72-75].

Two new syntheses of $\Delta^{1(6)}$-THC metabolite (L) from $\Delta^{1(7)}$-THC (XVII) have been reported. In one[76] osmium tetroxide was used to oxidise the exocyclic double bond as a key step.

Reagents: (1) Ac_2O/Pyridine; (2) OsO_4; (3) toluene-p-sulphonic acid; (4) Lithium aluminium hydride.

Weinhardt, Razdan and Dalzell[77] transformed $\Delta^{1(7)}$-THC acetate to 7-bromo-$\Delta^{1(6)}$-THC acetate in nearly quantitative yield with N-bromoacetamide-perchloric acid. Treatment of the product with silver acetate and hydrolysis with alkali gave the metabolite (L).

Recent studies by Merkus[78] have shown that substantial amounts of cannabidivarin and tetrahydrocannabivarin (the n-propyl analogues of cannabidiol and Δ^1-THC respectively; *see Figure 1*) are present in samples of Nepal hashish. In addition a new constituent, cannabivarin, which corresponds to the n-propyl analogue of cannabinol was isolated from these samples.

The presence of cannabidiol, Δ^1-THC and cannabinol on the one hand and their n-propyl analogues on the other, in the same samples of hashish, suggests that products are formed by two parallel biosynthetic pathways.

A detailed n.m.r. analysis of cannabicyclol has appeared[80].

REFERENCES

1. (a) SCHULTES, R. E., in *The Botany and Chemistry of Cannabis*, eds, Joyce, C. R. B. and Curry, S. H., J. & A. Churchill, London, 11 (1970); (b) KREJCI, Z., ibid., p. 49
2. MECHOULAM, R. and GAONI, Y., *Fortschr. Chem. org. NatStoffe.*, **25**, 175 (1967) and references cited therein
3. MECHOULAM, R., *Science*, **168**, 1159 (1970)
4. GAONI, Y. and MECHOULAM, R., *J. Am. chem. Soc.*, **93**, 217 (1971)
5. YAMAUCHI, T., SHOYAMA, Y., ARAMAKI, H., AZUMA, T. and NISHIOKA, I., *Chem. Pharm. Bull.*, **15**, 1075 (1967)
6. ARAMAKI, H., TOMIYASU, N., YOSHIMURA, H. and TSUKAMOTO, H., *Chem. Pharm. Bull.*, **16**, 822 (1968)
7. SHOYAMA, Y., FUJITA, T., YAMAUCHI, T. and NISHIOKA, I., *Chem. Pharm. Bull.*, **16**, 1157 (1968)
8. YAMAUCHI, T., SHOYAMA, Y., MATSUO, Y. and NISHIOKA, I., *Chem. Pharm. Bull.*, **16**, 1164 (1968)
9. VON SPULAK, F., CLAUSSEN, U., FEHLHABER, H. W. and KORTE, F., *Tetrahedron*, **24**, 5379 (1968)
10. VOLLNER, L., BIENIEK, D. and KORTE, F., *Tetrahedron Lett.*, 145 (1969)
11. TURK, R. F., FORNEY, R. B., KING, L. J. and RAMACHANDRAN, S., *J. Forensic Sci.*, **14**, 385 (1969)

12. JACOB, A. and TODD, A. R., *Nature*, **145**, 350 (1940)
13. COVELLO, M., *Il Farmaco, Sci. Tec.*, **3**, 8 (1948)
14. NIGAM, M. C., HANDA, K. L., NIGAM, I. C. and LEVI, L., *Can. J. Chem.*, **43**, 3372 (1965)
15. SALEMINK, C. A., VEEN, E. and DEKLOET, W. A., *Planta Med.*, **13**, 211 (1965)
16. HEGNAUER, R., *Chemotaxonomie der Pflanzen*, **3**, s. 350 (1964)
17. OBATA, Y. and ISHIKAWA, Y., *Agric. Biol. Chem. Japan*, **30**, 619 (1966)
18. JEN, T. Y., HUGHES, G. A. and SMITH, H., *J. Am. chem. Soc.*, **89**, 4551 (1967)
19. PETRZILKA, T., HAEFLIGER, W. and SIKEMEIER, C., *Helv. chim. Acta*, **52**, 1102 (1969)
20. RAZDAN, R. K., HANDRICK, G. R. and ZITKO, B. A., *Experientia*, **28**, 121 (1972)
21. FAHRENHOLTZ, K. E., LURIE, M. and KIERSTEAD, R. W., *J. Am. chem. Soc.*, **89**, 5934 (1967)
22. MECHOULAM, R., BRAUN, P. and GAONI, Y., *J. Am. chem. Soc.*, **89**, 4552 (1967)
23. RAZDAN, R. K. and HANDRICK, G. R., ibid., **92**, 6061 (1970)
24. KORTE, F., HAAG, M. and CLAUSSEN, U., *Angew. Chem. Int. Edn*, **4**, 872 (1965)
25. MECHOULAM, R., BEN-ZVI, Z., YAGNITINSKY, B. and SHANI, A., *Tetrahedron Lett.*, 2239 (1969)
26. MECHOULAM, R., SHANI, A., YAGNITINSKY, B., BEN-ZVI, Z., BRAUN, P. and GAONI, Y., in *The Botany and Chemistry of Cannabis*, eds, Joyce, C. R. B. and Curry, S. H., J. & A. Churchill, London, 93 (1970)
27. HAND, E. S. and HOROWITZ, R. M., *J. Am. chem. Soc.*, **86**, 2084 (1966)
28. MECHOULAM, R. and BEN-ZVI, Z., *Chem. Commun.*, 343 (1969)
29. CLAUSSEN, U., VON SPULAK, F. and KORTE, F., *Tetrahedron*, **24**, 1021 (1968)
30. KORTE, F. and SIEPER, H., *J. Chromat.*, **13**, 90 (1964)
31. (a) KANE, V. V. and RAZDAN, R. K., *J. Am. chem. Soc.*, **90**, 6551 (1968); (b) ibid., *Tetrahedron Lett.*, 591 (1969)
32. (a) CROMBIE, L. and PONSFORD, R., *Chem. Commun.*, 894 (1968); (b) idem, *Tetrahedron Lett.*, 4557 (1968); idem, *J. chem. Soc. C*, 796 (1971)
33. CROMBIE, L., PONSFORD, R., SHANI, A., YAGNITINSKY, B. and MECHOULAM, R., *Tetrahedron Lett.*, 5771 (1968)
34. YAGEN, B. and MECHOULAM, R., ibid., 5353 (1969)
35. BEGLEY, M. J., CLARKE, D. G., CROMBIE, L. and WHITING, D. A., *Chem. Commun.*, 1547 (1970)
36. MECHOULAM, R. and YAGEN, B., *Tetrahedron Lett.*, 5349 (1969)
37. GAONI, Y. and MECHOULAM, R., *Israel J. Chem.*, **6**, 679 (1968)
38. MECHOULAM, R., YAGNITINSKY, B. and GAONI, Y., *J. Am. chem. Soc.*, **90**, 2418 (1968)
39. CARDILLO, G., CRICCHIO, R. and MERLINI, L., *Tetrahedron*, **24**, 4825 (1968)
40. CLAUSSEN, U., VON SPULAK, F. and KORTE, F., ibid., **22**, 1477 (1966)
41. GAONI, Y. and MECHOULAM, R., *Chem. Commun.*, 20 (1966)
42. GILL, E. W., PATON, W. D. M. and PERTWEE, R. G., *Nature*, **228**, 134 (1970)
43. SHANI, A. and MECHOULAM, R., *Chem. Commun.*, 273 (1970)
44. RAZDAN, R. K., PUTTICK, A. J., ZITKO, B. A. and HANDRICK, G. R., *Experientia*, **28**, 121 (1972)
45. MECHOULAM, R., BEN-ZVI, Z. and GAONI, Y., *Tetrahedron*, **24**, 5615 (1968)
46. RAZDAN, R. K. and KANE, V. V., *J. Am. chem. Soc.*, **91**, 5190 (1969)
47. KOPECKY, K. R. and MUMFORD, C., 51st Annual Conference of the Chemical Institute of Canada, Vancouver, B.C., June 1968, Abstracts, p. 41; *Can. J. Chem.*, **47**, 709 (1969)
48. SCHAAP, A. P. and BARTLETT, P. D., *J. Am. chem. Soc.*, **92**, 3223 (1970)
49. GAONI, Y. and MECHOULAM, R., ibid., **88**, 5673 (1966)
50. RAZDAN, R. K. and ZITKO, B. A., *Tetrahedron Lett.*, 4947 (1969); see also ref. 34
51. JOHNSON, W. S., *Accts. Chem. Res.*, **1**, 1 (1968)
52. TAYLOR, E. C., LENARD, K. and LOEV, B., *Tetrahedron*, **23**, 77 (1967)
53. GOSH, R., TODD, A. R. and WRIGHT, D. C., *J. chem. Soc.*, 137 (1941)
54. CLAUSSEN, U., MUMMENHOFF, P. and KORTE, F., *Tetrahedron*, **24**, 2897 (1968)
55. CHAZAN, J. and OURISSON, G., *Bull. Soc. chim. Fr.*, 1374 (1968)
56. RAZDAN, R. K. and PARS, H. G., in *The Botany and Chemistry of Cannabis*, eds, Joyce, C. R. B. and Curry, S. H., J. & A. Churchill, London, 137 (1970)
57. ARCHER, R. A., BOYD, D. B., DEMARCO, P. V., TYMINSKI, I. J. and ALLINGER, N. L., *J. Am. chem. Soc.*, **92**, 5200 (1970)
58. GAONI, Y. and MECHOULAM, R., *Tetrahedron*, **22**, 1481 (1966)
59. BUDZIKIEWICZ, H., ALPIN, R. T., LIGHTNER, D. A., DJERASSI, C., MECHOULAM, R. and GAONI, Y., ibid., **21**, 1881 (1965)
60. CLAUSSEN, U. and KORTE, F., *Tetrahedron*, Suppl. No. 7, 87 (1966)
61. CLAUSSEN, U., FEHLHABER, H. W. and KORTE, F., *Tetrahedron*, **22**, 3535 (1966)

62. AGURELL, S. ref. 1, p. 57
63. LARS, J., NILSSON, G., NILSSON, I. M. and AGURELL, S., *Acta. chem. scand.*, **23**, 2209 (1969)
64. BURNSTEIN, S. H. and MECHOULAM, R., *J. Am. chem. Soc.*, **90**, 2420 (1968)
65. TIMMONS, M. L., PITT, C. G. and WALL, M. E., *Tetrahedron Lett.*, 3129 (1969)
66. (a) BURSTEIN, S. H., MENEZES, F., WILLIAMSON, E. and MECHOULAM, R., *Nature*, **225**, 87 (1970); (b) BEN-ZVI, Z., MECHOULAM, R. and BURSTEIN, S. H., *Tetrahedron Lett.*, 4495 (1970)
67. FOLTZ, R. L., FENTIONAN, A. F., LEIGHTY, E. G., WALTER, J. L., DREWES, H. R., SCHWARTZ, W. E., PAGE, T. F. and TRUITT, E. B., *Science*, **168**, 844 (1970)
68. WALL, M. E., BRINE, D. R., BRINE, G. A., PITT, C. G., FRUDENTHAL, R. I. and CHRISTENSEN, H. D., *J. Am. chem. Soc.*, **92**, 3466 (1970)
69. AGURELL, S., NILSSON, J. L. G., OHLSSON, A., SANDBERG, F. and WAHLQUIST, M., *Science*, **168**, 1228 (1970)
70. BEN-ZVI, Z., MECHOULAM, R. and BURSTEIN, S. H., *J. Am. chem. Soc.*, **92**, 3468 (1970)
71. WILDES, J. W., MARTIN, N. H., PITT, C. G. and WALL, M. E., *J. org. Chem.*, **36**, 721 (1971)
72. BEN-ZVI, Z., MECHOULAM, R., EDERY, H. and PORATH, G., *Science*, **174**, 951 (1971)
73. WALL, M. E., *Ann. N.Y. Acad. Sci.*, **191**, 23 (1971)
74. CHRISTENSEN, H. D., FREUDENTHAL, R. I., GIDLEY, J. T., ROSENFELD, R., BOEGLI, G., TESTINO, L., BRINE, D. R., PITT, C. G. and WALL, M. E., *Science*, **172**, 165 (1971)
75. MAYNARD, D. E., GURNY, O., PITCHER, R. G. and KIERSTEAD, R. W., *Experientia*, **27**, 1154 (1971)
76. NILSSON, J. L. G., NILSSON, I. M., AGURELL, S., AKERMARK, B. and LAGERLUND, I., *Acta Chem. Scand.*, **25**, 768 (1971)
77. WEINHARDT, K. K., RAZDAN, R. K. and DALZELL, H. C., *Tetrahedron Lett.*, 4827 (1971)
78. MERKUS, F. W. H. M., *Nature*, **232**, 579 (1971)
79. NEUMEYER, J. L. and SHAGOURY, R. A., *J. Pharm. Sci.*, **60**, 1433 (1971)
80. KANE, V. W., *Tetrahedron Lett.*, 4101 (1971)

4
RECENT PENICILLIN CHEMISTRY
R. J. Stoodley

INTRODUCTION	102
Nomenclature	102
General	103
MODE OF ACTION	104
CONFORMATION	104
STRUCTURE–ACTIVITY RELATIONSHIPS	105
PENICILLIN→CEPHALOSPORIN TRANSFORMATIONS	108
PENICILLIN SULPHOXIDES	109
SYNTHESES OF β-LACTAM-ANTIBIOTIC ANALOGUES	111
DERIVATIVES OF POTENTIAL VALUE IN THE CONSTRUCTION OF β-LACTAM-ANTIBIOTIC ANALOGUES	114
Sulphenic Acids	114
Anhydropenicillins	114
Cephams	115
Secopenicillins	116
EPIMERISATIONS	116
Thermodynamic Considerations	117
Mechanism	118
DEAMINATIONS	123
REARRANGEMENTS INVOLVING CLEAVAGE OF THE β-LACTAM RING	124
1,2-Bond Cleavages	124
1,5-Bond Cleavages	125
4,7-Bond Cleavages	125

INTRODUCTION

PENICILLINS, historically the first antibiotics, have played a vital role in man's quest to combat bacterial diseases. In spite of a gargantuan investment of effort which has spanned three decades, organic chemists continue to be fascinated and often bewildered by these molecules. Their chemistry has been reviewed on several occasions[1-6]; this chapter attempts to survey and assess the current achievements and objectives of penicillin research.

Nomenclature

Penicillins are a family of compounds possessing the general structure (I). In naming such substances the substituent, R, is employed as a prefix; thus, when R = PhCH$_2$ the derivative is known as benzylpenicillinic acid or benzylpenicillin.

Penicillins contain three asymmetric centres and can, therefore, exist in eight stereoisomeric forms. X-ray crystallographic studies[7] have established that natural penicillins possess the *S*-configuration at position 3 and the *R*-configuration at position 5 and 6 as shown in (II).

The terms 'penam' and 'penicillanic acid' denote the derivatives (III) and (IV), respectively. It is often convenient to designate the stereochemistry of a substituent at position 6 of (IV) by the α,β-notation; it is then inferred that (IV) possesses the 3*S*, 5*R*-configurations characteristic of natural penicillins. Thus, the 'penicillin nucleus' (V), which can be obtained from penicillins by selective chemical or enzymatic cleavage of the amido side-chain, is known as 6β-aminopenicillanic acid.

Cephalosporins (VI) differ from penicillins in that they contain the dihydro-1,3-thiazine ring in place of the thiazolidine ring. The bicyclic unit (VII) is called 'cepham' and the 'cephalosporin nucleus' (VIII) is named 7β-aminocephalosporanic acid.

General

The development of the chemistry of β-lactam antibiotics is a fascinating and exciting chapter in organic chemistry. Four important phases characterise the story up to the present time.

The first phase culminated in the structural determination of the penicillins and was the result of a monumental collaborative effort involving British and American scientists[2] during World War II.

The total synthesis of penicillins, accomplished by Sheehan and his collaborators[8] in an extensive and adept investigation, and the elaboration of methods[3] for their partial degradation to (V) are included in the second phase of the story. The latter achievement was particularly significant since a vast array of penicillins, which were unavailable by fermentation methods, could be assembled by chemical acylation of (V). Some of these semi-synthetic penicillins have improved pharmacological properties over natural penicillins (e.g. increased stability to acid and improved resistance to penicillinases) and they have acquired a prominent position in modern medicine[3-5].

The unfolding of the chemistry of cephalosporins, which has been reviewed[9-11] on numerous occasions, represents the third phase of the story. Cephalosporin C (VI; R = $(CH_2)_3CHNH_3^+CO_2^-$), although possessing only a relatively low order of antimicrobial activity, showed two promising properties. Firstly, it was effective against organisms which had developed resistance to penicillins and, secondly, its activity extended to many Gram-negative bacteria against which penicillins were ineffective. The elaboration of methods for the production of (VIII) and its subsequent acylation to give semi-synthetic cephalosporins was an inevitable consequence of the success of this approach in penicillin chemistry. This has led to the marketing of some exciting new antibiotics.

The antimicrobial properties of cephalosporins revealed that the thiazolidine ring of penicillins was not obligatory for biological function; in consequence, organic chemists have been encouraged to explore methods for the chemical synthesis of analogues of the β-lactam antibiotics. This research area, which is at present under vigorous investigation, corresponds to the fourth phase of the chapter; it has already yielded some promising results.

MODE OF ACTION

The cell wall is a rigid protective coating which envelops the bacterium and is essential to the organism's welfare. Its rigidity is attributable to the presence of a peptidoglycan network which consists of polysaccharide strands compaginated with peptide bridges. Penicillins and cephalosporins prevent cell-wall biosynthesis by inhibiting the cross-linking process[12,13].

At the molecular level the bridging reaction involves a transpeptidation in which the terminal R-alanine of a side chain ending in R-alanyl-R-alanine is displaced by the amino group of the terminal amino acid of a neighbouring side-chain. The displacement is probably triggered by the attack of a thiol group of the enzyme on the amido bond of the R-alanyl-R-alanine to give an intermediate thioester. Tipper and Strominger[14] have suggested that the β-lactam antibiotics are mimics of N-acyl derivatives of R-alanyl-R-alanine and that they block the transpeptidase. Persuasive evidence[15] for the formation of a penicillanoyl derivative of this enzyme has been recently presented.

CONFORMATION

An understanding of the mechanism of action of a penicillin at the molecular level must ultimately involve consideration of the molecule's conformation. Although knowledge in this area is meagre, a good deal is known about the preferred conformations of penicillanic acid derivatives.

X-ray crystallographic studies of penicillins[7,16] have indicated that the β-lactam ring is planar and the thiazolidine ring is puckered with the 2β-methyl group pressed close to the 6β-amido side-chain as in (IX). This conformational preference is not restricted to the solid state since methyl phenoxymethylpenicillinate[17] and methyl 6β-phthalimidopenicillanate[18] also adopt this conformation in non-polar solvents, the evidence being based on

Nuclear Overhauser Effects. Evidently, the alternative conformation (X) is inherently less stable in the above examples.

The conformation of a penicillanic acid derivative will be determined by the summation of a large number of interactions, and it is perhaps unwise to single out any one of them and suggest that its importance is overriding. Nevertheless, there are some features which are worthy of comment. In (IX) the 2β-methyl group and the 6β-side chain are compressed; however, even with the bulky phthalimido group this interaction is not serious enough[18] to change the conformation from (IX) to (X). Compared to (IX), conformation (X) contains an additional *syn*-clinal methyl–carboxy interaction and it is suggested that this feature ensures that conformation (IX) is favoured.

In marked contrast to the above examples, the thiazolidine rings of the R- and S-sulphoxides and sulphones of penicillanic acid derivatives[17-19] have the 2β-methyl groups directed away from the 6β-amido side-chains as in conformation (X). With the S-sulphoxides of penicillins (XI), X-ray crystallographic and n.m.r. spectroscopic studies have indicated that a strong intramolecular hydrogen bond exists between the amido hydrogen and the sulphoxide oxygen atoms; this is accommodated better in conformation (X) than in conformation (IX). Intramolecular hydrogen bonding is precluded in the R-sulphoxides (XII) and yet these derivatives still adopt conformation (X), possibly because in the alternative conformation an unfavourable 1,3-*syn* axial interaction involving the sulphoxide and carboxyl groups is present.

STRUCTURE–ACTIVITY RELATIONSHIPS

According to the biochemical evidence (*see* Section on Mode of action), penicillins are analogues of *N*-acyl derivatives of *R*-alanyl-*R*-alanine and they function as acylating agents. This hypothesis implies that the amido side-chain, the β-lactam ring and the 3-carboxyl group are essential structural features for biological action.

Chemical studies support the above description. Thus, (V) possesses only a low order of activity[20]. While the nature of the acyl side-chain may have important secondary effects, all penicillins show significant antibacterial activity[21]; however, N-alkyl derivatives of penicillins are inactive[22]. Epi-penicillins, in which the stereochemistry of the amido side-chain is inverted, are also biologically ineffective[23,24] (see Section on Epimerisations).

The β-lactam rings of penicillins are readily cleaved and the products are inactive. Indeed, a hydrolytic β-lactam fission, using the catalytic effect of penicillinases, has been adapted by many micro-organisms as a defence against penicillins.

Several chemical modifications of the carboxyl function of penicillins have been described, but only in the case of thioacids[25] is significant activity retained.

(XIII) (XIV) (XV)

The minimum structural requirement for activity which meets the biochemical evidence is depicted in (XIII). While (XIII) does not appear to have been reported, (XIV) is readily available from penicillins by Raney nickel desulphurisation; these substances lack antimicrobial activity[26] and, consequently, the sulphur atom is also necessary. However, the secopenicillin (XV) (see Secopenicillins on page 116) is also inactive[27]; this result suggests that the constraint of the thiazolidine ring is important, probably because it augments the β-lactam reactivity and possibly because it specifically orientates the carboxyl group. Biological activity is substantially reduced but not altogether eliminated in penicillin sulphoxides and sulphones[28], which also suggests that the sulphur atom does not play a direct role in the transpeptidase inhibition.

(XVI) (XVII) (XVIII)

The gem-dimethyl group at position 2 of a penicillin is not essential for biological effectiveness. Thus, both (XVI)[29] and (XVII)[30] (see Section on Syntheses of β-lactam-antibiotic analogues), display significant activity.

The structural features of penicillins which are considered to be necessary for biological function are duplicated in cephalosporins. However, they are also reproduced in the Δ²-cephalosporins (XVIII) and yet these derivatives are inactive[31]. Recent X-ray crystallographic work[32] has provided a possible explanation for the inactivity of (XVIII): the β-lactam nitrogen atom is essentially planar as in a monocyclic β-lactam. In contrast, the β-lactam nitrogen atom of a penicillin or a cephalosporin possesses pyramidal character. Evidently, the reactivity of the β-lactam linkage is suppressed in (XVIII) because of the more effective interaction of the nitrogen lone pair electrons with the carbonyl group.

Some recent work[33] from the Woodward Research Institute supports the above picture. Whereas (XIX) and (XX) are inactive, (XXI) and (XXII) (*see* Section on Syntheses of β-lactam-antibiotic analogues) possess significant antibacterial activity. It is probable, therefore, that the double bond at

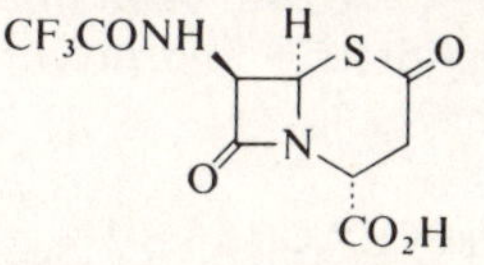

(XIX) 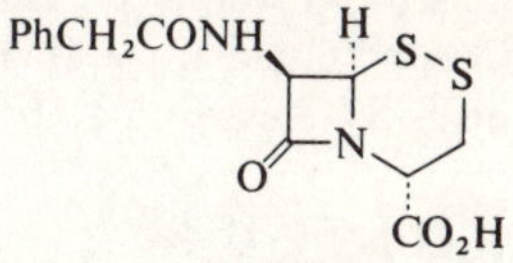   (XX)

position 3 introduces enamino character to the bridgehead nitrogen atom and increases the reactivity of the β-lactam bond.

If penicillins and cephalosporins function as acylating agents then an increase in the chemical reactivity of their β-lactam bonds may augment their biological effectiveness. There is some justification for this supposition in certain cases[31] where an approximate correlation has been noted between the i.r. frequency of the β-lactam carbonyl group and biological activity. However, in a recently reported study, Woodward and his co-workers[33]

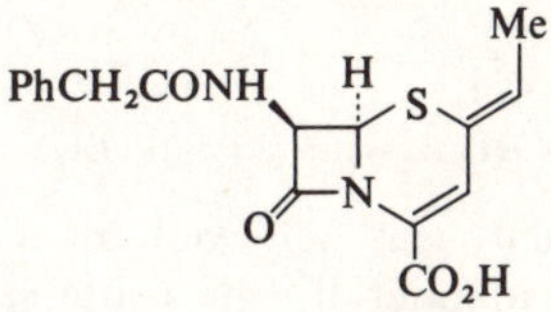

(XXI) 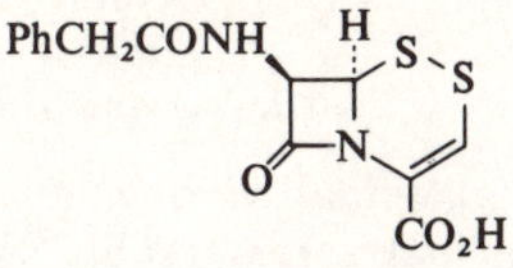   (XXII)

examined a series of biologically active derivatives (XXIII). The chemical reactivity of the β-lactam bond was varied by changing the substituent, R, but there was no correlation with biological activity.

It is evident from the foregoing discussion that neither the acetoxymethyl group at position 3 nor the methylene group at position 2 is an essential requirement for activity in cephalosporins. Indeed, (XXIV) (*see* Section on Penicillin→cephalosporin transformations) was the first unnatural β-lactam derivative to show significant activity[31].

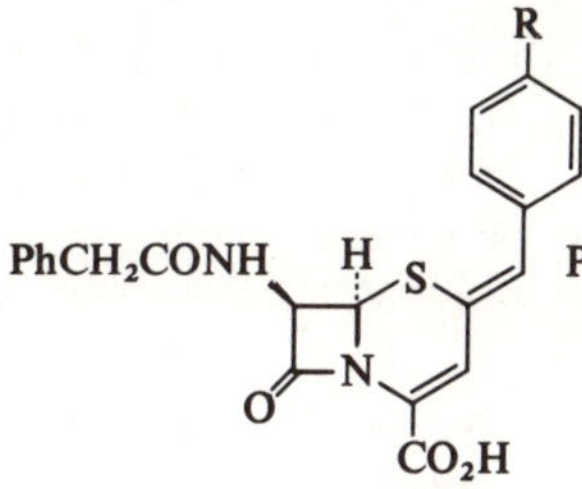

(XXIII) 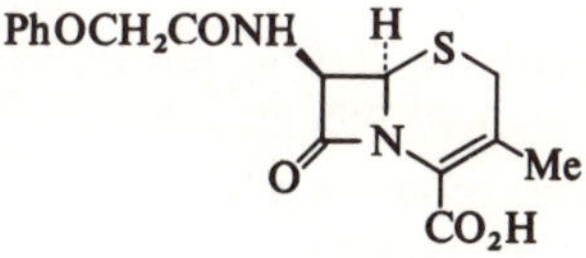   (XXIV)

PENICILLIN→CEPHALOSPORIN TRANSFORMATIONS

In essence, the transformation of a penicillin into a cephalosporin requires an oxidation of the methyl groups and a ring expansion. At the outset, this provided a problem of considerable challenge since there was no indication that the thiazolidine ring of a penicillin could be opened without disruption of the labile β-lactam linkage. Moreover, since penicillins are more readily available and cheaper to produce than cephalosporins, the conversion of the former antibiotics into the latter is of potential commercial importance. The problem has been solved through the considerable and inventive studies of workers at the Lilly Research Laboratories.

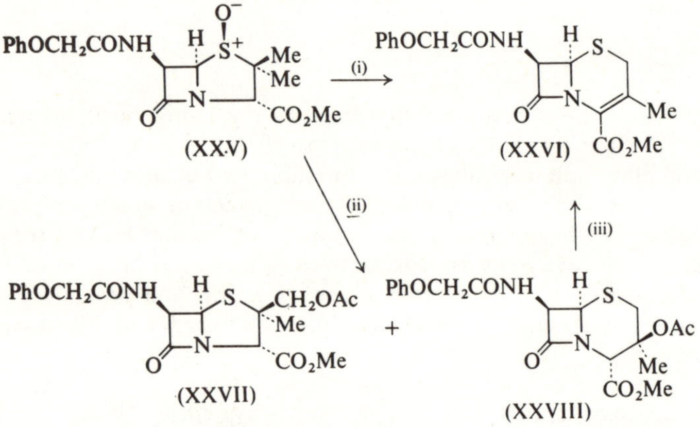

Chart 1. Reagents: (i) Δ/p-TsOH; (ii) Δ/Ac$_2$O; (iii) Et$_3$N

The vital oxidative ring enlargement was achieved by Morin and his co-workers[31] in a remarkably simple manner by the route summarised in Chart 1. It is noteworthy that (XXVI) provided the first direct chemical correlation between the penicillins and cephalosporins since it was also obtained from methyl phenoxymethylcephalosporinate by a palladium-catalysed hydrogenolysis.

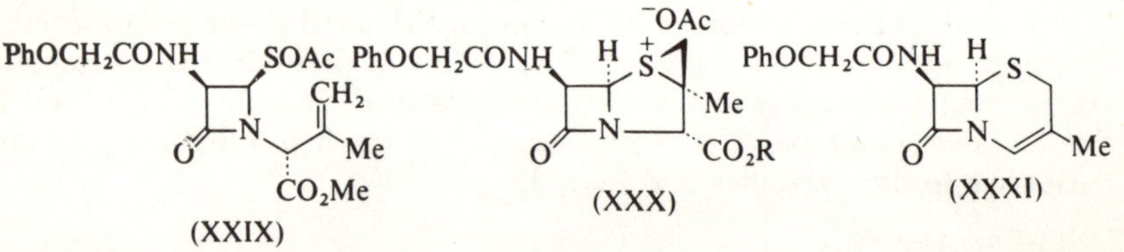

The stereospecific nature of the acetic anhydride-induced rearrangement of (XXV) implies that the ion pair (XXX; R = Me), which probably arises from the sulphenic anhydride (XXIX) (see p. 111), intervenes[31]. It is necessary to protect the carboxyl function during the rearrangement otherwise decarboxylation occurs, possibly at the ion-pair stage (XXX; R = H), to give (XXXI).

The conversion of (XXVI) into (VI; R = PhOCH$_2$) was achieved[34] in a circuitous but, nonetheless, elegant manner. The Δ^2-cephem (XXXII; R^1 = R^2 = H), obtained by the action of sodium hydroxide on (XXVI), was converted into the ester (XXXII; $R^1 = H$, $R^2 = p$-methoxybenzyl) which was oxidised with N-bromosuccinimide to (XXXII; $R^1 = Br$, $R^2 = p$-methoxybenzyl). The latter derivative afforded an equilibrium mixture of (XXXII; $R^1 = OAc$, $R^2 = p$-methoxybenzyl) and (XXXIII; $R = p$-methoxybenzyl) in the ratio of 7:3 with potassium acetate. Oxidation of this mixture with m-chloroperbenzoic acid yielded (XXXIV; $R^1 = p$-methoxybenzyl, $R^2 = PhOCH_2$) which was converted into (VI; $R = PhOCH_2$) by reduction and ester hydrolysis. The mild procedures which were developed for the reduction of the sulphoxide to the sulphide, are likely to be of general interest and application.

The transformation of (XXVII) into (VI; $R = PhOCH_2$) may, in principle, be achieved by the acetic anhydride-induced rearrangement of its sulphoxide. However, the stereochemical requirements of this rearrangement are rigorous and it appears necessary to have the sulphoxide function *cis* to the 2-methyl group and also in its thermodynamically stable configuration (*see* next Section). These conversions have been achieved[35] with (XXXV;

$R^1 = R^2 = Me$) and (XXXVI). The phthalimido group of the Δ^2-isomer of methyl 7β-phthalimidocephalosporanate can be removed by hydrazinolysis and, consequently, a method exists for the chemical transformation of (V) into (VIII).

PENICILLIN SULPHOXIDES

The oxidation of penicillin esters by a variety of methods, including sodium metaperiodate, m-chloroperbenzoic acid and hydrogen peroxide, has long been known to yield a single sulphoxide. Cooper and his co-workers[17], on the basis of n.m.r. spectroscopic evidence, have assigned the S-configuration (XI) to these derivatives. In the case of (XI; $R^1 = Me$, $R^2 = PhOCH_2$) the stereochemistry has been confirmed by an X-ray crystallographic study[17].

The preferential formation of the S-sulphoxides has been attributed[17] to the directing influence of the amido side-chain, which is considered to form a hydrogen bond with the oxidant. Some support for this speculation comes from the oxidation of the methyl ester[19] and the *N*-t-butylamide[36] of 6β-phthalimidopenicillanic acid; in each case the *R*-sulphoxide was produced.

It was first shown that both *R*- and *S*-sulphoxides were formed in the sodium metaperiodate oxidation of methyl-6,6-dibromopenicillanate[37]. Barton and his co-workers[36] also noted that iodobenzene dichloride in aqueous pyridine oxidised methyl benzylpenicillinate to a mixture containing equal amounts of the *R*- and *S*-sulphoxides. Moreover, Archer and DeMarco[19] discovered that the *S*-sulphoxides of methyl benzylpenicillinate and methyl methylpenicillinate isomerised to the *R*-derivatives when irradiated.

The *S*-sulphoxides of penicillins are thermodynamically more stable than

(XXXVII)   (XXXVIII)

the *R*-isomers and the latter derivatives isomerise to the former when heated[19,29,36]. The lower free energy of the *S*-sulphoxide is probably due to the presence of an intramolecular hydrogen bond between the amido hydrogen and the sulphoxide oxygen atoms; this interaction is absent in the *R*-isomer.

The isomerisation of the *R*-sulphoxide of β,β,β-trichloroethyl benzylpenicillinate in boiling *O*-deutero-t-butanol was accompanied by specific deuterium incorporation into the 2β-methyl group; under similar conditions the *S*-isomer did not undergo isotopic exchange[38]. However, under more forcing conditions the *S*-sulphoxide of methyl phenoxymethylpenicillinate also specifically exchanged the 2β-methyl hydrogen atoms[39]. The

(XXXIX)   (XL)

R-sulphoxide of methyl 6β-phthalimidopenicillanate is the thermodynamically stable isomer since, although it failed to isomerise to the *S*-isomer on heating, it underwent deuterium exchange in the 2α-methyl group[39].

The above results provide compelling support for the formation of a sulphenic acid (XXXVIII) during the thermolysis of a penicillanic acid ester sulphoxide. The specificity of the deuterium exchange implies that (XXXVIII) is probably formed by a concerted six-electron sigmatropic process, i.e. (XXXVII) involving the sulphoxide function and the 2-methyl group which is *cis* to it.

Authentication of this pathway comes from the observation that (XXXIX; $R^1 = CH_2CCl_3$, $R^2 = PhCH_2$) was converted into (XXXV;

$R^1 = CH_2CCl_3$, $R^2 = PhCH_2$)[29] and (XXXIX; $R^1 = R^2 = Me$) into (XXXV; $R^1 = R^2 = Me$)[35] on heating. Consequently, the isomerisation is accompanied by rotation about the 2,3-bond. The stereochemistry of the product is probably determined by the formation of a hydrogen bond between the sulphenic acid and the amide.

In the light of the foregoing observations the rearrangement of a penicillanic ester sulphoxide which occurs in boiling acetic anhydride (see previous Section) probably involves the initial formation of a sulphenic acid (XXXVIII) which is rapidly converted into the reactive sulphenic anhydride (XXIX).

The photoisomerisation of penicillin ester sulphoxides also involves 1,2-bond cleavage[35]; thus, both (XXXIX; $R^1 = R^2 = Me$) and (XL; $R^1 = R^2 = Me$) were formed when (XXXV; $R^1 = R^2 = Me$) was irradiated.

SYNTHESES OF β-LACTAM-ANTIBIOTIC ANALOGUES

The possibility of constructing β-lactam derivatives which are more effective than penicillins and cephalosporins is an alluring objective for organic chemists. From the foregoing evidence it is likely that such derivatives will incorporate the structural unit (XLI), where the bracket represents a five- or a six-membered ring.

The opportunity to assemble such substances was anticipated by Woodward and his co-workers in their total synthesis of cephalosporins[11,40]. The key intermediate (XLII), which was so ingeniously elaborated from R-cysteine, provided a masterly solution to the synthetic objective and, furthermore, it appeared to be a potentially valuable forerunner of unnatural β-lactam antibiotics. The degradation of (V) to (XLII), which has recently been described[33], provides a more convenient route to the intermediate.

The conversion of (V) into (XLII) must involve the elimination of C-3 and, therefore, cleavages of the 2,3- and 3,4-bonds of the thiazolidine ring. The latter bond fission was first reported in 1955 by Sheehan[41], who showed that Curtius degradation of 6β-phthalimidopenicillanic acid azide afforded the aldehyde (XLIII; R = CHO). Woodward and his co-workers[33] extended this approach to penicillins: the acid azide (XLIV; $R^1 = CON_3$) was heated in β,β,β-trichloroethanol to give (XLIV; $R^1 = NHCO_2CH_2CCl_3$) which afforded (XLV) on reduction with zinc in acetic acid. In contrast to (XLIII; R = CHO), the aldehydes derived from penicillins exist predominantly in the ring-closed form (XLV), suggesting that the equilibrium is sensitive to the size of the C-6 substituent (see p. 118).

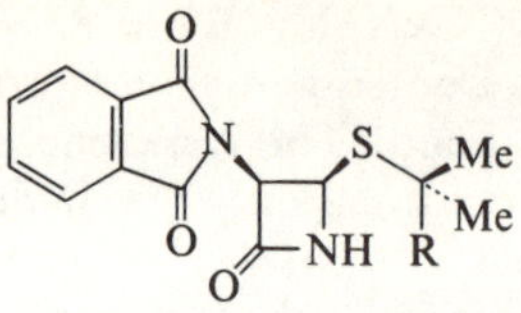

(XLIII) 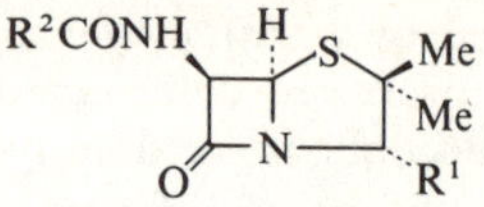   (XLIV)

The two methods by which (XLV; R = t-BuO) was transformed into (XLII) are summarised in *Chart 2*.

The Wittig reagent (XLVI), which was prepared from (XLII) by sequential reaction with a glyoxylic acid ester, thionyl chloride and triphenylphosphine, has played a key role in the synthesis of novel bicyclic β-lactams which incorporate the structural unit (XLI). Some examples are summarised in *Chart 3*. It should be pointed out that both olefins are formed in the reaction of (XLVI; R = t-Bu) with glyoxal derivatives although only the *trans*-isomer (XLVII) serves as a precursor of (XXIII). Moreover, the derivatives (XIX) and (XX) are accompanied by their C-4-stereoisomers. It is particularly encouraging, however, that (XXII) and (XXIII) show antibacterial activity.

The conversion of (XLIII; R = CHO) into (XLVIII) via a Curtius rearrangement of the acid azide (XLIII; R = CON_3) has been reported recently by Sheehan[30]. In the presence of benzyl pyruvate (XLVIII) afforded (XLIX) which, after removal of the protecting groups and phenylacetylation, yielded (XVII). The latter derivative displayed antibacterial

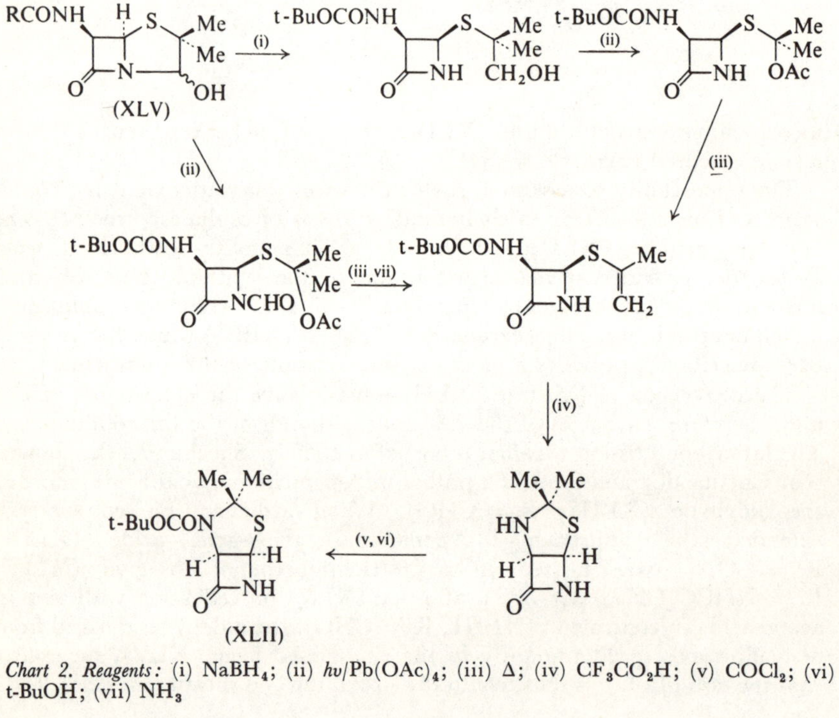

Chart 2. Reagents: (i) $NaBH_4$; (ii) $hv/Pb(OAc)_4$; (iii) Δ; (iv) CF_3CO_2H; (v) $COCl_2$; (vi) t-BuOH; (vii) NH_3

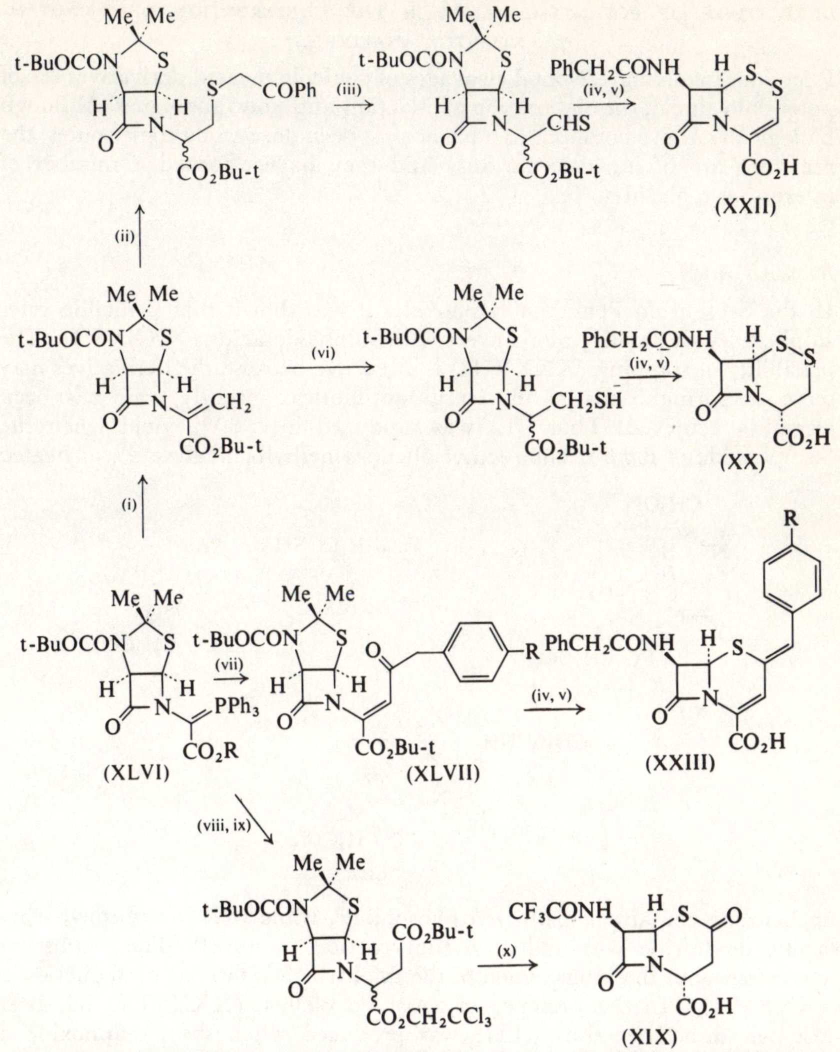

Chart 3. Reagents: (i) CH_2O; (ii) $PhCOCH_2SH$; (iii) $h\nu$; (iv) CF_3CO_2H; (v) $PhCH_2COCl$; (vi) H_2S; (vii) *p*-$RC_6H_4CH_2COCHO$; (viii) t-BuO_2CCHO; (ix) H_2/Pd; (x) CF_3CO_2H $(CF_3CO)_2O$

activity and, consequently, it seems likely that it possesses the *S*-configuration at position 3. It should be noted that the reversal of the anhydropenicillin rearrangement (*see* p. 115) shows a similar specificity.

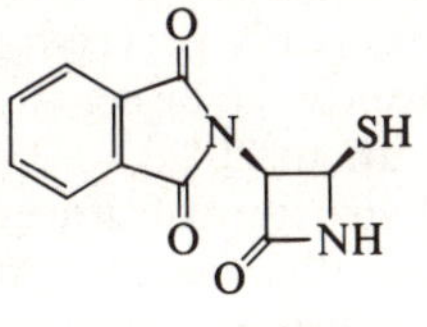

(XLVIII)

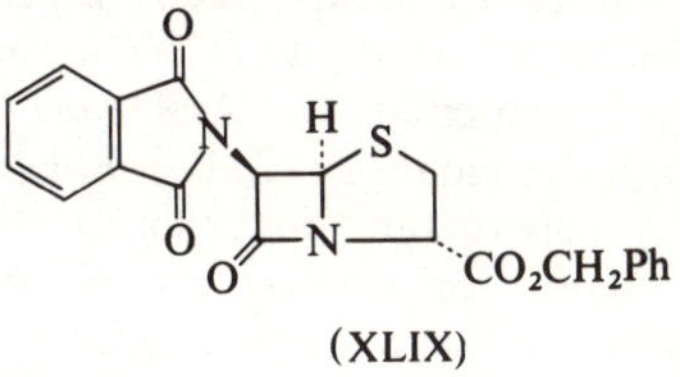

(XLIX)

DERIVATIVES OF POTENTIAL VALUE IN THE CONSTRUCTION OF β-LACTAM-ANTIBIOTIC ANALOGUES

Reactions involving 1,2-bond cleavages of penicillanic acid derivatives are of potential value in the elaboration of β-lactam-antibiotic analogues. Although biologically active compounds have not yet been derived by these routes, the reactions are of intrinsic curiosity and they have afforded a number of interesting molecules.

Sulphenic Acids

In the Section on Penicillin sulphoxides it was shown that penicillin ester sulphoxides thermally interconvert with sulphenic acids (XXXVIII). The possibility of trapping (XXXVIII) is attractive because the derivatives may serve as springboards to unnatural antibiotics; recently, this has been elegantly achieved. Thus, (L) was produced in >80% yield when the S-sulphoxide of β,β,β-trichloroethyl phenoxymethylpenicillinate was heated

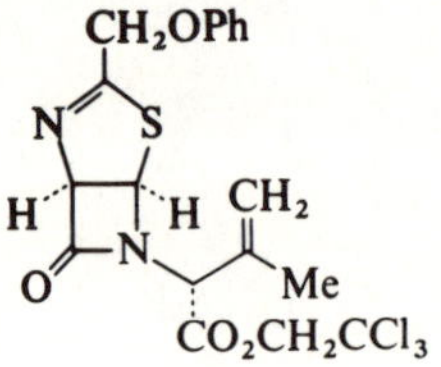

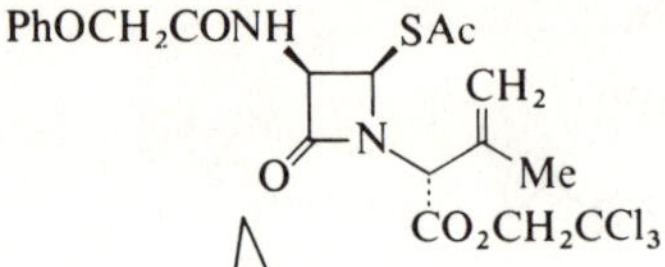

in benzene containing trimethyl phosphite[42], while (LI) was formed when acetic anyhdride was added to the reaction mixture[43]. The results are consistent with the intervention of the thiol which is formed by reduction of (XXXVIII). In the presence of activated olefins, (XXXVIII) has been trapped directly[29]; thus, (LII) was produced when the S-sulphoxide of β,β,β-trichloroethyl benzylpenicillinate was heated in benzene containing norbornadiene.

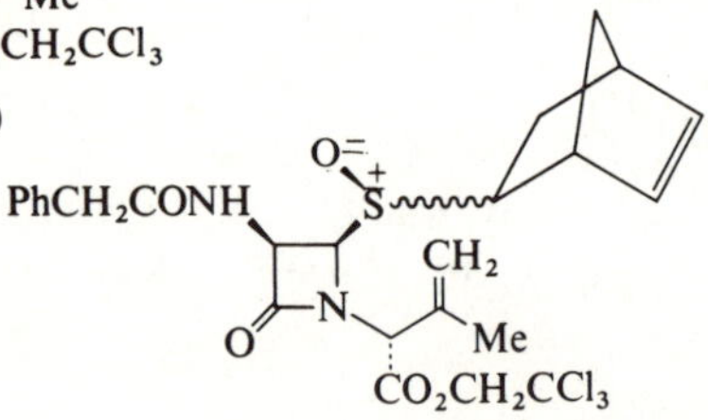

Anhydropenicillins

In 1963 workers at Bristol Laboratories[44] reported an intriguing triethylamine-induced rearrangement of penicillinic acid chlorides and anhydrides to anhydropenicillins (LIII). The reaction, which probably proceeds via the thiolate intermediate (LIV; R^1 = Cl or OCO_2R) appears to be a general one although the yield of (LIII) is not high (generally about 20%).

Anhydropenicillins incorporate some interesting structural features; thus, in principle, the methyl groups are activated for allylic attack and the thiolactone linkage is open to hydrolysis. The latter reaction occurred with

(LV) under carefully controlled conditions[45] to give (II; R = o-carboxyphenyl); significantly, the S-configuration at position 3 is regenerated in the process.

The promise which anhydropenicillins appeared to hold as precursors for novel β-lactam antibiotics has, unfortunately, not been fulfilled. In particular, the allylic methyl groups are inert to the action of lead tetraacetate and selenium dioxide[46]. Experiments designed to trap (LIV) do not appear to have been explored. Although an external reagent might not be expected to compete with the intramolecular acylation, (LVI) has been

isolated from the reaction of 6,6-dibromopenicillanic acid mixed anhydride with triethylamine[37].

Cephams

The principle embodied in the anhydropenicillin rearrangement has been extended recently to the preparation of the cepham derivatives (LVII)[47]. Thus, 6β-phthalimidopenicillanoylchloromethane (LVIII; R^1 = Cl, R^2 = phthalimido) yielded (LVII; R = phthalimido) in the presence of a suitable base. However, the reaction was accompanied by the formation of the 7-epimer of (LVII; R = phthalimido). This arose from 6α-phthalimidopenicillanoylchloromethane which was formed from (LVIII; R^1 = Cl, R^2 = phthalimido) by a competitive base-catalysed epimerisation (*see* next Section). The extent of the epimerisation was sensitive to the solvent, the base and the leaving group; it was minimised when 6β-phthalimidopenicillanoyliodomethane (LVIII; R^1 = I, R^2 = phthalimido) was treated with 1,5-diazabicyclo[4.3.0]non-5-ene in dimethyl sulphoxide. Deuterium labelling experiments have established that the abstraction of the C-3 proton of (LVIII; R^1 = Cl, R^2 = phthalimido) is the rate-controlling step in the rearrangement.

The reorganisation appears to be a general one and is limited only by the availability of the haloketone. Moreover, in the case of (LVIII; R^1 = Cl, R^2 = PhOCH$_2$CONH) the reaction to yield (LVII; R = PhOCH$_2$CONH) was not accompanied by any epimerisation[48].

Secopenicillins

Recently, workers at Beecham Research Laboratories[27] have discovered that esters of 6β-triphenylmethylaminopenicillanic acid react with sodium

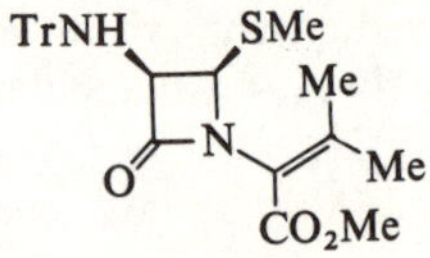

(LIX)

hydride and methyl iodide in tetrahydrofuran to give (LIX). The mechanism of this interesting reaction has not been fully established although it seems likely that the alkylating agent is intimately involved in the 1,2-bond cleavage. Replacement of the trityl group of (LIX) by an acyl substituent and subsequent ester hydrolysis has yielded derivatives, e.g. (XV), which have been termed 'secopenicillins'.

EPIMERISATIONS

Penicillins are considered to be structural analogues of R-alanyl-R-alanine (*see* Section on Mode of action). At the molecular level this description is not

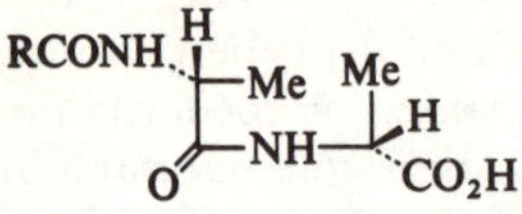

(LX) 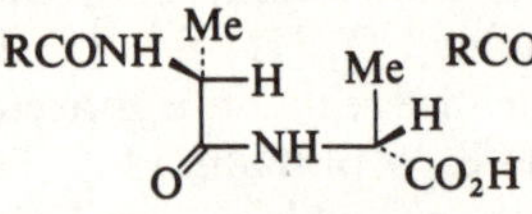   (LXI) 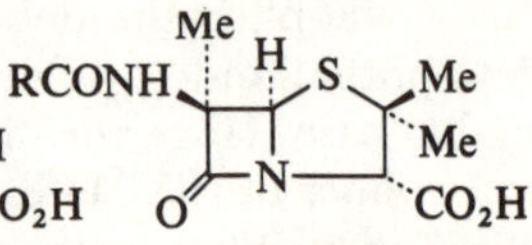   (LXII)

so convincing as the organic chemist might like. Thus, if (LX) approximates to the biologically active conformation of R-alanyl-R-alanine, penicillins apparently possess the incorrect stereochemistry at position 6. It is highly unlikely, however, that antibiotics which are as effective as the penicillins could have such a gross stereochemical discrepancy. Consequently, (LXI) is probably a more apt representation of the biologically active conformation. In such an event it is conceivable that a penicillin which possessed a methyl group in the 6α-position, i.e. (LXII), might be a more effective analogue and, therefore, a better antibiotic[48a].

The above considerations provide an incentive for the investigation of epimerisations and alkylations of penicillins at position 6. In principle, such reactions may be achieved directly on suitable penicillanic acid derivatives since the β-lactam carbonyl group should provide the appropriate activation.

Wolfe and Lee first reported[49] that methyl 6β-phthalimidopenicillanate underwent epimerisation at position 6 in the presence of potassium t-butoxide in t-butyl alcohol, sodium hydride in tetrahydrofuran or triethylamine in dichloromethane. Almost simultaneously, Johnson and his co-workers[23] described the epimerisation of hetacillin (LXIII) in aqueous solution at pH 11·5. Hetacillin, whilst not incorporating the features for activity, readily loses acetone under the conditions of testing to provide (II; $R = PhCHNH_2$) which is a very important antibiotic known as ampicillin. However, under corresponding conditions epihetacillin shows no significant activity. Removal of the side-chain from epihetacillin yielded 6α-aminopenicillanic acid, from which benzylepipenicillin was synthesised[24]; it also exhibits negligible antibacterial activity. Consequently, the R-configuration at position 6 of a penicillin is an essential feature for biological effectiveness.

Several further examples of epimerisations have subsequently been encountered. Nayler and his co-workers[50] have demonstrated that the electronegativity of the 6-substituent is an important factor in the isomerisation process. Thus, 6β-aminopenicillanic acid, 6β-N,N-dimethylaminopenicillanic acid, 6β-triphenylmethylaminopenicillanic acid and various penicillins failed to epimerise at pH 11, while penicillanic acid itself showed no tendency to undergo deuterium exchange. However, the betaine (LXIV)

(LXIII)   (LXIV)

readily epimerised to the 6α-isomer at neutral pH, and 6α-bromopenicillanic acid underwent deuterium exchange at position 6 at pH 11.

The sulphoxide function dramatically lowers the free energy of activation for the epimerisation process. Thus, (XI; $R^1 = CH_2CCl_3$, $R^2 = PhOCH_2$) was isomerised[51] under essentially neutral conditions with N,O-bis-(trimethylsilyl)acetamide, although under similar conditions cephalosporin ester sulphoxides were unaffected. However, with a weak base even the latter derivatives underwent epimerisation[52]; thus (XXXIV; R^1 = 9-fluorenyl, R^2 = 2-thienylmethyl) isomerised in the presence of triethylamine in dimethyl sulphoxide.

Thermodynamic Considerations

In almost all of the epimerisations so far considered overwhelming preferences for the 6α-substituted penicillanic acid derivatives have been noted. The higher free energy of the 6β-derivative may be attributed to the presence of a *cis*-interaction between the C-6 substituent and the sulphur atom and to a compressional interaction involving the former and the 2β-methyl group. In the epimerisation of methoxymethyl 6β-phthalimidopenicillanate, less than 1% of the 6β-isomer was present at equilibrium[53].

A variation in the steric requirement of the C-6 substituent is expected to alter the position of the equilibrium. In the case of simple β-lactams such an effect has been observed. Thus, although cis-1,4-diphenyl-3-phthalimido-azetidin-2-one underwent complete epimerisation to the trans-isomer, cis- or trans-3-bromo-1,4-diphenylazetidin-2-one was converted into an equilibrium mixture containing 30% of the cis-isomer[54] in the presence of 1,5-diazabicyclo[4.3.0]non-5-ene in benzene solution. However, 6α-bromopenicillanic acid underwent base-catalysed deuterium exchange at position 6 without any detectable formation of the 6β-isomer[50]. This observation offers little hope that the equilibrium concentration of a 6β-derivative could become significant. Nevertheless, in the case of (XI; $R^1 = CH_2CCl_3$, $R^2 = PhOCH_2$) the equilibrium mixture contained 20% of the 6β-derivative[51]. It may be argued, however, that this is a special case in which the 6β-derivative is stabilised by the presence of an intramolecular hydrogen bond between the acylamino hydrogen and the sulphoxide oxygen atoms (see p. 105).

Recently, the reactions of the Schiff base derivatives (LXV) with a trace of 1,5-diazabicyclo[4.3.0]non-5-ene in dichloromethane have been examined[53]. Epimerisation rapidly occurred at position 6 but in each case the starting material was readily detected by n.m.r. spectroscopy at equilibrium; thus, (LXV; R = p-nitrophenyl) and (LXV; R = 2-hydroxy-1-naphthyl) comprised 19% and 39% of the respective equilibrium mixtures. The equilibrium constant is solvent sensitive, and in benzene solution the

$$RCH=N\underset{O}{\overset{H}{\underset{|}{\diagdown}}}\overset{S}{\underset{N}{\diagdown}}\overset{Me}{\underset{CO_2CH_2OMe}{\diagdown}}$$

(LXV)

amounts of (LXV; R = p-nitrophenyl) and (LXV; R = 2-hydroxy-1-naphthyl) were increased to 30% and 47%, respectively.

The above results are significant in two respects. Firstly, the aldimino group is probably a more reasonable steric model for the acylamino group than are the diacylamino, acylalkylamino or trialkylammonium substituents. Consequently, the overwhelming thermodynamic preference which has been noted in the latter cases should not be extrapolated to penicillins. Secondly, they provide a method for inverting the stereochemistry at position 6 of methyl 6α-aminopenicillanate, which has been prepared by Bose and his co-workers[55], enabling a total chemical synthesis of penicillins to be achieved by this route.

Mechanism

The mechanistic aspects of the epimerisation reaction will be considered in detail, partly because there is some controversy concerning the nature of the intermediates involved and also because some of the interpretations are relevant to the general theme of carbon acid-proton transfers.

Wolfe and Lee[49] noted that the conversion of methyl 6β-phthalimido-penicillanate into the 6α-isomer with potassium t-butoxide in t-BuOD was accompanied by deuterium exchange at position 6, although when the reaction was interrupted the recovered starting material contained no isotope. Under similar conditions the 6α-derivative was unchanged and incorporated no deuterium. On the basis of this evidence it was tentatively suggested that the enethiolate (LXVI; $R^1 = CO_2Me$, $R^2 =$ phthalimido) provided a better model for the reaction transition state than the enolate (LXVII; $R^1 = CO_2Me$, $R^2 =$ phthalimido).

Nayler and his co-workers[50] investigated the epimerisation of hetacillin (LXIII) in alkaline deuterium oxide solution. The formation of epihetacillin, which was essentially complete, was accompanied by isotope incorporation at position 6. In contrast to methyl 6α-phthalimidopenicillanate, epihetacillin exchanged deuterium at a rate which was not markedly slower than the

(LXVI)   (LXVII)

original isomerisation. However, a kinetic preference for *endo*-protonation of the intermediate was again observed. The enolate (LXVII; $R^1 = CO_2^-$, $R^2 =$ 2,2-dimethyl-5-oxo-4-phenylimidazolidin-1-yl) was proposed to account for the results.

The reaction of methyl 6β-phthalimidopenicillanate with triethylamine in dichloromethane was re-examined by Sjoberg and his collaborators[56] and they isolated the 1,4-thiazepine (LXVIII; $R^1 = CO_2Me$, $R^2 =$ phthalimido) in addition to the 6α-isomer. Under similar conditions (LXV; R = *p*-nitrophenyl) was converted[57] into (LXVIII; $R^1 = CO_2CH_2OMe$, $R^2 = p$-nitrobenzylidenimino). Thiazepine formation necessitates the cleavage of the 1,5- and 4,7-bonds of the penicillanic acid precursor, bond formation between

(LXVIII)   (LXIX)   (LXX)

the sulphur atom and the C-7 carbonyl group and a proton transfer from position 6 to the nitrogen atom. The rearrangement may be triggered by a 1,5-bond fission involving the intermediacy of the enethiolate (LXVI); alternatively, the ylide (LXIX), derived by nucleophilic rupture of the β-lactam linkage may be the thiazepine forerunner.

In methanol containing a trace of sodium methoxide (LXV; R = *p*-nitrophenyl) rapidly equilibrated with the 6α-isomer and the mixture was quantitatively transformed into (LXVIII; $R^1 = CO_2CH_2OMe$, $R^2 = p$-nitrobenzylidenimino)[58]. The enethiolate is implicated under these conditions since the thiazepine is not likely to arise from (LXX; $R^1 = CH_2OMe$, $R^2 = p$-nitrobenzylidenimino). Wolfe and his co-workers[59] have further examined the reaction of methyl 6β-phthalimidopenicillanate with triethylamine in a mixture of chloroform and t-butyl alcohol; equal amounts of the 6α-derivative and thiazepine were produced at 298 K and 323 K implicating a common rate-determining step which is considered to involve enethiolate formation.

It is evident from the foregoing discussion that the enethiolate intervenes in thiazepine formation. This result is of interest because it provides the first illustration that a penicillanic acid derivative can undergo 1,5-bond cleavage. Moreover, the enethiolate is a curious species since the azetinone ring, being isoelectronic with cyclobutadiene, is formally antiaromatic. However, the enethiolate is not necessarily the common intermediate which is involved in the formation of the C-6 epimer and the thiazepine. Indeed, a possible objection to this description is that, since methyl 6α-phthalimidopenicillanate is produced, it is necessary to postulate that the chiral centre at position 3 of (LXVI) directs the reprotonation to regenerate exclusively the R-configuration at position 5 (*see* p. 123 for chemical evidence in support of this stereochemical assignment).

Recently, the reactions of methyl 6β-phthalimidohomopenicillanate (LXXI) with a variety of organic bases have been scrutinised[60] and some of the results are presented in *Table 4.1*.

Table 4.1. THE REACTION OF METHYL 6β-PHTHALIMIDOHOMOPENICILLANATE (LXXI) WITH ORGANIC BASES IN $CDCl_3$ SOLUTION AT 306 K

Base	pK_a	Relative rate of disappearance of (LXXI)	Product composition	
			6α-Isomer, %	Thiazepine, %
1,5-Diazabicyclo[4.3.0]non-5-ene	—	40.5×10^3	>99	<1
N,N,N',N'-Tetramethylguanidine	13.6	3.4×10^3	>99	<1
1,2,2,6,6-Pentamethylpiperidine	11.2	1.1	80	20
Triethylamine	10.8	8.2	50	50
1-Methylpiperidine	10.1	14	35	65
1-Methylmorpholine	7.4	1.0	15	85

With a strong base, such as 1,5-diazabicyclo[4.3.0]non-5-ene, (LXXI) was converted only into methyl 6α-phthalimidohomopenicillanate. Moreover, with the same base in pyridine containing deuterium oxide the 6α-derivative underwent deuterium exchange at position 6 approximately 14-times slower than (LXXI) epimerised. Therefore, the ground-state free energies of (LXXI) and its 6α-isomer are reflected in the energies of the transition states leading to the proton transfers (*see Figure 4.1a*). Since the

(LXXI)   (LXXII)   (LXXIII)

large free-energy difference between the α- and β-isomers of a 6-phthalimidopenicillanic acid derivative is probably due to the steric effect of the phthalimido group (see p. 117), this effect must still be important at the transition states leading to the removal of the C-6 protons. The ion pairs (LXXII) and (LXXIII) appear to be satisfactory models for these transition states.

It is apparent from *Table 4.1* that the formation of (LXVIII; $R^1 = CH_2CO_2Me$, $R^2 =$ phthalimido) becomes important when a weak base is employed and that its yield increases as the strength of the base diminishes. With a weak base the reaction is essentially irreversible; thus, in the case of 1-methylpiperidine it has been estimated[48] that the conversion of methyl 6α-phthalimidohomopenicillanate into the thiazepine is approximately 300-times slower than the transformation of (LXXI) into the 6α-isomer and the thiazepine. Evidently, the transition states involving the proton transfers are much closer in energy when a weak base is involved (see *Figure 4.1b*). Therefore, the intermediates which intervene must differ

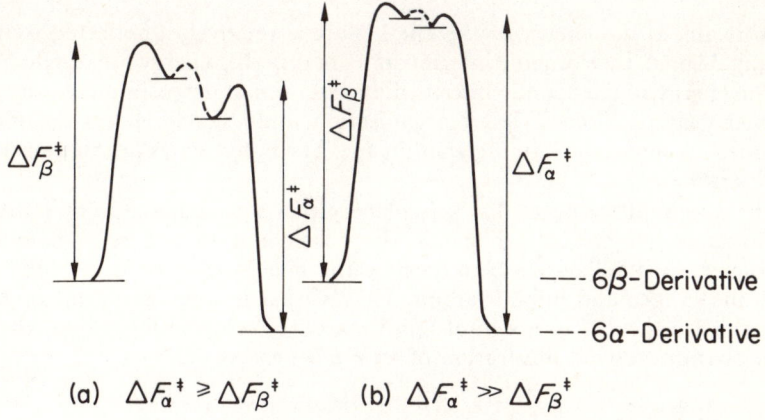

Figure 4.1. Free-energy profiles for the proton transfers involved in the reaction of methyl 6β- and 6α-phthalimidopenicillanic acid derivatives with (a) a strong base and (b) a weak base

significantly from the ion pairs which have been invoked in the strong base-promoted reactions. It is postulated that this difference is mainly one of geometry and that as the base strength diminishes the intermediates develop increasing sp^2 character at position 6.

The ratios of the 6α-isomer to the thiazepine which are formed from (LXXI) (see *Table 4.1*) reflect the relative rate constants for protonation and rearrangement of the common intermediate. The transition state of the reaction leading from the intermediate to the 6α-isomer almost certainly involves a proton-transfer process and, therefore, the free energy of activation of this process is expected to diminish as the strength of the conjugate acid increases. More thiazepine is formed as the base strength decreases and, consequently, the free energy of activation for thiazepine formation decreases to a greater extent than that for α-isomer formation. The slow step in thiazepine formation, which is unlikely to involve a proton-transfer process, is postulated to involve conversion of the anionic intermediate into the

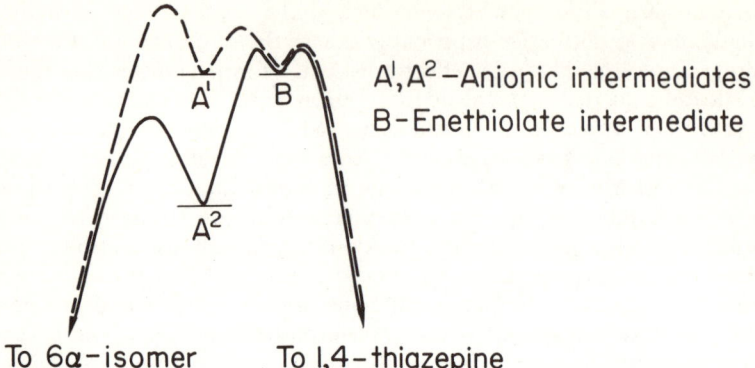

Figure 4.2. *Free-energy profiles for the formation of a 6α-phthalimidopenicillanic acid derivative and a 1,4-thiazepine in the presence of a weak base (broken line) and a strong base (unbroken line)*

enethiolate. If the energy of the enethiolate is relatively unaffected by the conjugate acid, then anionic intermediates of differing energy are implicated. As the energy of the anionic intermediate rises (i.e. it develops increasing sp^2 geometry at position 6) less reorganisation and, therefore, less energy is required to convert it into the enethiolate. These postulates are summarised in *Figure 4.2*.

The precise description that one places upon the common anionic intermediates is not critical: they may be considered to possess enolate-like character, enethiolate-like character or the charge may be delocalised on both the oxygen and sulphur atoms. However, it is unnecessary to suppose that the 1,5-bond is ever completely broken. Thiazepine formation, therefore, provides a good illustration of an *E1cB* process [60a].

(LXXIV)   (LXXV)

It has already been pointed out (p. 116) that penicillins with a 6α-alkyl substituent, e.g. (LXII), may be of biological interest. In principle, such substances might be derived by performing the epimerisation in the presence of an electrophilic carbon source. Attempts[49] to trap the anionic intermediate involved in the isomerisation of methyl 6β-phthalimidopenicillanate have been unsuccessful. However, Reiner and Zeller[61] isolated (LXXIV) from the reaction of (V) with benzaldehyde at pH 7·5; the stereochemistry at position 6 of (LXXIV) was not established. The imino linkage of (LXXIV) was cleaved by acid and the derived amine was converted by phenylacetylation into (LXXV), which was biologically inactive.

DEAMINATIONS

The replacement of the acylamino group of a penicillin by an acyloxy group appears to be a modest chemical objective and the derivative may possibly be of biological interest. Consequently, a number of workers have investigated the deamination of (V).

Testa and his co-workers[62] first reported that (V) was converted into 6-chloro- and 6-bromo-penicillanic acid in the presence of sodium nitrite and the hydrohalic acid. The smaller coupling constant (J 1·5–2·0 Hz) of the β-lactam protons of these derivatives[50,63] compared to those of penicillins[64] (J 4·0–4·5 Hz) indicates that they possess the α-configuration at position 6. When the deamination was performed in the absence of a strongly nucleophilic acid, 6α-hydroxypenicillanic acid was the predominant product. Not surprisingly, its O-acyl derivatives are devoid of biological activity.

There was no evidence to suggest that the 6β-derivatives were formed in the deaminations and, consequently, these reactions proceeded with inversion of configuration at position 6. Deuteration studies[63] have revealed that 6-azopenicillanic acid (LXXVI; R = H) is a likely intermediate; this species, therefore, determines the stereochemical outcome of the reaction. Furthermore, under carefully controlled conditions, benzyl 6-azopenicillanate (LXXVI; R = PhCH$_2$) was isolated[63] from the reaction of benzyl 6β-aminopenicillanate with nitrous acid. This substance has also been implicated[65]

(LXXVI)

in the conversion of benzyl phenoxymethylpenicillinate into benzyl 6α-phenoxyacetyloxypenicillanate induced by dinitrogen tetroxide.

Further support for the intervention of (LXXVI; R = H) comes from the reaction of (V) with sodium nitrite and hydrobromic acid[37], when 6,6-dibromopenicillanic acid was found to be a minor product; its yield was improved when bromine was added to the reaction medium.

Recently, the deaminations of methoxymethyl 6β- and 6α-aminopenicillanates have been studied[58]. In each case the major β-lactam-containing product was methoxymethyl 6α-chloropenicillanate. Methoxymethyl 6α-aminopenicillanate was prepared by an acid-catalysed hydrolysis of methoxymethyl 6α-(2-hydroxy)naphthylideniminopenicillanate and, with N-ethoxycarbonylphthalimide, it afforded methoxymethyl 6α-phthalimidopenicillanate. The latter material was identical with the substance obtained from the epimerisation of methoxymethyl 6β-phthalimidopenicillanate and, consequently, the R-configuration at position 5 of a penicillanic acid derivative is preserved during epimerisation (see p. 120).

In attempts to prepare 6β-substituted penicillanic acids, the reactions of 6α-halo-[63], 6α-mesyloxy- and 6α-brosyloxy-penicillanic acid derivatives[65] with nucleophiles have been studied. The objective was not realised although rearranged products were isolated in certain cases[63] (see next Section).

REARRANGEMENTS INVOLVING CLEAVAGE OF THE β-LACTAM RING

An intriguing facet of the character of penicillanic acid derivatives is their exceptional propensity to undergo rearrangement reactions. Consequently, the chemical modification of these molecules in a predictable manner may be a perplexing exercise: even banal functional-group transformations can be fraught with unexpected side reactions. These reorganisations are usually of sufficient intrinsic interest to warrant investigation and comprehension; furthermore, the information gleaned may ultimately aid the chemist who seeks to control the molecule's reactivity. Rearrangements occurring

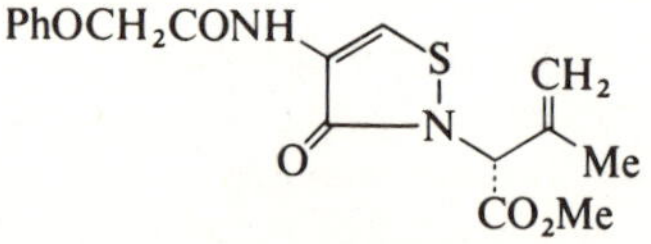

(LXXVII) 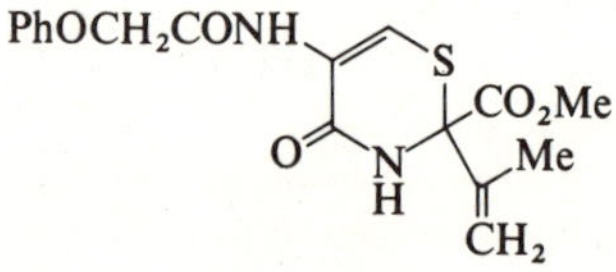   (LXXVIII)

without disruption of the β-lactam ring have been discussed (*see* pp. 108–111, 114–116). Examples in which the azetidinone ring is destroyed are considered in this Section: they are classified according to the bond-breaking process which is considered to trigger the reaction.

1,2-Bond Cleavages

The thermal sigmatropic 1,2-bond cleavage of penicillin sulphoxides to give sulphenic acids and the reactions of these intermediates to give β-lactam-containing products have been considered (*see* pp. 108–111 and 114). Both

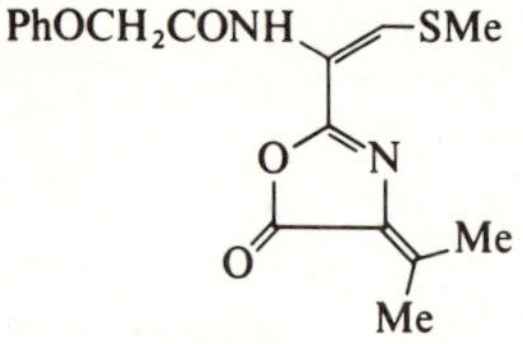

(LXXIX) 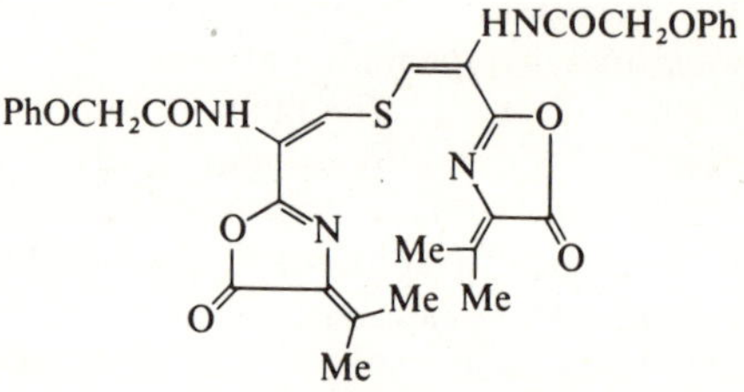   (LXXX)

(LXXVII) and (LXXVIII) have been isolated as by-products from the reaction of (XXV) with acetic anhydride[31], which suggests that the sulphenic anhydride (XXIX) can undergo ring expansion.

The derivatives (LXXIX) and (LXXX) have been isolated from the reaction of phenoxymethylpenicillinic acid with methyl chloroformate and triethylamine in dimethylformamide[66], conditions similar to those employed in the anhydropenicillin rearrangement (*see* p. 114). It is likely that the thiolate (LIV; $R^1 = OCO_2Me$, $R^2 = PhOCH_2$) is a precursor of (LXXIX) and (LXXX).

1,5-Bond Cleavages

The rearrangement of a penicillanic acid derivative into a 1,4-thiazepine (LXVIII), which has been fully considered in the Section on Epimerisations, is triggered by a 1,5-bond cleavage and involves the intermediacy of the enethiolate (LXVI).

In dichloromethane containing triethylamine, methyl 6β-phthalimidopenicillanate was converted into a mixture containing equal amounts of (LXVIII; $R^1 = CO_2Me$, $R^2 =$ phthalimido) and methyl 6α-phthalimidopenicillanate. However, under similar conditions, methoxymethyl 6β-phthalimidopenicillanate afforded the 6α-derivative in about 30% yield

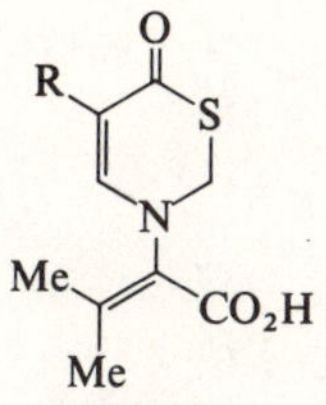

(LXXXI)

and the acid (LXXXI; R = phthalimido) in approximately 70% yield[58]. The conversion of methoxymethyl 6β-phthalimidopenicillanate into (LXXXI; R = phthalimido) constitutes a deep-seated skeletal change and dramatically illustrates how an apparently trivial modification to the ester group can have a profound effect upon the reaction outcome. The formation of (LXXXI; R = phthalimido) almost certainly involves the intermediacy of (LXVIII; $R^1 = CO_2CH_2OMe$, $R^2 =$ phthalimido); in the rearrangement of (LXV; R = p-nitrophenyl) to (LXXXI; R = p-nitrobenzylidenimino) the intermediate 1,4-thiazepine (LXVIII; $R^1 = CO_2CH_2OMe$, $R^2 = p$-nitrobenzylidenimino) has been isolated[57].

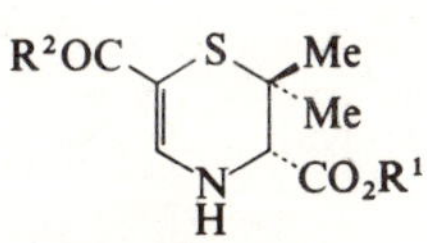

(LXXXII) 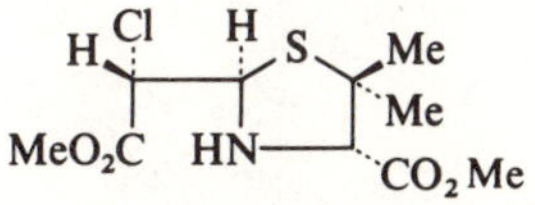 (LXXXIII) 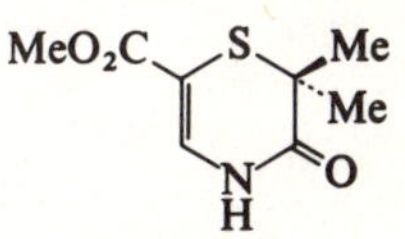 (LXXXIV)

The rearrangement of methyl penicillanates to the corresponding thiazepines has also been induced[67] by Lewis acids, e.g. antimony pentachloride.

4,7-Bond Cleavages

The conversions of penicillins into penillic acids, penillonic acids, penicillenic acids and azlactones are long-established examples[2] of rearrangements which are probably initiated by cleavage of the β-lactam bond.

The reaction of 6α-chloropenicillanic acid with sodium azide in dimethylformamide, which was originally studied in the hope of preparing 6β-azidopenicillanic acid, afforded (LXXXII; $R^1 = H$, $R^2 = N_3$)[63]. Corresponding rearrangements have been induced with other nucleophiles; the reactions are probably triggered by cleavage of the β-lactam bond since (LXXXIII) has been shown to be an intermediate in the sodium methoxide-promoted conversion of methyl 6α-chloropenicillanate into (LXXXII; $R^1 = Me$, $R^2 = OMe$)

The reaction of (V) with sodium nitrite and hydrochloric acid is solvent dependent: 6α-chloropenicillanic acid was the major product in water and in 70% aqueous methanol, while in methanol (LXXXIV) was formed in about 30% yield and no 6α-chloropenicillanic acid was detected[68]. The acid

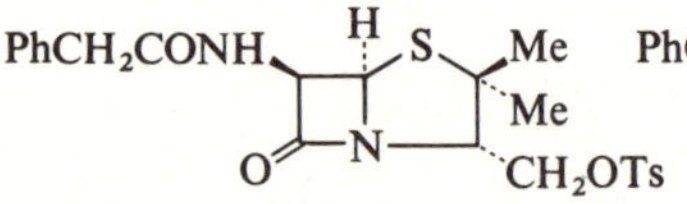

(LXXXV)

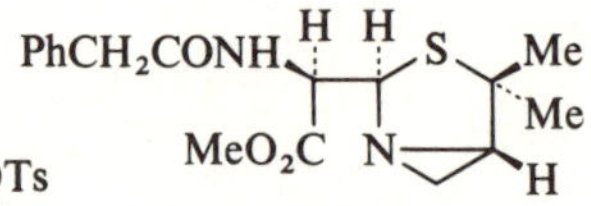

(LXXXVI)

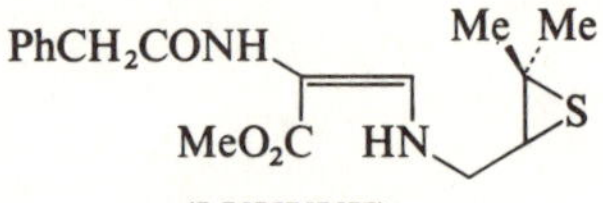

(LXXXVII)

(LXXXII; $R^1 = H$, $R^2 = OMe$) is a likely precursor of (LXXXIV) and it has been established that the rearrangement is probably initiated by methanolysis of the β-lactam bond of (V) to give (LXX; $R^1 = H$, $R^2 = NH_2$).

A related rearrangement is provided by the reaction of (LXXXV) with sodium hydrogen carbonate in methanol in which (LXXXVI) was formed[69]. Interestingly, the thiazolidine nitrogen atom acted as the nucleophile in preference to the sulphur atom although, under more forcing conditions, (LXXXVI) was converted into (LXXXVII).

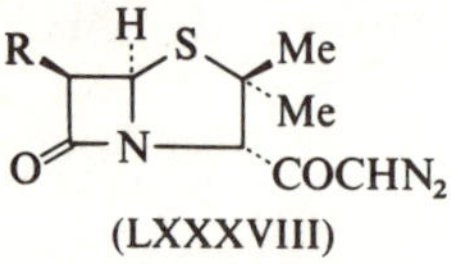

(LXXXVIII)

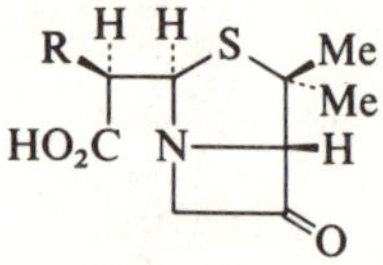

(LXXXIX)

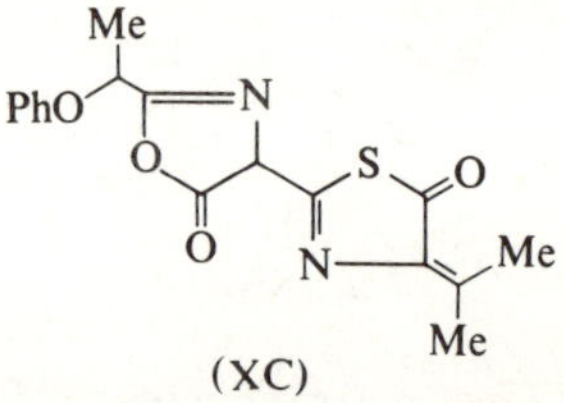

(XC)

Irradiation of the diazoketone (LXXXVIII; R = phthalimido) in aqueous dioxan yields (LXXXIX; R = phthalimido) in addition to 6β-phthalimidohomopenicillanic acid[70]. Hydrolysis of the 4,7-bond of the diazoketone is probably triggered by the interaction of the bridgehead nitrogen atom with an electrophilic intermediate such as the oxocarbene. A similar product (LXXXIX; R = PhOCH$_2$CONH) accompanied the formation of (LVIII; $R^1 = Cl$, $R^2 = PhOCH_2CONH$) in the reaction of (LXXXVIII; R = PhOCH$_2$CONH) with hydrochloric acid[48].

Although anhydropenicillins (LIII) are inert to the action of lead tetraacetate and selenium dioxide, they undergo an oxidative rearrangement in the presence of mercuric acetate in benzene to form substances with antibacterial activity, which have been termed anhydropenicillenes[46]. Thus, anhydro-α-phenoxyethylpenicillin (LIII; R = PhOCHCH$_3$) afforded (XC); the reaction is probably initiated by a mercuric acetate-catalysed cleavage of the β-lactam bond.

Manuscript received April 1971.

REFERENCES

1. COOK, A. H., *Q. Rev.*, **2**, 103 (1948)
2. CLARKE, H. T., JOHNSON, J. R. and ROBINSON, R., eds, *The Chemistry of Penicillin*, Princeton University Press (1949)
3. DOYLE, F. P. and NAYLER, J. H. C., *Adv. Drug Res.*, **1**, 1 (1964)
4. SHEEHAN, J. C. in 'Molecular Modification in Drug Design', *Adv. Chem. Series No. 45*, Am. Chem. Soc., Washington, 15 (1964)
5. SHEEHAN, J. C., *Ann. N.Y. Acad. Sci.*, **145**, 216 (1967)
6. MANHAS, M. S. and BOSE, A. K., *Synthesis of Penicillin, Cephalosporin C and Analogues*, Dekker, New York (1969)
7. PITT, G. J., *Acta Crystallogr.*, **5**, 770 (1952)
8. SHEEHAN, J. C. and HENERY-LOGAN, K. R., *J. Am. chem. Soc.*, **81**, 5838 (1959); ibid., **84**, 2983 (1962)
9. VAN HEYNINGEN, E., *Adv. Drug Res.*, **4**, 1 (1967)
10. ABRAHAM, E. P., *Q. Rev.*, **21**, 231 (1967); *Top. Pharm. Sci.*, **1**, 1 (1968)
11. HEUSLER, K., *Top. Pharm. Sci.*, **1**, 33 (1968)
12. STROMINGER, J. L., in *Antibiotics I. Mechanism of Action*, ed, Gottlieb, D. and Shaw, P. D., Springer-Verlag, New York (1967)
13. NAYLER, J. H. C., *Rep. Prog. appl. Chem.*, **54**, 194 (1969)
14. TIPPER, D. J. and STROMINGER, J. L., *Proc. natn. Acad. Sci. U.S.A.*, **54**, 1133 (1965); *J. biol. Chem.*, **243**, 3169 (1968)
15. LAWRENCE, P. J. and STROMINGER, J. L., *J. biol. Chem.*, **245**, 3653, 3660 (1970)
16. CROWFOOT, D., BUNN, C. W., ROGERS-LOW, B. W. and TURNER-JONES, A., in *The Chemistry of Penicillin*, eds, Clarke, H. T., Johnson, J. R. and Robinson, R., Princeton University Press, 310 (1949)
17. COOPER, R. D. G., DEMARCO, P. V., CHENG, J. C. and JONES, N. D., *J. Am. chem. Soc.*, **91**, 1408 (1969)
18. COOPER, R. D. G., DEMARCO, P. V. and SPRY, D. O., *J. Am. chem. Soc.*, **91**, 1528 (1969)
19. ARCHER, R. A. and DEMARCO, P. V., *J. Am. chem. Soc.*, **91**, 1530 (1969)
20. ROLINSON, G. N. and STEVENS, S., *Proc. R. Soc. B*, **154**, 509 (1961)
21. DOYLE, F. P. and NAYLER, J. H. C., *Adv. Drug Res.*, **1**, 42 (1964)
22. LEIGH, T., *J. chem. Soc.*, 3616 (1965)
23. JOHNSON, D. A., MANIA, D., PANETTA, C. A. and SILVESTRI, H. H., *Tetrahedron Lett.*, 1903 (1968)
24. JOHNSON, D. A. and MANIA, D., *Tetrahedron Lett.*, 267 (1969)
25. GOTTSTEIN, W. J., BABEL, R. B., CRAST, L. B., ESSERY, J. M., FRASER, R. R., GODFREY, J. C., HOLDREGE, C. T., MINOR, W. F., NEUBERT, M. E., PANETTA, C. A. and CHENEY, L. C., *J. Med. Chem.*, **8**, 794 (1965)
26. KACZKA, E. and FOLKERS, K., in *The Chemistry of Penicillin*, eds, Clarke, H. T., Johnson, J. R. and Robinson, R., Princeton University Press, 243 (1949)
27. CLAYTON, J. P., NAYLER, J. H. C., SOUTHGATE, R. and TOLLIDAY, P., *Chem. Commun.*, 590 (1971)
28. GUDDAL, E., MORCH, P. and TYBRING, L., *Tetrahedron Lett.*, 381 (1962)
29. BARTON, D. H. R., GREIG, D. G. T., LUCENTE, G., SAMMES, P. G., TAYLOR, M. V., COOPER, C. M., HEWITT, G. and UNDERWOOD, W. G. E., *Chem. Commun.*, 1683 (1970)
30. SHEEHAN, J. C. U.S. Pat. 3 487 074; 3 487 079; *Chem. Abstr.*, **72**, 66933, 100686 (1970)
31. MORIN, R. B., JACKSON, B. G., MUELLER, R. A., LAVAGNINO, E. R., SCANLON, W. B. and ANDREWS, S. L., *J. Am. chem. Soc.*, **85**, 1896 (1963); ibid., **91**, 1401 (1969)
32. SWEET, R. M. and DAHL, L. F., *J. Am. chem. Soc.*, **92**, 5489 (1970)

33. HEUSLER, K. and WOODWARD, R. B., Ger. Offen 1 935 459; 1 935 640; *Chem. Abstr.*, **72**, 100689, 100690 (1970)
34. WEBBER, J. A., VAN HEYNINGEN, E. M. and VASILEFF, R. T., *J. Am. chem. Soc.*, **91**, 5674 (1969)
35. SPRY, D. O., *J. Am. chem. Soc.*, **92**, 5006 (1970)
36. BARTON, D. H. R., COMER, F. and SAMMES, P. G., *J. Am. chem. Soc.*, **91**, 1529 (1969)
37. CLAYTON, J. P., *J. chem. Soc. C*, 2123 (1969)
38. BARTON, D. H. R., COMER, F., GREIG, D. G. T., LUCENTE, G., SAMMES, P. G. and UNDERWOOD, W. G. E., *Chem. Commun.*, 1059 (1970)
39. COOPER, R. D. G., *J. Am. chem. Soc.*, **92**, 5010 (1970)
40. WOODWARD, R. B., *Science*, **153**, 487 (1966)
41. SHEEHAN, J. C. and BRANDT, K. G., *J. Am. chem. Soc.*, **87**, 5468 (1965)
42. COOPER, R. D. G. and JOSE, F. L., *J. Am. chem. Soc.*, **92**, 2575 (1970)
43. HATFIELD, L. D., FISCHER, J., JOSE, F. L. and COOPER, R. D. G., *Tetrahedron Lett.*, 4897 (1970)
44. WOLFE, S., GODFREY, J. C., HOLDREGE, C. T. and PERRON, Y. G., *J. Am. chem. Soc.*, **85**, 643 (1963); *Can. J. Chem.*, **46**, 2549 (1968)
45. WOLFE, S., BASSETT, R. N., CALDWELL, S. M. and WASSON, F. I., *J. Am. chem. Soc.*, **91**, 7205 (1969)
46. WOLFE, S., FERRARI, C. and LEE, W. S., *Tetrahedron Lett.*, 3385 (1969)
47. RAMSAY, B. G. and STOODLEY, R. J., *Chem. Commun.*, 1517 (1970)
48. RAMSAY, B. G., Ph.D. Thesis, University of Newcastle upon Tyne (1970)
48a. STROMINGER, J. L. and TIPPER, D. J., *Am. J. Med.*, **39**, 708 (1965)
49. WOLFE, S. and LEE, W. S., *Chem. Commun.*, 242 (1968)
50. CLAYTON, J. P., NAYLER, J. H. C., SOUTHGATE, R. and STOVE, E. R., *Chem. Commun.*, 129 (1969)
51. GUTOWSKI, G. E., *Tetrahedron Lett.*, 1779 (1970)
52. SASSIVER, M. L. and SHEPHERD, R. G., *Tetrahedron Lett.*, 3993 (1969)
53. JACKSON, J. R. and STOODLEY, R. J., *Chem. Commun.*, 647 (1971)
54. BOSE, A. K., NARAYANAN, C. S. and MANHAS, M. S., *Chem. Commun.*, 975 (1970)
55. BOSE, A. K., SPIEGELMAN, G. and MANHAS, M. S., *J. Am. chem. Soc.*, **90**, 4506 (1968)
56. KOVACS, O. K. J., EKSTROM, B. and SJOBERG, B., *Tetrahedron Lett.*, 1863 (1969)
57. JACKSON, J. R. and STOODLEY, R. J., *Chem. Commun.*, 14 (1970)
58. JACKSON, J. R., Ph.D. Thesis, University of Newcastle upon Tyne (1971)
59. WOLFE, S., LEE, W. S. and MISRA, R., *Chem. Commun.*, 1067 (1970)
60. RAMSAY, B. G. and STOODLEY, R. J., *Chem. Commun.* 450 (1971); (a) MCLENNAN, D. J., *Q. Rev.* **21**, 490 (1967)
61. REINER, R. and ZELLER, P. *Helv. chim. Acta*, **51**, 1905 (1968)
62. CIGNARELLA, G., PIFFERI, G. and TESTA, E., *J. org. Chem.*, **27**, 2668 (1962)
63. MCMILLAN, I. and STOODLEY, R. J., *Tetrahedron Lett.*, 1205 (1966); *J. chem. Soc. C*, 2533 (1968)
64. GREEN, G. F. H., PAGE, J. E. and STANIFORTH, S. E., *Chem. Commun.*, 597 (1966)
65. HAUSER, D. and SIGG, H. P., *Helv. chim. Acta*, **50**, 1327 (1967)
66. KUKOLJA, S., COOPER, R. D. G. and MORIN, R. B., *Tetrahedron Lett.*, 3381 (1969)
67. CLAYTON, J. P., SOUTHGATE, R., RAMSAY, B. G. and STOODLEY, R. J., *J. chem. Soc. C*, 2089 (1970)
68. STOODLEY, R. J., *Tetrahedron Lett.*, 941 (1967); *J. chem. Soc. C*, 2891 (1968)
69. BELL, M. R. and OESTERLIN, R., *Tetrahedron Lett.*, 4975 (1968); BELL, M. R., CLEMANS, S. D. and OESTERLIN, R., *J. Med. Chem.*, **13**, 389 (1970)
70. RAMSAY, B. G. and STOODLEY, R. J., *J. chem. Soc. C*, 1319 (1969)

5
RECENT ADVANCES IN THE CHEMISTRY OF CYCLIC PEPTIDES

P. M. Hardy and B. Ridge

INTRODUCTION	130
2,5-DIOXOPIPERAZINES	131
2,5-Dioxopiperazine Metabolites	131
The Preparation of Dioxopiperazines	134
The Conformation of Dioxopiperazines	135
Spectroscopic Studies	137
cyclo-TRIPEPTIDES	140
cyclo-TETRAPEPTIDES	141
cyclo-PENTAPEPTIDES	146
cyclo-HEXAPEPTIDES	148
The Conformation of *cyclo*-Hexapeptides	148
Configurational Stereochemistry	152
Spectroscopic Studies on Cyclic Peptides	154
Enzyme Substrates and Enzyme Models	155
Cyclodimerisation	156
Enniatins	159
Ferrichromes	163
cyclo-HEPTAPEPTIDES	168
cyclo-DECAPEPTIDES	169
Gramicidin S	169
The Conformation of Gramicidin S	174
Tyrocidines	180
HIGHER *cyclo*-PEPTIDES	182
Valinomycin	182
Mycobacillin	186
CYCLIC PEPTIDES CONTAINING UNUSUAL AMINO-ACIDS	187
Alamethicin	187
Monamycin	189
Cycloheptamycin	191
Viomycin	191
Stendomycin	193
Telomycin	193
Peptide Antibiotics containing 3 Hydroxypicolinic Acid	193
Quinoxaline Antibiotics	196
Actinomycin	197
Thiostrepton	201
Bacitracin	203
Cyclic Peptides from Amanita Fungi	204
Peptide Alkaloids	210

CYCLIC PEPTIDES CONTAINING CYSTINE	212
Malformin	213
Neurohypophyseal Hormones	213
Calcitonin	215
Oxidative Studies on Model Cysteine Peptides	216
Ribonuclease	217
Insulin	217
Pro-insulin	220
Immunoglobulins	221

INTRODUCTION

NATURALLY occurring cyclic peptides which possess biological activity fall into three main groups. The hormones are of animal origin, the toxins are found in plants and fungi, and the antibiotics are extracted from fungal and bacterial sources. If those cyclic globular proteins which have been structurally characterised are also included, then the enzymes constitute a fourth group. The peptide hormones and enzymes consist solely of α-linked amino-acids whose rings are closed by disulphide bridges. The antibiotics and toxins, in contrast, show more structural diversity; they contain non-protein, dehydro-, and D-amino-acids, α-hydroxyacids, links through side-chain functional groups, and various fatty acid and heterocyclic acid components.

Interest in the synthesis of cyclic peptides has not been confined to the construction of naturally occurring compounds or their structural analogues. Rings varying in size from 2,5-dioxopiperazines to *cyclo*-hexapeptides have been widely explored as models for spectroscopic and conformational work, and other synthetic studies have been concerned with models of the active sites of enzymes and enzyme substrates. As in many other areas of organic chemistry, the number of publications in this field has increased rapidly in recent years. It is the purpose of this review to outline the advances which have been made in cyclic peptide chemistry over the past six years. We have chosen to concentrate on 1965 and the subsequent quinquennium as Schröder and Lubke[1] have conveniently covered the literature of much of this field (with the exception of 2,5-dioxopiperazines and the ferrichromes) up to the end of 1964. In general the material presented here abuts the account of Schröder and Lubke; considerations of space limit reference to work published prior to 1965 to such background as is pertinent to the understanding of the more recent advances. This approach has, of course, resulted in the exclusion of much of the classical work in the field, and the structural determination of many of the cyclic peptides discussed is not mentioned. Compounds which on acid hydrolysis split off relatively small proportions of their structures as amino-acids are only briefly treated, and penicillin, the cephalosporins, the polymyxins, and the circulins have been omitted entirely. Proteins containing cross-links other than disulphide bridges have been ignored, and even those proteins which contain disulphide bridges have been considered from a narrow viewpoint. Only the relation of the disulphide bridges to the conformation and biological activity is touched upon; the significance of amino-acid sequences and other considerations such as their mechanism of action are excluded. The synthesis of peptides has not been treated except in some illustrative examples, and only brief reference is made to the extensive studies which have been made of the structure–activity relationships of the cyclic peptide hormones and antibiotics.

In the earlier sections of this review cyclic peptides are discussed in order of their increasing complexity. Cyclodepsipeptides have been integrated into the text according to their ring size and have not been separated from the homodetic cyclic peptides. In the later section on peptides containing unusual amino-acids it has proved convenient to ignore ring size and the order of presentation has been arbitrarily chosen to facilitate structural comparisons. Standard abbreviations for peptide chemistry have been used; these are conveniently outlined on pp. 226–243 of Volume 2 of the Chemical Society Specialist Periodical Report on Amino-acids, Peptides and Proteins.

2,5-DIOXOPIPERAZINES

In recent years much interest has centred on the 2,5-dioxopiperazines, partly because a number of them have been identified as fungal metabolites, but mainly because they serve as accessible models of larger peptides. Extensive studies have been made of their conformations (the methodology of these being clearly of relevance to the conformational study of higher cyclic peptides), and they have also been utilised to evaluate the contribution of peptide chromophores (albeit *cis*) and aromatic side-chains of amino-acids to the optical behaviour of proteins.

2,5-*Dioxopiperazine Metabolites*

2,5-Dioxopiperazines occur as metabolites both in the simple form and as parts of more complex structures. An example of the former type is *cyclo*-(L-alanyl-L-leucyl), isolated from *Aspergillus niger*[2]. In some instances modified systems occur. Albonoursin, present in culture filtrates[3] of *Streptomyces albulus* and *Streptomyces noursei*, is known to be a bis-dehydro-derivative, 3-isobutylidene-6-benzylidene-2,5-dioxopiperazine[4] (I). Its synthesis[5] has been reported recently, and is outlined in Scheme 1. Although the structure of echinulin[6] (II) looks quite complex, it is basically a *cyclo*-(alanyl-tryptophyl) in which alkylation has taken place on three positions of the tryptophan residue. Hydrogenation of this crystalline metabolite of *Aspergillus echinulatus*[7] gives hexahydroechinulin, which on acid hydrolysis yields L-alanine and a new amino-acid hexahydroechinin (III). Hydrolysis of echinulin itself does not liberate echinin. Echinulin and hexahydroechinulin show the same broad positive Cotton effect[8] in the 220–300 nm region as *cyclo*-(L-alanyl-L-tryptophyl) (although in the latter case the extremum is at shorter wavelength) indicating that the echinin moiety possesses the L-configuration. This formulation of echinulin as the *cis*-2,5-dioxopiperazine (II) is in agreement with further o.r.d. studies[9] using the synthetic compound (IV) and also with studies on the behaviour of diastereoisomeric 2,5-dioxopiperazines[10] on thin layer chromatography.

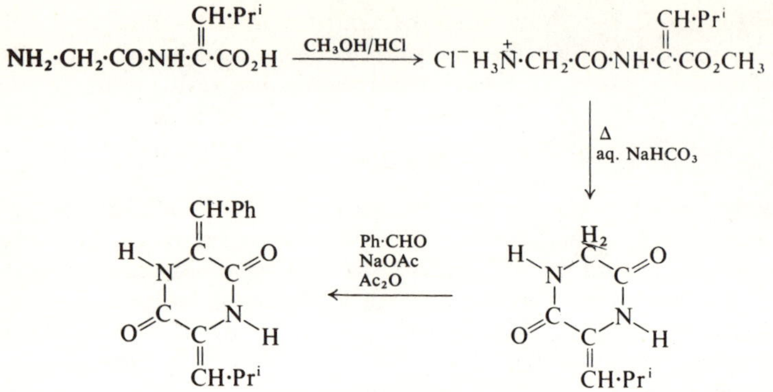

Scheme 1. *The synthesis of albonoursin*[5]

A number of fungi produce 2,5-dioxopiperazine metabolites with two features in common. They contain a disulphide bridge across the dioxopiperazine ring, and at least one of the two amino-acid constituents of the ring is a heterocyclic derivative. The simplest compound of this type is gliotoxin[11] (V), an antibiotic from the fungus *Gliocardium fimbriatum*; its absolute configuration has been determined by X-ray diffraction analysis[12,13]. A group of compounds called sporidesmins have been isolated[14] from *Pithomyces chartrum*. There are five members in this group, sporidesmin (VI; R = OH), and sporidesmin B (VI; R = H), C[15], D[16] (VII), and

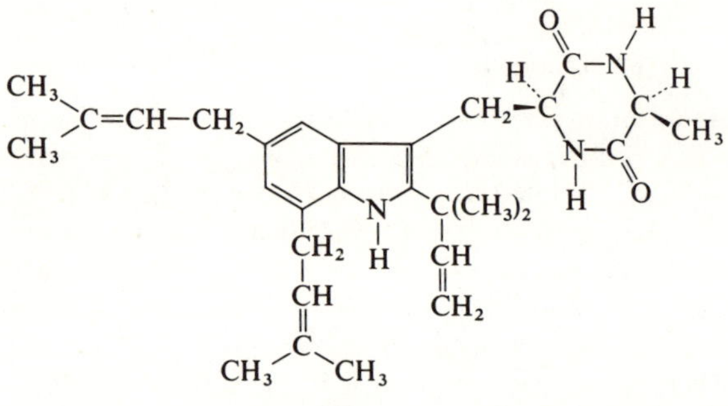

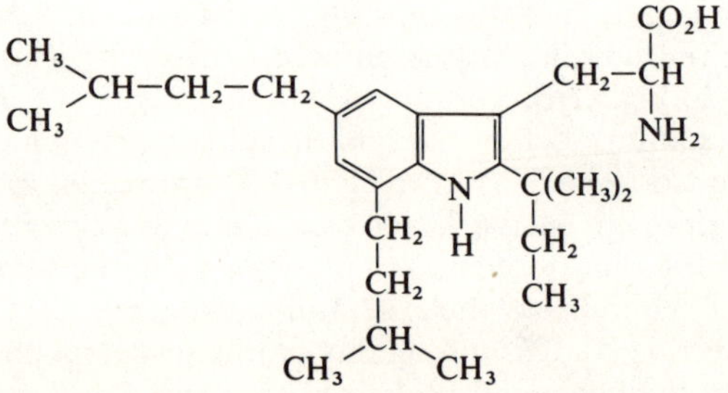

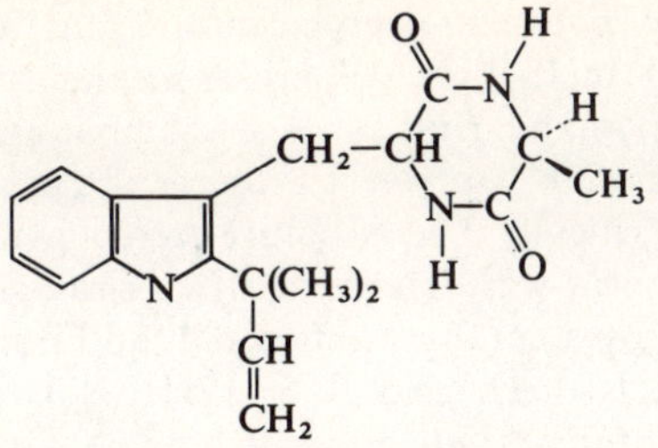

(IV)

E[17] (VIII). Sporidesmin is the causative agent of facial eczema, an hepatotoxic condition of sheep. The structures of sporidesmin and sporidesmin B were determined by chemical and spectroscopic methods[18], including mass spectrometry. The c.d. spectra of these compounds in comparison with that of gliotoxin showed that the common structural unit possesses the same absolute configuration[19]. X-ray crystal structure analyses of sporidesmin

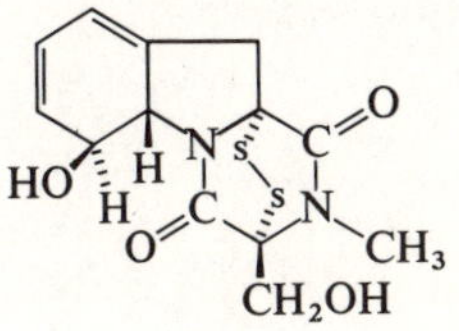

(V) 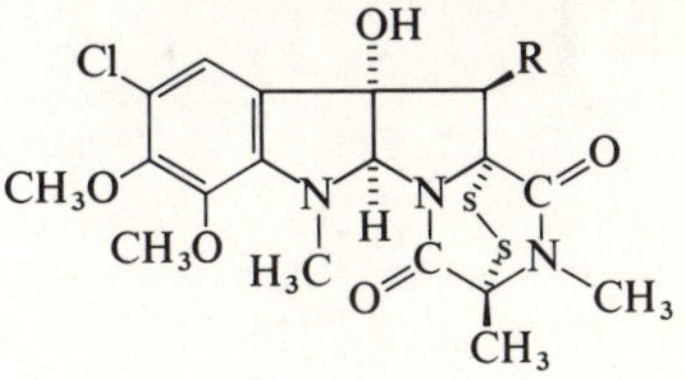　(VI)

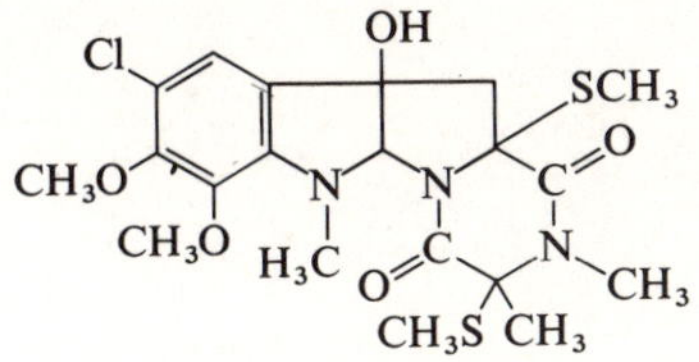

(VII) 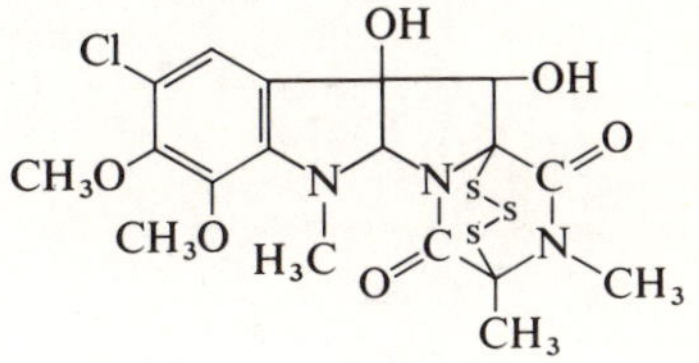　(VIII)

revealed the relative and later the absolute configuration[12,20] and confirmed the circular dichroism studies. It is interesting to note that the —S—S—dihedral angle in these compounds is not the usual value of ~ 100° but rather ~ 10–20°. The structural relationship of the other sporidesmins was shown by the conversion of sporidesmin E to sporidesmin using the thiophile triphenylphosphine[17]. Sporidesmin can be converted to sporidesmin D. Recently it was suggested on the basis of n.m.r. studies that the trithia-bridge in sporidesmin E can exist in solution in two distinct conformations; these conformations show a differential reactivity towards the thiophile[21].

A fungus which shows antiviral activity (*Arachniotus aureus*) produces a rather complex symmetrically substituted 2,5-dioxopiperazine, aranotin[22] (IX; R = H). Apoaranotin (X), acetylaranotin (IX; R = Ac) and its bisdethiodi(methythio)derivative (XI) co-occur as metabolites. The structural relationship between acetylaranotin and apoaranotin was determined by physical methods[23]. Whereas in the sporidesmins and the gliotoxins the dioxopiperazine ring exists in the boat form, in acetylaranotin the skew-boat conformation is preferred due to the extra strain imposed by the fused pyrrolidine rings[24]. The absolute stereochemistry of bisdethiodi-(methythio)acetylaranotin was determined by X-ray crystallography[25], and that of acetylaranotin by a combination of c.d. and n.m.r. spectroscopy[26].

Verticillin A (XII; R = H, and R' = OH), isolated from a species of *Verticillium*[27], exerts antimicrobial activity against Gram-positive bacteria and mycobacteria, but not against Gram-negative bacteria and fungi[27].

(IX) (X) (XI) (XII)

Degradative studies[27] again revealed the presence of the dithiodioxopiperazine cage-structure. A closely related structure, chaetocin (XII; R = OH, and R' = H), a metabolite of the fungus *Chaetomium minutum*, possesses antibacterial and cytostatic activity[28]. Its structure was determined by a combination of X-ray crystallography and n.m.r. spectroscopy[28]. The c.d. spectra of verticillin A and chaetocin are antipodal to those of gliotoxin, sporidesmin, and aranotin, thus establishing the absolute configuration of verticillin A.

The Preparation of Dioxopiperazines

The standard method for the preparation of cyclic dipeptides involves the spontaneous cyclisation of free dipeptide alkyl esters. The latter can either be formed by the action of ammonia on the corresponding amine salts[29,30] or by the hydrogenation of N-benzyloxycarbonyl-dipeptide alkyl esters in neutral solution[31]. Two useful methods giving rise to optically homogeneous dioxopiperazines have recently been described. The heating of a free dipeptide or its salt in molten phenol effects ring closure[32] (this is similar to, although cleaner than, an older method in which β-naphthol was used as solvent[33]). Dipeptide methyl ester formate salts on being heated in neutral solution also give rise to good yields of dioxopiperazines[34]. The formate salt can be prepared directly from a t-butyloxycarbonyl-dipeptide alkyl ester by fission of the N-protecting group with formic acid.

The Conformation of Dioxopiperazines

Attempts have been made to compute the stable conformations of the side-chains in substituted dioxopiperazines using Van der Waals contact criteria and also energy minimisation calculations[35], and quantum mechanical calculations using the PCILO method[36]. However, in these approaches the dioxopiperazine ring system was assumed to be planar (see later). In the molecular orbital approach[36], the most stable arrangements of the side-chains in cyclo-(glycyl-phenylalanyl) and cyclo-(glycyl-valyl) were found to be the gauche conformers (XIIIa) and (XIV) respectively (see later).

(XIII)

(XIV)

The first X-ray diffraction analysis of a cyclo-dipeptide, cyclo-(glycyl-glycyl) (XV), led to the conclusion that the dioxopiperazine ring is planar[37,38]. However, similar studies of cyclo-(L-alanyl-L-alanyl)[39,40] (cis-form, (XVI)) and cyclo-(L-alanyl-D-alanyl)[39,41] (trans-form, (XVII)) indicate that while the trans-compound possesses an essentially planar ring, in the cis-compound the planes containing the cis-amide linkages are folded away[39] from the side-chains through an angle of 26°. Sletten[40] describes his model of the cis-compound as a skewed boat conformation with the side-chain methyl groups located quasi-equatorially. These two conformations are essentially similar, but there are small differences in some bond parameters. Benedetti et al.[42] suggest that in the cyclo-(alanyl-alanyl) system small bond angle deformations can be magnified to produce large variations in the internal torsion angles leading to large variations in ring conformation.

From an X-ray diffraction analysis of *cyclo*-(sarcosyl-sarcosyl) (XVIII) Groth[43] concluded that the ring system deviates from planarity, adopting a flattened chair conformation.

Kopple and Marr[44] studied the 60 MHz n.m.r. spectrum of tyrosine-containing dioxopiperazines such as *cyclo*-(glycyl-L-tyrosyl) (XIX; R = OH) and compared the chemical shifts obtained with those of model compounds

(XV)   (XVI)   (XVII)   (XVIII)

which did not contain the aromatic side-chain. It was observed that the *p*-hydroxybenzyl side-chain causes the *cis*-hydrogen of the glycyl methylene group to resonate 1–1·5 ppm upfield of its normal position. Similar observations have been made in independent 100 MHz n.m.r. studies on dioxopiperazines containing one residue of glycine together with either phenylalanine[45] (XIX; R = H), histidine[46] (XX), or tryptophan[46] (XXI).

(XIX)   (XX)

These observations have been interpreted in terms of a preferential location of the aromatic side-chain in a sandwich conformation (XIIIa), thus subjecting the glycyl methylene hydrogens to aromatic ring-current shielding. On the assumption that the dioxopiperazine ring is planar, the theoretical shielding of the glycyl methylene hydrogens by the aromatic ring in this conformation was calculated[44,45] from the data of Johnson and Bovey[47]. Kopple[44], in the case of the tyrosine-containing dioxopiperazine, noted that the observed magnetic shielding effects were as large as those calculated. A study of the temperature dependence of the chemical shifts enabled the equilibrium constant between folded (XIIIa) and unfolded forms (XIIIb and c) to be calculated. Kenner and Sheppard[45], however, in the case of the phenylalanine dioxopiperazine, observed that the shielding found was too large even for the conformation shown to be completely dominant. The implication here is that the dioxopiperazine ring is buckled into a boat conformation (the planes containing the *cis*-amide bonds being buckled away from the benzyl substituent by 5 ∼ 10° [cf. the earlier discussion of the X-ray work on *cyclo*-(L-alanyl-L-alanyl)]) bringing H_A nearer to the aromatic nucleus (XXII). This interpretation was corroborated by an analysis of the coupling constants of the ABX pattern of the phenylalanine side-chain and also by the larger coupling of the *trans*-hydrogen, H_B, with the adjacent NH group (XXIII). The benzyl group thus takes up a 'flagpole' orientation in agreement with the results of the molecular orbital calculations mentioned earlier. Using similar arguments the rings of *cyclo*-(glycyl-L-tyrosyl,-L-histidyl, and -L-tryptophyl) have been shown to be non-planar[46] in dimethyl sulphoxide. In the case of *cyclo*-(L-alanyl-L-phenylalanyl) the shielding of the methyl group is only slightly less than that calculated (0·86 ppm as opposed to 0·96 ppm). It was concluded that in this case the molecule possesses a more nearly planar dioxopiperazine ring since the two 'flagpole' substituents interact sterically and reduce the angle of buckling.

(XXI)   (XXII)   (XXIII)

Experiments with *N*-methylated dioxopiperazines[45], which are more soluble in less polar solvents than are the unsubstituted dioxopiperazines, show that large shielding effects persist in media of low dielectric constant. This indicates that although hydrophobic effects may contribute to the dominance of the folded conformation, especially in more polar solvents, a direct interaction between the amide groups and the aromatic ring is clearly important.

Spectroscopic Studies

In the o.r.d. spectra of cyclic dipeptides multiple Cotton effects have been observed whose position depends on solvent composition[48,49]. The o.r.d. and c.d. spectra of simple dioxopiperazines have been interpreted in terms of an n—π* transition, and exciton interactions resulting in a splitting of the π—π* transition of the *cis*-amide chromophore[48,49]. This effect could be attributed to non-planarity of the dioxopiperazine ring system. According to Schellman and Nielson[50] the optical properties of *cyclo*-(L-alanyl-L-alanyl) (XVI) can be interpreted in terms of a folding of the dioxopiperazine ring away from the methyl groups. This folding allows overlap to occur between the orbitals of the two amide chromophores, which produces the observed exciton splitting effect. The c.d. absorption spectrum of the phenol and indole chromophores in the dioxopiperazines of tyrosine and tryptophan have been determined in the near ultraviolet in dioxan and dimethyl sulphoxide. The observed molecular ellipticities are several times larger than those of equivalent linear dipeptides, and this enhancement is ascribed to the rigid conformation of these dioxopiperazines[51].

In the mass spectrometer simple dioxopiperazines give rise to a highly characteristic fragmentation pattern[52]. The most probable molecular ion is

(XXIV). The three most characteristic fragmentations, deduced from studies with deuterated compounds, are listed below as (i), (ii), and (iii). The more complex dioxopiperazines derived by desulphurisation of acetylaranotin (IX; R = Ac) gliotoxin (V), etc., show a more complex fragmentation pattern[53], which was fully utilised in the structural studies of these compounds. The four general modes of cleavage revealed are illustrated for a tetrasubstituted proline dioxopiperazine (XXV, a, b, and c) and are listed below as (i), (ii), (iv), and (v).

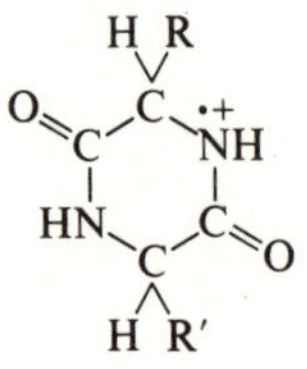

(XXIV)

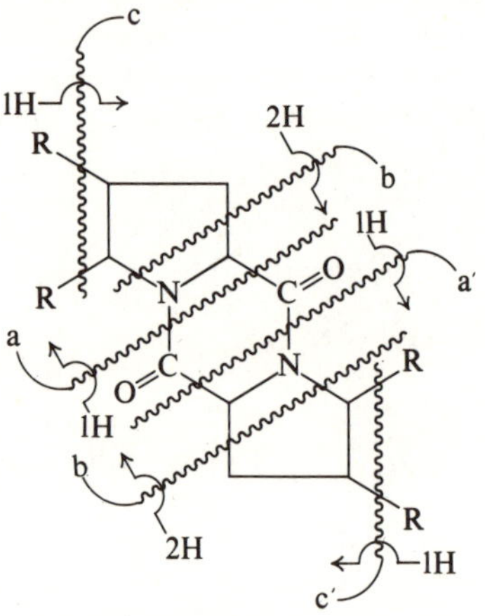

(XXV)

(i) Elimination of CO or CHO (*Scheme 2*).
(ii) Amine fragmentation (*Scheme 3*, and XXVa,a′).
    The amine fragment which arises from cleavage a,a′ constitutes the most important diagnostic evidence for structure elucidation of dioxopiperazines. This fragmentation occurs with the transfer of one hydrogen from the rest of the molecule.
(iii) Elimination of CONH (*Scheme 4*).
(iv) Dioxopiperazine fragmentation (XXVb,b′).
    This fragmentation occurs with the transfer of two hydrogens from the rest of the molecule.
(v) Elimination of the ring adjacent to the proline moiety (XXVc, c′).
    This fragmentation occurs with transfer of one hydrogen from the rest of the molecule.

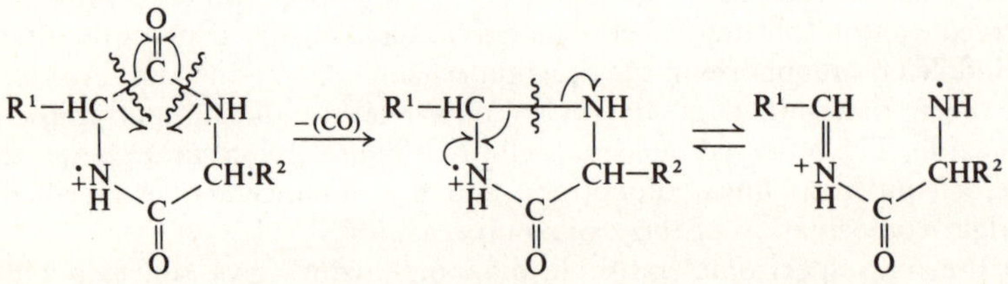

Scheme 2

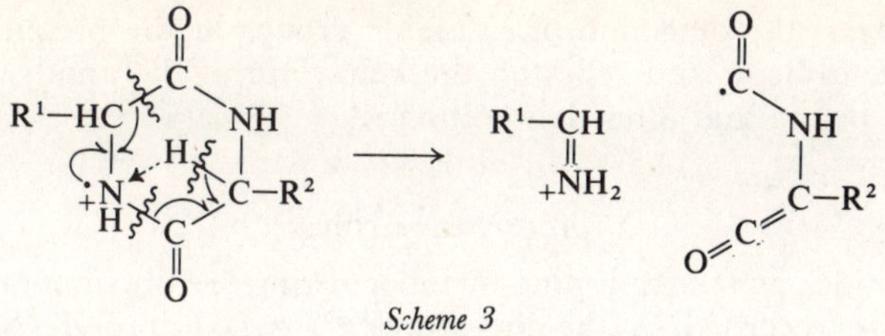

Scheme 3

The absence of an amide II band at 1550 cm^{-1} in the i.r. spectrum of a secondary amide is taken as negative evidence for the *cis*-amide conformation[54], a band in this region being regarded as characteristic of the *trans*-link. Bláha, Smolíková, and Vítek[55] have studied the i.r. spectra of a wide range of 3-, and 3,6-substituted 2,5-dioxopiperazines as nujol or fluorolube mulls and

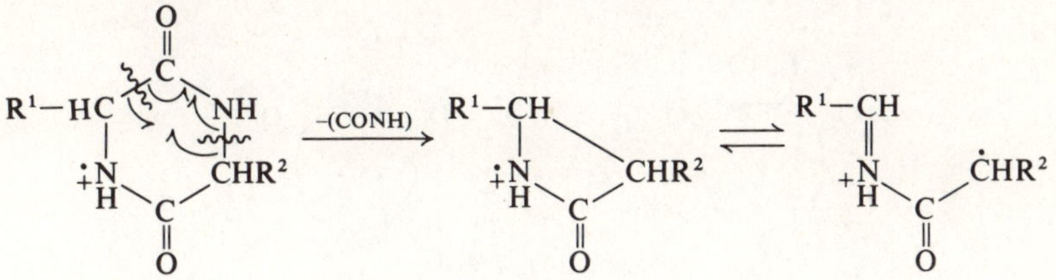

Scheme 4

identified the frequencies characteristic of the *cis*-conformation of the amide group, using the correlations of Miyazawa[56], which arose from a study of dioxopiperazine itself. The situation is summarised in *Table 5.1*. The detection of *cis*-amide bonds alone clearly does not present difficulty, for the amide A, amide I, amide II (no band at 1 550 cm^{-1}), and amide III bands can be

Table 5.1. CHARACTERISTIC I.R. FREQUENCIES (cm^{-1}) SHOWN BY *cis*-SECONDARY AMIDES (DIOXOPIPERAZINES) COMPARED WITH THOSE FOR SIMPLE *trans*-AMIDES*

Band	cis-Amide			trans-Amide	
	Frequency	Frequency on deuteration	Vibrational modes	Frequency	Frequency on deuteration
Amide A	3 180–3 195	2 250–2 350	ν N-H - - -	~3 300	~2 450
Amide B	3 040–3 080 (3 040–3 055)	—	ν N-H - - -	~3 100	
Amide I	1 670–1 690		$\begin{cases} \nu \text{ C=O} \\ \nu \text{ C-N} \\ \delta \text{ N-H in plane} \end{cases}$	~1 650	
Amide II	1 440–1 455	1 230–1 250	δ N-H in plane	~1 550	1 450
Amide III	1 305–1 345	No change	ν C-N	$\begin{cases} \nu \text{ C-N} \\ \delta \text{ N-H} \\ \text{in} \\ \text{plane} \end{cases}$	

*With oligopeptides, however, the frequencies observed depend on their secondary structures[57].

used. However, the detection of *cis*-amide groups in the presence of *trans*-ones is more difficult and relies on the behaviour of the amide II and III bands both before and after deuteration[55].

CYCLO-TRIPEPTIDES

cyclo-Tripeptides possessing a nine-membered ring are uncommon; only two examples are recorded in the literature[58,59]. *cyclo*-Triprolyl (XXVI) was prepared by Rothe and co-workers[58] by cyclising diprolylproline in dilute solution either as its *p*-nitrophenyl active ester or by using tetraethyl pyrophosphite as the condensing agent. The sublimable product (obtained in excellent yield) was characterised by partial hydrolysis and molecular weight determination (Rast method and mass spectrometric method). *cyclo*-Trisarcosyl (XXVII) was prepared by Dale and Titlestad[59] by the

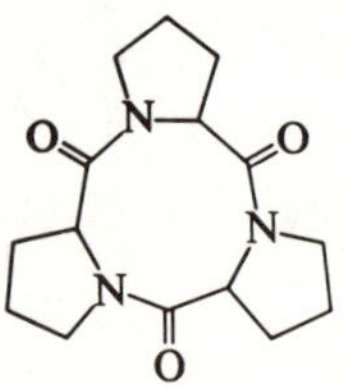

(XXVI) 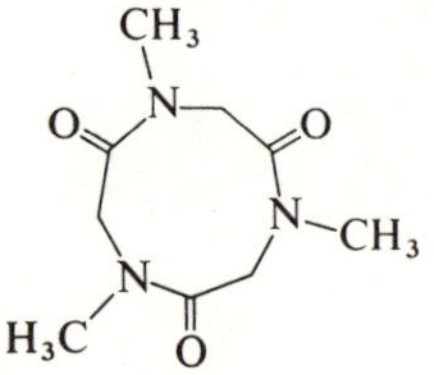   (XXVII)

high dilution cyclisation of 2,4,5-trichlorophenyl disarcosylsarcosinate. The molecular weight was determined by mass spectrometry. The n.m.r. spectrum possessed a singlet for the N—CH_3 groups and a single AB pattern for the methylene hydrogens. A high temperature was required to cause the AB pattern to collapse, and the molecule possessed a high dipole moment. The only conformation consistent with this data is a nine-membered ring possessing *cis*-peptide units with all the carbonyl groups located on the same side of the ring-plane (XXVIII). Venkatachalam[60] showed that a model of *cyclo*-triprolyl could only be constructed with *cis*-amide bonds. Stereochemical studies using contact criteria showed that a satisfactory structure (XXIX) could be generated only if the *cis*-peptide units were distorted from planarity; this is similar to Dale's proposed structure (XXVIII) for *cyclo*-trisarcosyl.

Only one *cyclo*-tripeptide has been isolated from natural sources, albeit one with a twelve-membered ring. This compound, aspochracin[61], was

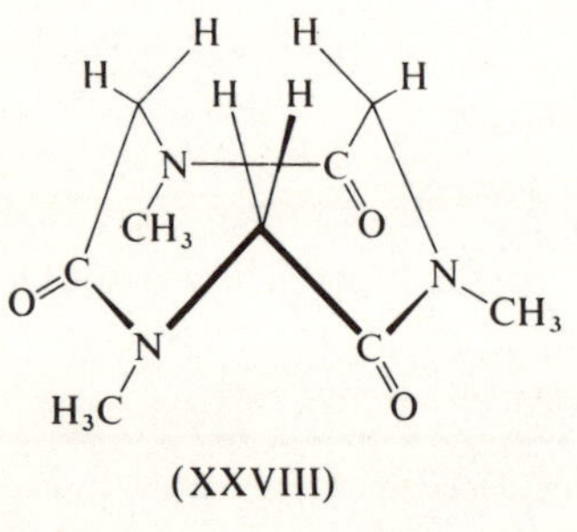

(XXVIII) 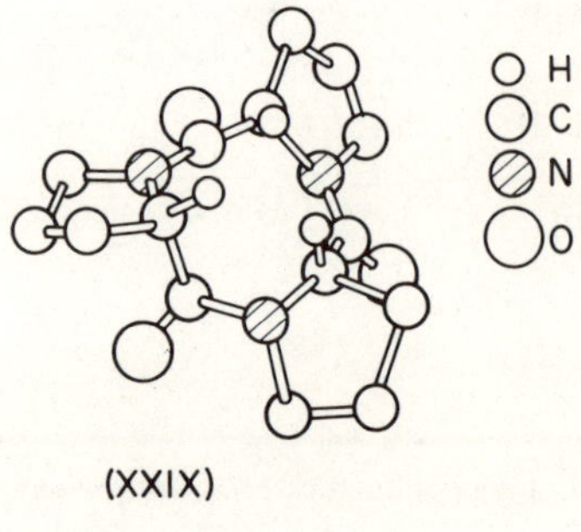   (XXIX)

isolated from the culture medium of a muscardine fungus, *Aspergillus ochraceus*. It is an insecticidal pathogen, in particular causing paralysis and death to silkworm larvae. Aspochracin possesses a λ_{max}^{MeOH} at 297 nm, but its hydrogenation product, hexahydroaspochracin, possesses only end absorption. Total acid hydrolysis of the latter compound yielded *N*-methyl-L-valine, L-ornithine, *N*-methyl-L-alanine, and caprylic acid. Identification of the latter compound indicated that the chromophore in aspochracin was probably an octatrienecarboxamide. Partial hydrolysis of hexahydroaspochracin with baryta yielded a peptide with the acid and amino-acid content of hexahydroaspochracin, which was shown to contain *N*-terminal ornithine and *C*-terminal *N*-methyl-alanine. The δ-amino group of ornithine is presumably the *N*-terminus.* The structures proposed for aspochracin (XXX) and hexahydroaspochracin (XXXI) are consistent with their spectroscopic properties. Hexahydroaspochracin has been synthesised in

(XXX, R = $CH_3-(CH=CH)_3-$)
(XXXI, R = $CH_3-(CH_2)_6-$)

(XXXII)

poor yield from *N*-methyl-L-valyl-*N*-methyl-L-alanyl-α-octanoyl-L-ornithine (XXXII) by cyclisation with half an equivalent of dicyclohexylcarbodiimide.

CYCLO-TETRAPEPTIDES

Only a few cyclic tetrapeptides are known. The first to be studied was the simplest, *cyclo*-tetraglycyl, which was prepared by high dilution cyclisation of triglycylglycine cyanomethyl or thioglycollic acid active esters[62]. A tentative conformation for *cyclo*-tetraglycyl (XXXIII) incorporating four *trans*-amide bonds has been suggested, but this structure involves rather distorted bond parameters; i.r. evidence is in accord with *trans*-peptide bonds. Theoretical calculations of the conformation of *cyclo*-tetra-L-alanyl (XXXIV; R = R¹ = CH_3, and R² = R³ = H) based on stereochemical criteria (contact criteria) and energy considerations have been made[63], and in

* Ref. 61, however, cites the α-amino group, which is inconsistent with the structure proposed, and probably represents a typographical error.

order to get acceptable N—C$^\alpha$—C' bond angles it was necessary to distort the *trans*-amide units from planarity (XXXV).

Most cyclic peptides have been prepared by a variation of the active ester method under conditions of high dilution. A novel variation of this method in which the phenolic component of the active ester is a high molecular weight insoluble polymer has been used to prepare a number of *cyclo*-tetrapeptides including *cyclo*-tetra-L-alanyl[63a]. The appropriate N-protected tetrapeptide is esterified to poly-4-hydroxy-3-nitrostyrene, and subsequent removal of the N-protecting group then allows intramolecular aminolysis of the polymeric active ester, releasing the cyclic product. Since

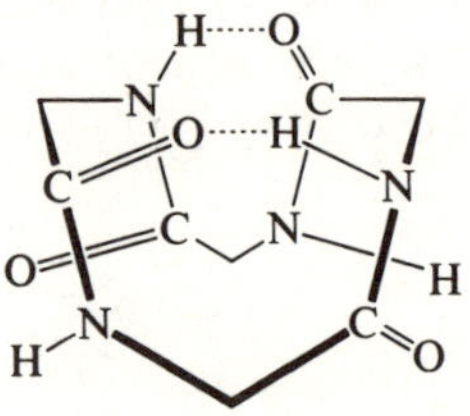

(XXXIII)

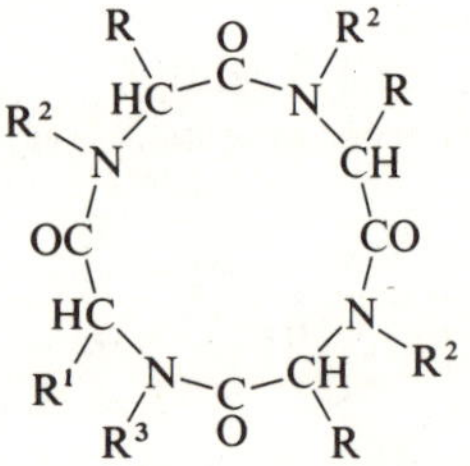

(XXXIV)

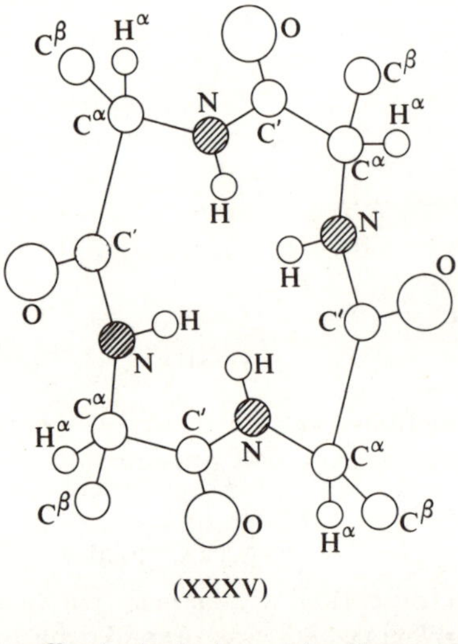

(XXXV)

the polymeric carrier effectively isolates the linear peptide chains, intermolecular side-reactions such as cyclodimerisation are reduced.

Dale and Titlestad[59] prepared *cyclo*-tetrasarcosyl (XXXIV; R = R^1 = H, and R^2 = R^3 = CH$_3$) and found its n.m.r. spectrum to consist of two N—CH$_3$ signals (intensity 6:6) and two methylene AB patterns (intensity 4:4). Symmetry considerations suggest that the molecule must be centrosymmetric with the four amide groups in the sequence *cis*-, *trans*-, *cis*-, *trans*-. The high temperature required to bring about coalescence of the AB pattern suggests that strong transannular interactions occur rendering inversion of the nonplanar twelve-membered ring difficult. The conformation postulated for this compound in solution was similar to but not identical with the conformation (XXXVI) deduced from X-ray crystallography[64]. The widely spaced AB pattern in the n.m.r. spectrum[65] ($\Delta\nu = 2\cdot1$ ppm) was assigned to the cyclohexane type hydrogens at the 2- and 8-positions, and the other AB pattern ($\Delta\nu = 0\cdot8$ ppm) to the extended chain type hydrogens at the 5- and 11-positions. This assignment then enabled a comparison to be made with the n.m.r. spectrum of synthetic cyclo-(trisarcosyl-N-methyl-L-alanyl) (XXXIV; R = H, and $R^1 = R^2 = R^3 = CH_3$) which showed that the α-methyl group occupies a quasi-equatorial orientation at the 2 (or 8) position. The interpretation of the n.m.r. spectrum of synthetic cyclo-

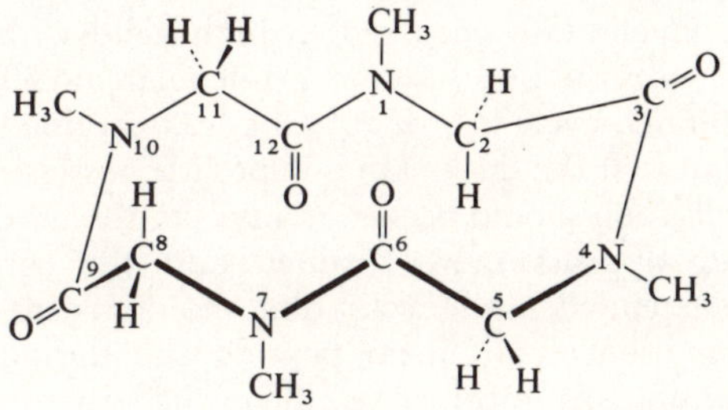

(XXXVI)

(trisarcosyl-L-alanyl) (XXXIV; $R = R^3 = H$, and $R^1 = R^2 = CH_3$) is consistent with the N—H group being part of a *trans*-system, which is confirmed by the large vicinal NH—CH coupling[66] ($^3J_{NH-CH} = 10$ Hz). Since the secondary amide linkage is *trans*-(i.r. evidence) the α-methyl group is similarly located at the 2-position. Synthetic cyclo-(glycyl-trisarcosyl) (XXXIV; $R = R^1 = R^3 = H$, and $R^2 = CH_3$) and cyclo-(glycyl-sarcosyl-glycyl-sarcosyl) show similar n.m.r. spectra, strongly suggesting that these two compounds also adopt a conformation of the type described above. In this conformation the stereochemical sequence LLDD would be necessary if a cyclo-tetrapeptide were to have its side-chains located in external (quasi-equatorial) positions.

The cyclic tetra-depsipeptide cyclo-(D-hydroxyisovaleryl-N-methyl-L-isoleucyl-D-hydroxyisovaleryl-N-methyl-L-leucyl) (XXXVII) was shown to adopt a different conformation[67] in the crystal, presumably because the stereochemical sequence is different to the one mentioned above, i.e. DLDL.

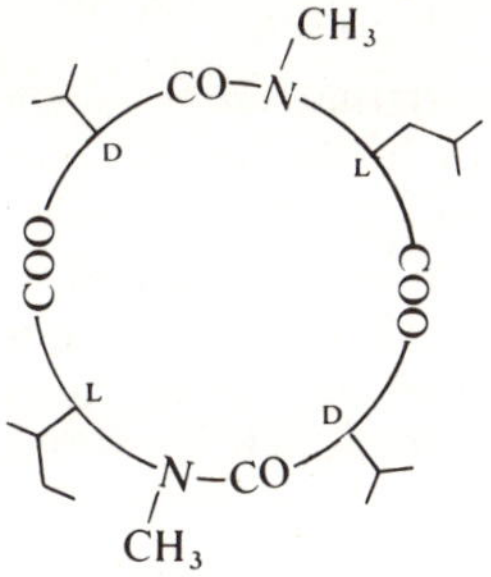

(XXXVII) 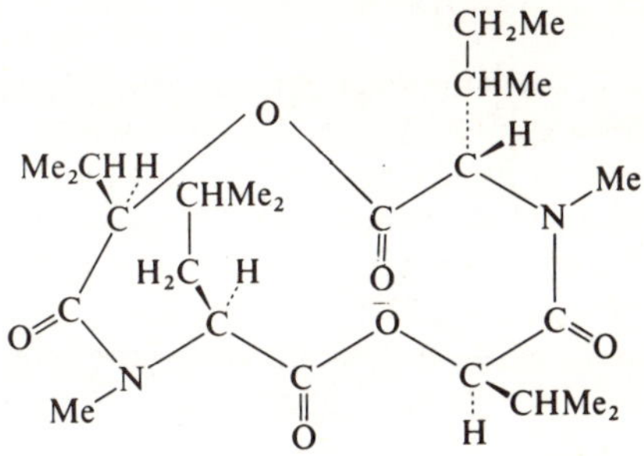   (XXXVIII)

143

The conformation (XXXVIII) contains two ester units in the *trans*-arrangement and two *cis*-N-methyl peptide units. All the carbonyl groups point to one side of the ring-plane while side-chains extend on the opposite side.

A comparative study of the ease of cyclisation of a series of diastereoisomeric linear tetradepsipeptides (XXXIX) related to the enniatins showed that the yield of *cyclo*-tetradepsipeptide obtained depends on the sequence of

(XXXIX)

configurations[68]. Cyclisation occurred in high yield with the LDLD-isomer, and in low yield with the DDDD-isomer. Molecular models of these peptides indicate that the LDLD-cycle is probably the least strained (cf. conformation (XXXVIII)) and that the linear LDLD-peptolide possesses a bent conformation[69], hence cyclisation should occur readily. On the other hand, the model of the DDDD-ring appears more strained and the corresponding linear peptolide more extended, and cyclisation should occur less readily. The preferred conformation of the linear peptide and the degree of conformational or transannular strain present in the cyclic structure to be formed are the two major factors influencing the ease of cyclisation.

The only totally α-linked cyclic peptide isolated from natural sources is fungisporin. On destructive distillation spores of *Penicillium* and *Aspergillus* yield this crystalline sublimate which was originally assigned the *cyclo*-octapeptide structure[70], *cyclo*-(D-Val-L-Val-D-Phe-L-Phe)$_2$. Redetermination of the molecular weight by mass spectrometry indicated the *cyclo*-tetrapeptide structure[71] (XL) which has been confirmed by a conventional synthesis[71].

(XL)   (XLI)   (XLII)

Two other cyclic tetrapeptides isolated from natural sources contain β-linkages. The lichen *Roccella canariensis* contains a cyclic peptide which on total acid hydrolysis gave proline and (+)-β-amino-β-phenylpropionic acid. High resolution mass spectrometry was consistent with its formulation as a cyclic tetrapeptide, and the n.m.r. spectrum indicated that it was symmetrical. Structure (XLI) appears to be the most probable, but (XLII) cannot be entirely ruled out[72]. Serratamolide is a cyclic tetradepsipeptide which has been more fully studied. The structure (XLIII) of this antibiotic was elucidated by Wasserman[1]. More recently a homologue, serratamolide B (XLIV), from the mycelium of *Serratia marcescens* has been characterised[73].

Serratamolide and some of its analogues have been synthesised by Shemyakin and his co-workers[74].

In a parallel study, the synthesis of the structurally related fourteen membered *cyclo*-depsipeptide, *cyclo*-di-(β-seryloxypropionyl) (XLV) has been carried out[75,76]. The ring was generated by cyclodimerisation of β-(O-t-butylseryloxy)propionyl chloride (XLVI) using triethylamine in

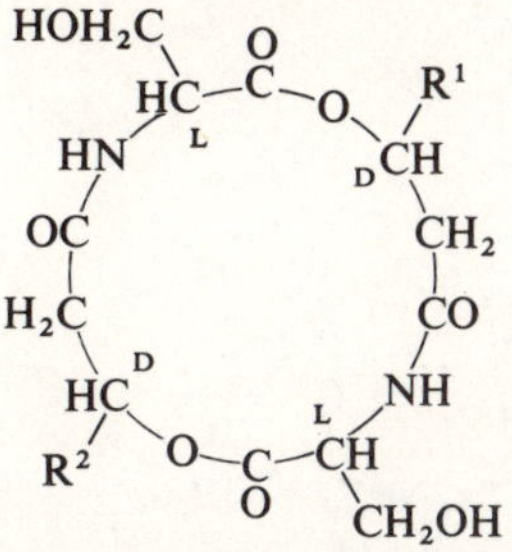

(XLIII, $R^1 = R^2 = (CH_2)_6CH_3$)
(XLIV, $R^1 = (CH_2)_8CH_3$
  $R_2 = (CH_2)_6CH_3$)
(XLV, $R^1 = R^2 = H$)

benzene. In the case of the cyclisation of β-(O-t-butyl-DL-seryloxy)propionyl chloride, two isomeric cyclic fourteen-membered rings were obtained[75]. The major product (present to 60%) was proved to be the DL (*meso*) form by an unambiguous cyclisation[76] of β-(O-t-butyl-L-seryloxy)propionyl-β-(O-t-butyl-D-seryloxy)propionyl chloride (XLVII). The minor product was the racemic macrocycle; this was proved by synthesis[76] of L,L and D,D-*cyclo*-di-

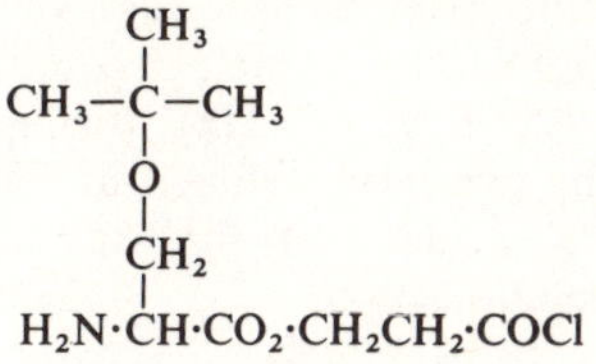

$$CH_3-\underset{\underset{O}{|}}{\overset{\overset{CH_3}{|}}{C}}-CH_3$$
$$|$$
$$CH_2$$
$$|$$
$$H_2N \cdot CH \cdot CO_2 \cdot CH_2CH_2 \cdot COCl$$

(XLVI)

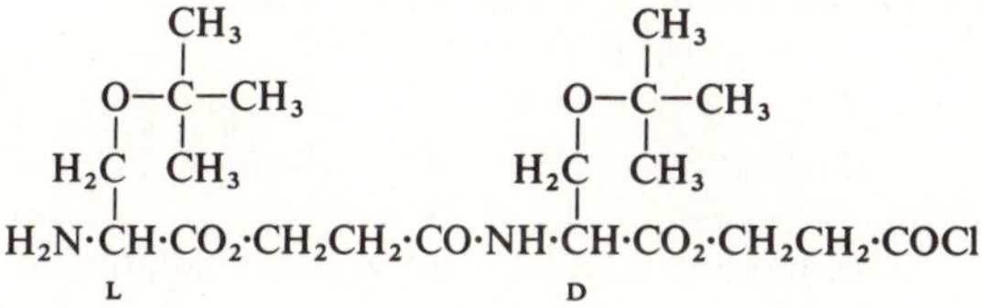

(XLVII)

(β-seryloxypropionyl) by twinning the appropriate optically homogeneous monomer. Further examples of the twinning of a racemic monomer giving rise to a *meso*-dimer will be encountered later. One would suggest on the basis of the cyclisation experiments of Kenner[69] (see later) and Shemyakin[68] that in all probability either the presumed intermediate in the twinning process (the partially protected β-L-seryloxypropionyl-β-D-seryloxypropionyl-derivative) possesses a conformation more favourable to cyclisation than the all L or all D diastereomer, or, alternatively, the conformational or transannular strain present in the *meso*-macrocycle may be less than in the all-L or all-D isomer. A conformation (XLVIII) was postulated for these rings[75,76]

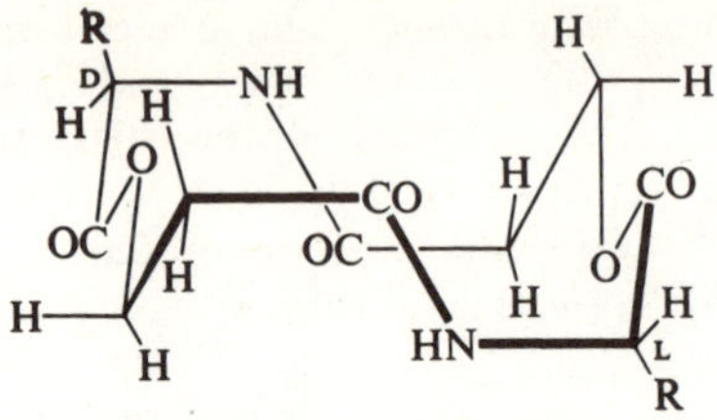

(XLVIII, R = CH$_2$—OH)

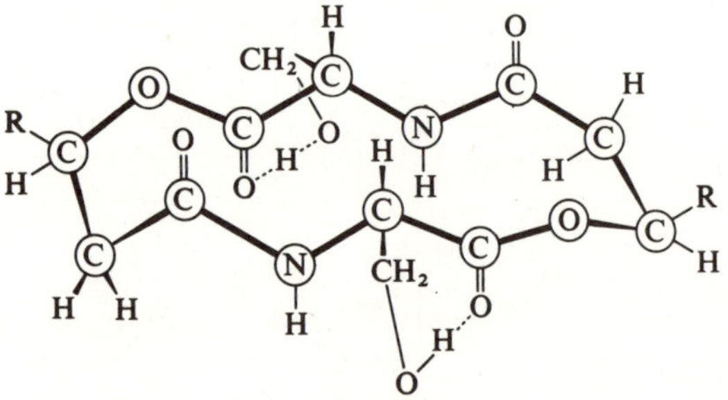

(XLIX)

based on the conformation of *cyclo*-tetradecane[77] and of a 1,8-diazacyclo-tetradecane derivative[78]. Recently Hassall[79] has rejected this type of conformation for serratamolide itself on the basis of n.m.r. studies and now favours the conformation (XLIX), which accounts for the observed large but identical couplings ($^3J_{NH-CH}$ = 9 Hz) arising from the pair of *trans*-NH—CH hydrogens of the two seryl residues, and also the small and similar couplings ($^3J_{CH-CH_2}$ = 2 ~ 5 Hz) of the ring methylene and methine hydrogens in a gauche relationship.

CYCLO-PENTAPEPTIDES

No cyclic peptides containing five common amino-acids in the ring system have been isolated from natural sources. The first *cyclo*-pentapeptides to be synthesised[80] were two diastereoisomers of *cyclo*-(glycyl-leucyl-glycylglycyl-leucyl); a conventional high dilution cyclisation of the *p*-nitrophenylthiol ester was the final step. The yield of the LD-isomer was higher than that of the LL-form. Subsequent studies on the cyclisation of glycyl-leucyl-leucyl-glycylglycine (LD and LL) using two methods indicate that movement of the leucyl-residues to adjacent positions in the peptide chain accentuates the difference in yields on cyclisation[69]. The dielectric increment of a linear peptide (the increase in dielectric constant of water on dissolving in it the zwitterionic peptide) is related to its dipole moment, and can be used to provide information about its length and hence conformational preferences. Dielectric increment measurements on oligopeptides, including the linear precursors of the *cyclo*-pentapeptides[69,80] discussed above, show that there is a smaller average distance between the ends of the chain in the DL-diastereoisomers compared to the corresponding LL-forms. Cyclisation would therefore be more favoured in the former cases.

The n.m.r. spectrum of *cyclo*-(pentasarcosyl) (L) contains five N—CH$_3$

singlets (intensity 3:3:3:3:3) and five recognisable methylene AB quartets (intensity 2:2:2:2:2). It is not until a temperature of 130°C that the spectrum collapses to a pair of singlets (N—CH$_3$ and CH$_2$), indicating that this conformationally homogeneous cycle possesses a high resistance to ring inversion. This suggests strong transannular interactions between the amide groups[59].

Theoretical calculations have been made to compute the conformations[81,82] of cyclic pentapeptides. In the case of *cyclo*-(glycylglycylglycylprolyl-prolyl)

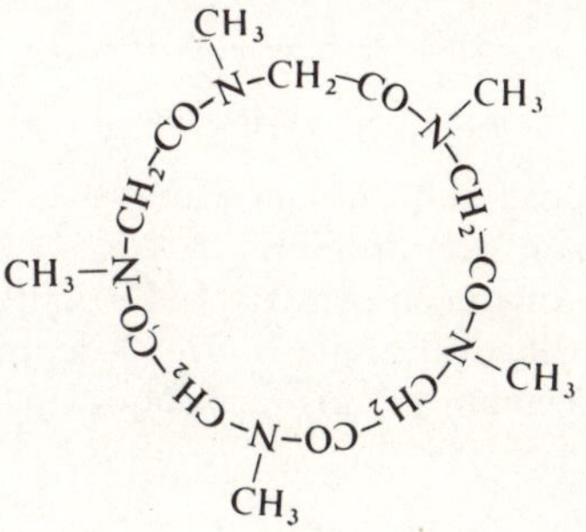

(L)

three conformations of minimum energy have been identified (LI; a, b, and c; the values of minimum energy associated with each conformation are given). It is suggested that conformations b, and c are the predominant forms co-existing in solution.

An attempt has been made to devise a synthetic model system that is capable of carrying out energy-coupled transport of cations across a cell-membrane[83]. The model was designed to complex with hydrated A-cations, which are located in the aqueous phase on one side of a lipophilic membrane,

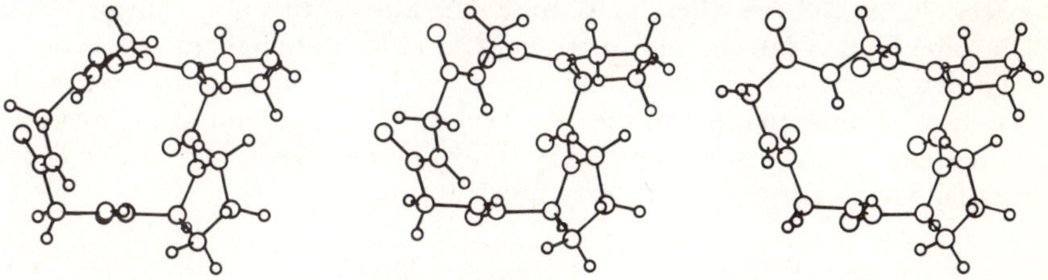

(a)   2·09 kcal   (b)   1·27 kcal   (c)   1·35 kcal

(LI)

and transport them through the membrane against a concentration gradient to the aqueous phase on the other side. The work done against the gradient of osmotic pressure could be compensated for by the energy released by a thiol–disulphide redox system[83]. The model compound synthesised[84] was the cyclic thiol, *cyclo*-(glycyl-L-cysteinyl-glycylglycyl-L-prolyl), which was oxidised to S,S'-bis-*cyclo*-(glycyl-L-hemicystinyl-glycylglycyl-L-prolyl) (LII), the first material to be prepared containing two homodetic peptide rings linked together. This compound is rather insoluble in water unless lithium, sodium, or potassium salts are present. Preliminary experiments on cation

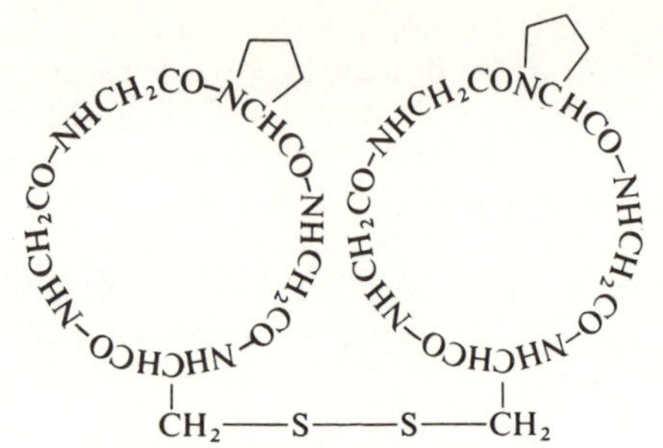

(LII)

transport using synthetic membranes show that the compound (LII) possesses a specificity for cations in the order $K^+ > Na^+ > Li^+ > Ca^{++}$. A positive c.d. band at 264 nm in uncomplexed (LII) points to a conformation in which the disulphide dihedral angle is 75–90 degrees with a right-handed helicity. The o.r.d. spectrum of this compound in water shows a negative

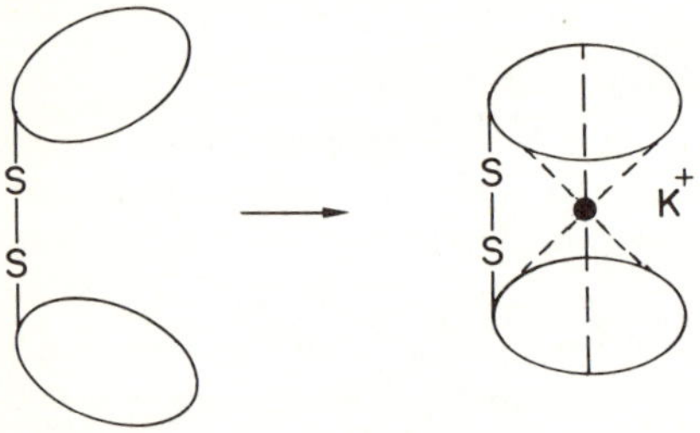

(LIII) 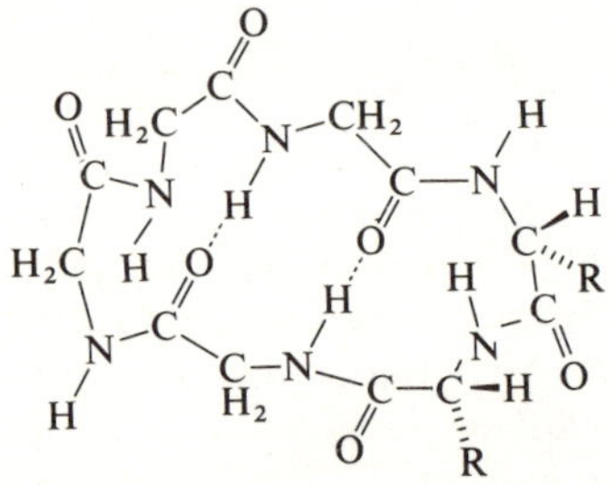   (LIV)

extremum at 233 nm whose amplitude increases on titration with potassium chloride. This is interpreted in terms of a conformational transition during complex formation which causes the carbonyl groups of the *trans*-amide bonds to become parallel to the ring axis and hence surround the metal ion with a cage of oxygen atoms (LIII). Thus the bicyclo-system is able to fold around some cations, particularly potassium ions.

CYCLO-HEXAPEPTIDES

The Conformation of cyclo-Hexapeptides

cyclo-Hexapeptides have been examined more thoroughly than any other type of cyclic peptide. Two X-ray diffraction studies have been reported. In cyclo-(hexaglycyl) hemihydrate[8b] eight molecules occupy the unit cell in four distinct conformations with *trans*-amide bonds. The predominant conformer (50% of the molecules exist in this form) contains two trans-annular hydrogen bonds (LIV; R = H) while the other conformers, which lack transannular hydrogen bonds, are hydrogen-bonded to interleaving water molecules. The lack of side-chains in this cycle allows enough conformational freedom for several conformer-types to be packed together in the crystal. In contrast, the single crystal X-ray diffraction analysis of *cyclo*-(glycylglycyl-D-alanyl-D-alanyl-glycylglycyl) trihydrate[86] shows only one conformer to be present; this corresponds to the intramolecularly hydrogen-bonded pleated sheet structure (LIV; $R = CH_3$), in which the methyl groups of the D-alanyl residues are orientated *quasi*-axially at corner positions in the cycle.

The knowledge gained of the conformation of cyclic hexapeptides in the crystalline phase together with improvements in technique and instrumentation of n.m.r. spectroscopy has given impetus to the study of such peptides in solution, to gather information not only on the shape of the peptide backbone, but also on the orientation of the side-chains. The n.m.r. spectra of two diastereoisomers of *cyclo*-(glycyl-phenylalanyl-glycylglycyl-phenylalanyl-glycyl) (LL and LD) in trifluoroacetic acid show a complex band of signals in the NH region[87]. However, *cyclo*-(glycyl-L-prolyl-glycylglycyl-L-prolyl-glycyl) shows only two distinct NH signals, which have been tentatively assigned as arising from solvated (exposed) and non-solvated (i.e. transannularly hydrogen bridged) peptide NH hydrogens[87]. Studies of the

(LV)

solvent dependence of the chemical shifts[88] confirm that the band at higher field arises from solvent shielded NH groups. On passing from d⁶-dimethyl sulphoxide to trifluoroacetic acid, the solvent exposed NH hydrogens become more shielded than the intramolecularly hydrogen-bonded ones. Moreover, the NH group occurring at low field undergoes deuterium exchange more rapidly than the NH group responsible for the high field signal. A model has been proposed for the conformation of the *cyclo*-hexapeptide (LV) taking into account that two types of glycyl methylene hydrogens are evident in the n.m.r. spectrum in d⁶-dimethyl sulphoxide.

Kopple and co-workers have studied the n.m.r. spectra of several *cyclo*-hexapeptides (LVI a–e). The amide hydrogen resonances were related to their parent α-CH resonances on the basis of spin-decoupling experiments, a comparison of deuterated and undeuterated derivatives, and by their apparent multiplicities. In d⁶-dimethyl sulphoxide each compound showed two distinct groups of amide hydrogen resonances; the high field group integrates for two and the low field group integrates for four hydrogens. It was established that the high field group should be assigned to transannularly hydrogen-bonded NH hydrogens by studying the solvent dependence and the temperature dependence of the amide hydrogen chemical shifts[89,90].

The mean chemical shift of the high field group was much less solvent dependent than the low field group, whose behaviour closely resembled that of linear peptides. In d⁶-dimethyl sulphoxide/water mixtures the chemical shifts of the low field amide hydrogens are at least twice as temperature dependent as the high field ones (e.g. in the case of LVIe the temperature

(LVI, a[89]: X = R¹ = H, R² = CH$_2$(C$_6$H$_4$)OHp
b[89]: X = D, R¹ = H, R² = CH$_2$(C$_6$H$_4$)OHp
c[90]: X = R¹ = H, R² = CH$_2$—CH(CH$_3$)$_2$

d[90]: X = H, R¹ = —CH$_2$—[imidazole],
R² = CH$_2$(C$_6$H$_4$)OHp

e[90]: X = D, R¹ = —CH$_2$—[imidazole],
R² = CH$_2$(C$_6$H$_4$)OHp)

coefficients are 0·61 Hz/deg. for the low field set and 0·12 Hz/deg. for the high field set). It is interesting that in this work, in contrast to the work with *cyclo*-(glycyl-L-prolyl-glycylglycyl-L-prolylglycyl) just discussed, exchange of all NH hydrogens with deuterium oxide proceeds at about the same rate. In d⁶-dimethyl sulphoxide the spin-decoupling experiments mentioned earlier also indicate that the α-hydrogens of histidine, leucine, or tyrosine (in the appropriate peptides (LVI a–e)) are vicinal to a low field amide hydrogen, thus establishing that these residues must be located at corner positions in the cycles[89,90]. At 220 MHz the resonances of the side-chains of the residues of histidine, leucine, and tyrosine were sufficiently resolved for the coupling constants to be measured. The values obtained favour a *trans*-relationship of the vicinal hydrogens on the α- and β-carbon atoms[90] (i.e. $x_1 = 180$ degrees (LVII)). A glycyl residue in partially deuterated *cyclo*-(pentaglycyl-L-tyrosyl) (LVIb) and *cyclo*-(glycylglycyl-L-histidyl-glycyl-glycyl-L-tyrosyl) (LVIe) shows a marked magnetic non-equivalence of its α-hydrogens[90]. Since in *cyclo*-(pentaglycyl-L-leucyl) (LVIc) the magnetic non-equivalence of a similar glycyl residue is much less, the tyrosyl residue is tentatively assigned as being responsible for this effect in the former compound.

Thus the conformation proposed for these cyclic hexapeptides (LVIII) in dimethyl sulphoxide and aqueous solutions, taking into account all the

(LVII)

data discussed, consists of two antiparallel extended peptide segments joined together by two intra-chain hydrogen bonds. These small pieces of β-structure are connected by two *trans*-peptide bonds which lie in planes perpendicular to the average plane of the hexapeptide ring. It is this conformation which has enabled Kopple[90] to offer a rationalisation for the

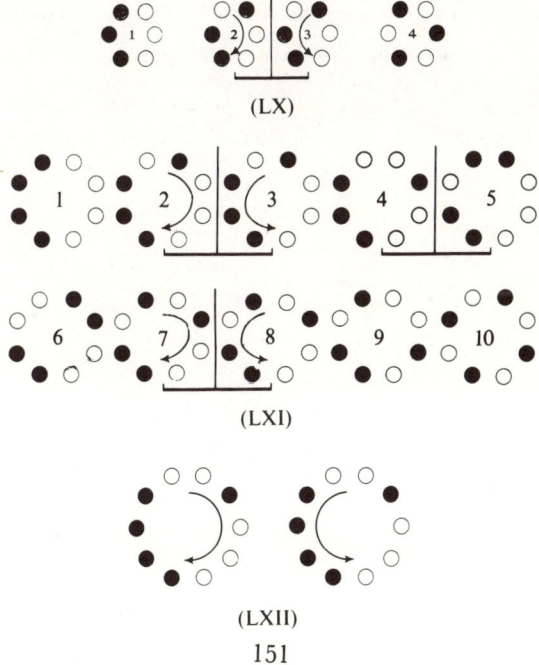

upfield shift of the internal amide hydrogens. If these hydrogens lie directly above and close to the planes of the two end amide groups, then they are in a region of diamagnetic shielding by the amide π-electron system and should resonate at unusually high field (LIX).

Configurational Stereochemistry

Prelog and Gerlach[91] have examined the stereochemical consequences of the introduction of an equal number of enantiomeric chiral centres into a ring which possesses the quality of direction. *cyclo*-Polypeptides are compounds of this type. Apart from optically inactive *meso*-forms and conventional enantiomers, two new classes of stereoisomer were noted, *cyclo*enantiomers and *cyclo*diastereoisomers. A *cyclo*enantiomeric relationship exists between two structures when there is only a change of ring direction between object and its non-superimposable mirror image about a reflection plane, the arrangement of centres of chirality not being altered. With a conventional enantiomeric relationship a different arrangement of the chirality centres is found. *Cyclo*diastereoisomers possess the same arrangement of chiral centres but a different ring direction, the arrangement of chiral centres being such as to preclude the existence of enantiomers. As an illustration of this classification consider the isomers of a cyclic hexapeptide composed of six similar chiral centres (depicted in LX 1–4), there being three centres with R-configuration (symbolised ●) and three centres with S-configuration (symbolised ○); there are two *meso*-forms (1 and 4), and a *cyclo*enantiomeric pair (2 and 3). The case of isomeric *cyclo*-octapeptides containing four R-chiral centres (●) and four S-centres (○) is depicted in (LXI; 1–10). In this case there are four *meso*-forms (1, 6, 9 and 10), one pair of conventional enantiomers (4 and 5), and two pairs of *cyclo*enantiomers (2 and 3, and 7 and 8). An example of cyclodiastereoisomerism is depicted using a pair of hypothetical decapeptides (LXII).

Gerlach, Ovchinnikov, and Prelog[92] synthesised the *cyclo*enantiomeric *cyclo*-hexa-alanyls shown in *Scheme 5*. The linear hexapeptides L-alanyl-D-alanyl-D-alanyl-L-alanyl-D-alanyl-L-alanine (LXIII) and L-alanyl-D-alanyl-L-alanyl-D-alanyl-D-alanyl-L-alanine (LXIV) were prepared in a form

L_1-Ala-D_2-Ala-D_3-Ala-L_4-Ala-D_5-Ala-L_6-Ala

(LXIII)

↓   (LXV, configurations written outside ring)

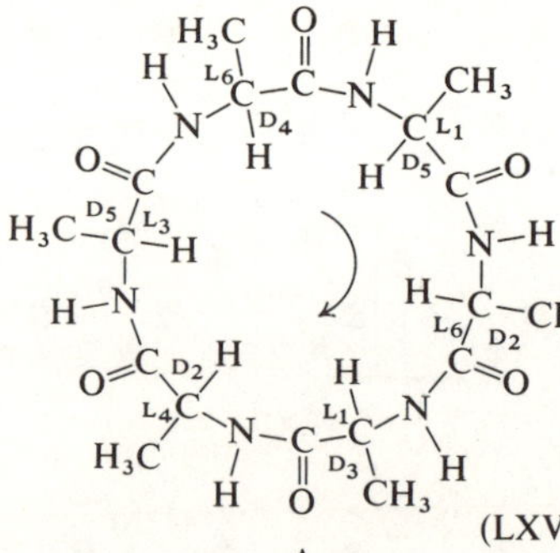

(LXVI, configurations written inside ring)

↑

L_1-Ala-D_2-Ala-L_3-Ala-D_4-Ala-D_5-Ala-L_6-Ala

(LXIV)

Scheme 5

suitable for subsequent azide cyclisation to the *cyclo*-hexapeptides (LXV) and (LXVI) respectively. The o.r.d. spectra of the two cyclic peptides were antipodal, with an extremum at 230 nm. The 100 MHz n.m.r. spectra taken in methanol/deuterium oxide showed ten signals corresponding to the C-methyl groups of alanine, which arise from a slight overlapping of the six methyl doublets. Equimolar admixture of the two *cyclo*enantiomers (LXV) and (LXVI) gave rise to a crystalline racemate. The structural isomers

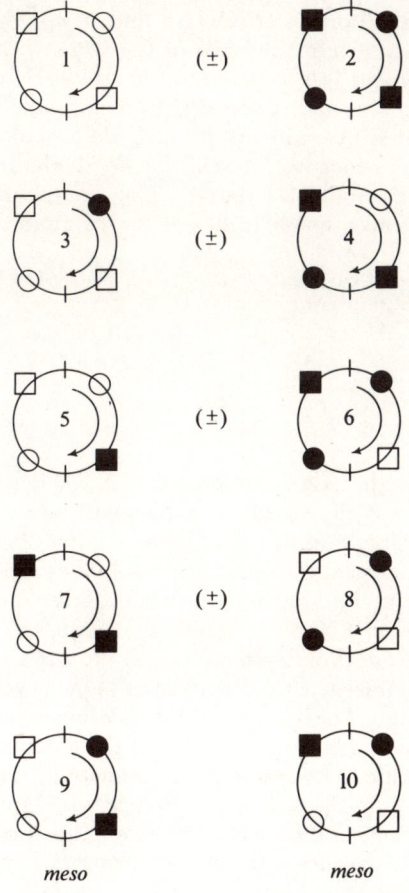

Scheme 6.. Diagrammatic representation of the stereoisomers of cyclo-(glycylphenylalanyl-leucyl-glycyl-phenylalanyl-leucyl)
L - Phe = ○,   D-Phe = ●,   L-Leu = □,   D-Leu = ■,   Gly = ı

cyclo-(L-alanyl-L-alanyl-glycylglycyl-L-alanyl-glycyl) and *cyclo*-(L-alanyl-L-alanyl-glycyl-L-alanyl-glycylglycyl)[93] and a series of cyclic peptides of glycine, D-alanine, and L-alanine[94] have been prepared. Ten stereoisomeric forms of the cyclic hexapeptide *cyclo*-(glycyl-phenylalanyl-leucyl-glycyl-phenylalanyl-leucyl) can exist and are represented diagrammatically in *Scheme 6*. One member of each of the four pairs of enantiomers (compounds 1, 3, 5, and 8) and one of the two *meso*-forms (compound 10) have been prepared[95,96].

153

Spectroscopic Studies on Cyclic Peptides
Some unexpected results have been obtained from a study of the i.r. spectra of a series of diastereoisomeric *cyclo*-hexapeptides[55,96] with the sequence *cyclo*-(glycyl-phenylalanyl-leucyl-glycyl-phenylalanyl-leucyl) (LXVII a–e) and the corresponding *retro*-compounds (LXVIII, a and b) in nujol or fluorolube mulls. All peptides show a strong series of absorptions characteristic of the *trans*-amide link. On deuteration the amide-II band was in general displaced to 1470 cm^{-1}, the region of CH$_2$ deformation vibrations, but in the case of the isomers (LXVII a and b) a new isolated band also appeared at 1235 cm^{-1}. On the basis of the work with dioxopiperazines[55] (*see Table 5.1*) this band was assigned to the amide-II band of a deuterated *cis*-amide grouping. This absorption is not recognisable in the undeuterated *cis*-amide because it is overlaid by the CH$_2$ deformation band. Further, a band at 1340 cm^{-1} which was insensitive to deuteration was tentatively assigned to the *cis*-amide III band. The compounds (LXVIIc) and (LXVIIIb) also showed some evidence for *cis*-amide bonds in their i.r.

┌─Gly-Phe-Leu-Gly-Phe-Leu─┐   ┌─Gly-Leu-Phe-Gly-Leu-Phe─┐
                  ←   ←

(LXVII, a^{95}: L   L   L   L   (LXVIII, a^{96}: L   D   L   D
       b^{95}: D   L   L   L   b^{96}: L   D   L   L).
       c^{95}: L   D   L   L
       d^{95}: D   L   D   L
       e^{96}: D   L   L   D).

spectra. However, in the case of compounds (LXVII d and e) and (LXVIIIa) no such evidence was discerned. A semi-quantitative estimate of the *cis*-amide content was made from the integrated intensity of the *cis*-amide-III band using the bands of the aromatic out-of-plane-bending deformations at 700 and 745 cm^{-1} as internal reference standards. The *cis*-amide content increases in the series (LXVII d, b and a, i.e. D,L,D,L; D,L,L,L; and L,L,L,L respectively) in the approximate ratio 0:1:2. *Cis*-amide bonds must be considered as a possible structural feature of higher cyclic peptides[97] as well as in the lower ones. The *cyclo*-peptides mentioned above which contain two amino-acid residues of the D- and two of the L-configuration show no i.r. evidence characteristic of *cis*-amide bonds, unlike the compounds containing three or four residues of the L-configuration[55,96]. It is interesting to note that the yields of azide cyclisation products from the linear diastereoisomeric peptides decrease in parallel with the increasing *cis*-content of amide-bonds in the cyclic products[95].

The o.r.d. and c.d. spectra of the series of diastereoisomeric *cyclo*-hexapeptides (LXVII a–e, and LXVIII a and b) just discussed have been measured to ascertain the effect of changing the absolute configuration of the amino-acid residues on optical behaviour[98,99]. Multiple Cotton effects at 250–270 nm are ascribed to the $B_{2u} \leftarrow A_{1g}$ transition of the aromatic chromophores, and two Cotton effects which generally occur at 209–233 nm have been related to the n–π* transition of the amide chromophore; a further two Cotton effects appearing at 200 nm and below are due to the perpendicular and parallel exciton bands of the π–π* transition of the amide chromophore. The multiple Cotton effect above 250 nm is evident only in those cases where the phenylalanine residues possess the same configuration (i.e. LXVII a, c and d, and LXVIIIa). Inspection of molecular models shows that only in these isomers can the aromatic chromophores approach one another closely enough for them to interact or undergo 'stacking'. There is no obvious correlation between the signs of the aromatic Cotton effects and the absolute configuration of phenylalanine. The two Cotton effects at 209–218 nm and 225–233 nm were assigned to the n–π* transition of the amide chromophore on the basis of their positions and solvent dependencies. The sign of the Cotton effect at shorter wavelength is determined solely by the configuration of phenylalanine, while the sign of the longer wavelength Cotton effect is determined by the absolute configuration and sequential arrangement of residues. The two Cotton effects at 190 and 200 nm arise from a splitting of the amide π–π* transition into two bands of opposite sign. The sign of these c.d. bands depends on the sequential arrangement of amino-acid side-chains of particular configuration, regardless of their character, since the signs of these bands are reversed when going from a normal *cyclo*-peptide sequence to the analogous *retro*-sequence.

Enzyme Substrates and Enzyme Models

The cleavage of *cyclo*-peptides by endopeptidases has not been studied systematically. It appears, however, from the fragmentary evidence which is available that while *cyclo*-hexapeptides containing lysine show a tendency to be cleaved by trypsin, those of smaller ring size do not. Ohno and Izumiya[100] found that the *cyclo*-hexapeptide, *cyclo*-(pentaglycyl-L-lysyl) was readily hydrolysed by trypsin to pentaglycyl-L-lysyl, but *cyclo*-(diglycyl-L-lysyl-diglycyl-L-lysyl) was only slowly cleaved[101]. Thus at least some *cyclo*-hexapeptides can form productive enzyme–substrate complexes with trypsin. *cyclo*-Pentapeptides and smaller rings on the other hand do not seem able to attain the necessary conformation in the complex. Kenner and Laird[102] found that *cyclo*-(glycyl-L-lysyl-glycyl-L-lysyl-glycyl) was resistant to trypsin under conditions which led to complete hydrolysis of the corresponding linear sequence. Izumiya[103] and co-workers failed to prepare L-phenylalanyl-L-lysine and L-lysyl-L-phenylalanine from *cyclo*-(L-phenylalanyl-L-lysyl) by treatment with trypsin and chymotrypsin respectively, but it is not surprising that a substrate with such a small ring system is not attacked by either enzyme. Even the naturally occurring *cyclo*-heptapeptide evolidine is not cleaved by enzymes[104].

cyclo-Peptides possess greater conformational rigidity than the corresponding linear peptides and are thus attractive structures to support suitable side-chain groups in order to look for co-operative interactions such as occur between amino-acid side-chains found at the catalytic sites of certain enzymes. The orientation of the side-chains is clearly important, necessitating a study of the conformation of the synthetic model. Such a *cyclo*-peptide system possesses the disadvantage, however, that the precise micro-environment of the enzyme cannot be reproduced. A series of cyclic peptides (LXIX a–e) containing histidine and tyrosine have been studied as models for the co-operative catalysis of ester hydrolysis[105,106]. The pKa values of the imidazolium and phenolic groups of these peptides do not appreciably differ from those of simple linear peptides containing histidine and tyrosine, leading to

a[105] : cis   ⌐L-His-L-Tyr⌐
b[106] : trans ⌐L-His-D-Tyr⌐
c[106] : cis   ⌐Gly-L-His-Gly-Gly-L-Tyr-Gly⌐
d[106] : trans ⌐Gly-L-His-Gly-Gly-D-Tyr-Gly⌐
e[105] : cis   ⌐Gly-L-His-Gly-Gly-Gly-L-Tyr⌐

(LXIX)

the conclusion that no significant hydrogen-bonding occurs between these two types of side-chain. A kinetic study revealed that the *cyclo*-peptides were no more active than imidazole in catalysing the hydrolysis of substituted phenyl acetates[105,106]. However, these phenolic groups undergo acetylation. The rate of acylation is greater for the *cis*-compounds than the *trans*-ones, and furthermore the rate of acylation is greater for the *cyclo*-dipeptide than the corresponding hexapeptides[106]. It has been suggested that either an *N*-acetyl-imidazole is formed which is then able to transfer the acetyl group to tyrosine (LXX) or that the imidazole acts as a general base (LXXI); the normality of the pKa values renders the latter mechanism less likely.

(LXX)   (LXXI)

Sheehan and McGregor[107] investigated *cyclo*-(glycyl-L-histidyl-L-seryl-glycyl-L-histidyl-L-seryl) as an esterase model. No significant activity above that expected for a histidine containing peptide was observed in the catalysis of the hydrolysis of *p*-nitrophenyl acetate, but the linear tripeptide was found to decompose in aqueous solution under extremely mild conditions giving rise to free serine and *cyclo*-(glycyl-L-histidyl). It appears that the tripeptide can catalyse its own destruction.

Cyclodimerisation

Tripeptide and pentapeptide monomers show a marked tendency to undergo dimerisation during cyclisation[54,108], e.g. diglycylglycine cyanomethyl ester yields *cyclo*-hexaglycyl[62]. Schwyzer originally suggested that two molecules of peptide derivative aggregate in an antiparallel fashion[54,109] (corresponding to the antiparallel pleated sheet of Pauling and Corey[110]), thus facilitating a double condensation (LXXII). On this basis peptide monomers containing an odd number of amino-acid residues might be expected to be especially prone to twinning because the monomer chains can be aligned in this antiparallel pleated sheet more efficiently[109]. More recent work has led to a

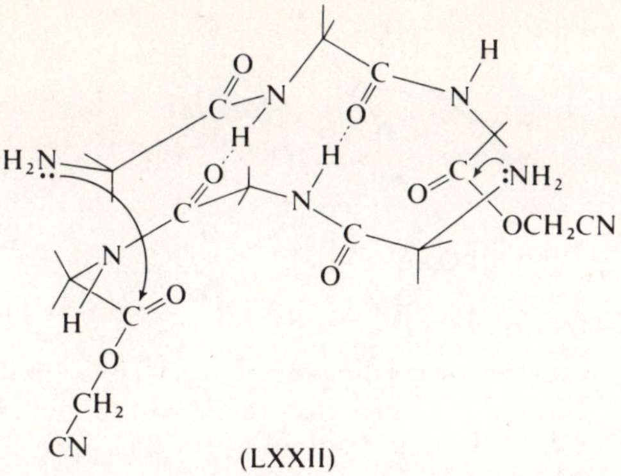

(LXXII)

modification of these ideas. A number of monomers with an even number of residues[59,111] have been shown to undergo cyclic dimerisation (*Scheme 7*). The degree of aggregation of peptide monomers by hydrogen bonding should be dependent on dilution and on the nature of the cyclising medium. Solvents which do not compete for peptide–peptide hydrogen bonds should facilitate cyclodimerisation. However, it has been shown with simple model amides that aggregation by hydrogen bonding is insignificant in aqueous

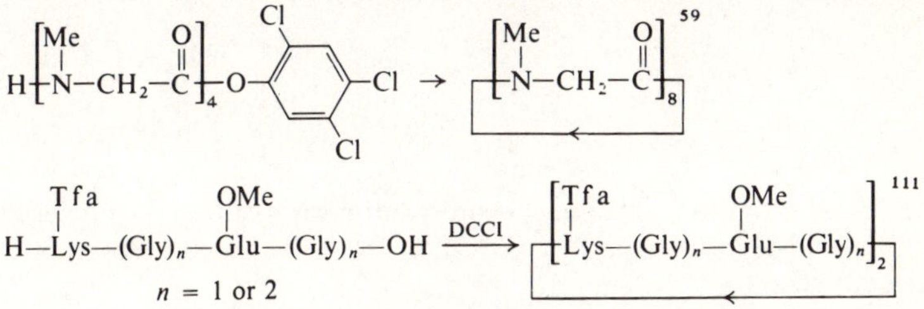

Scheme 7

solution[112] and weak in tetrahydrofuran and dioxan, whereas cyclodimerisation can occur in water and methanol[68]. Furthermore, a number of peptides which cannot associate according to (LXXII) undergo twinning. Glycyl-L-prolyl-glycine[87,113] (LXXIII), for example, gives rise to *cyclo*-(glycyl-L-prolyl-glycylglycyl-L-prolyl-glycyl) (LXXIV) as readily as diglycyl-L-proline[113] (LXXV) although the prolyl unit carries no NH group with which to associate by hydrogen bonding. The monomer unit (LXXIII)

H·Gly·L-Pro·Gly·OX   ⟶   (LXXIV)   ⟵   H·Gly·Gly·L-Pro·OX
(LXXIII)   (LXXV)

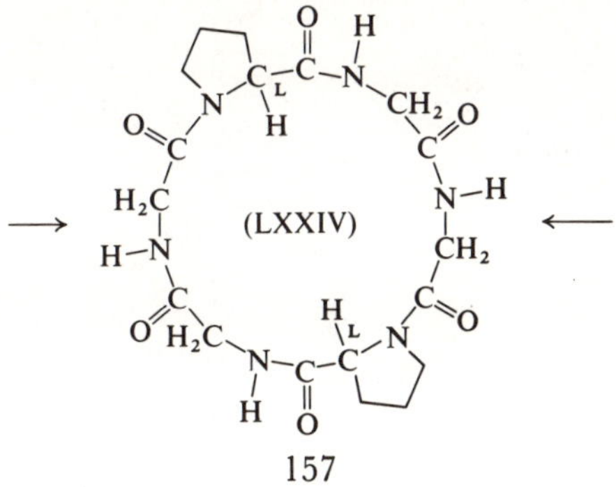

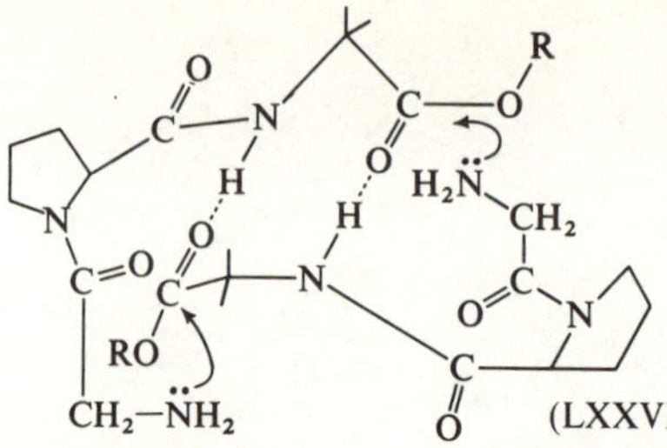

could associate, however, according to (LXXVI)[87]. Similarly L-prolyl-L-prolyl-glycine (LXXVII) yields *cyclo*-(L-prolyl-L-prolyl-glycyl-L-prolyl-L-prolyl-glycyl) (LXXVIII) on cyclisation[114]. The depsipeptide, glycyl-glycolyl-glycine (LXXIX), which has an ester linkage instead of

H·L-Pro·L-Pro·Gly·OX
(LXXVII)

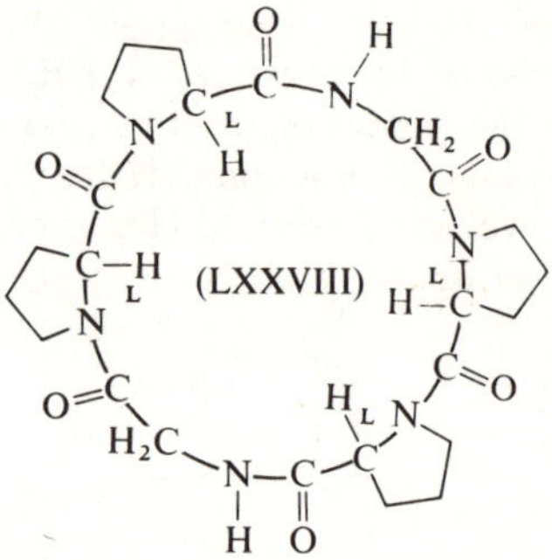

one of the amide linkages, readily dimerises to *cyclo*-(glycyl-glycolyl-glycylglycyl-glycolyl-glycyl[87]) (LXXX). *N*-Substituted depsipeptides also cyclodimerise[68]

NH$_2$CH$_2$CO—OCH$_2$CO—NHCH$_2$CO$_2$Np
(LXXIX)
↓
HNCH$_2$CO—OCH$_2$CO—NHCH$_2$CO
| |
OCCH$_2$NH — COCH$_2$O—COCH$_2$NH
(LXXX)

H·Gly·Gly·L-Phe·OX
H·Gly·Gly·D-Phe·OX
(LXXXI)

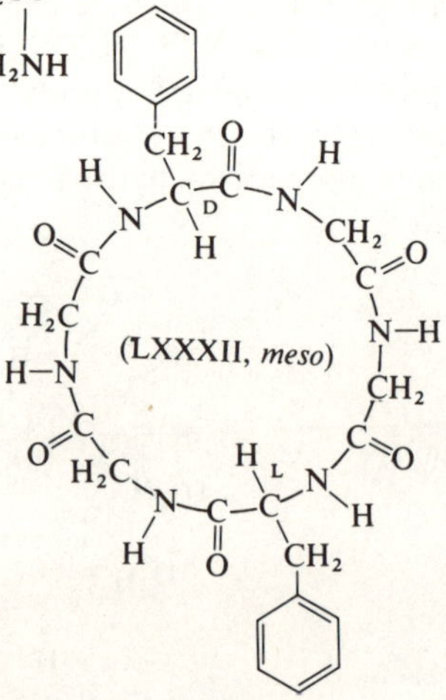

(LXXXII, *meso*)

H·Gly·Gly·Phe·Gly·Gly·Phe·N$_3$
(LXXXIII)

Cyclodimerisation of tripeptides probably occurs because there is prohibitive conformational, angular, or transannular strain present in the transition state leading to *cyclo*-tripeptide formation. With larger peptides dimerisation presumably occurs because association in the transition state is particularly favoured. Such association could be the result of hydrophobic forces. The preferred conformation of the linear monomer also has a role to play. If the spatial separation of the ends of the monomer chain is small (i.e. a bent chain conformation), then the tendency to form cyclic monomer will be greater than if the separation is large.

In a number of cases it has been found that cyclodimerisation is stereospecific. The cyclic hexapeptide obtained from glycylglycyl-DL-phenylalanine (LXXXI) by twinning has been shown by an independent synthesis to be exclusively the *meso*-form (LXXXII), i.e. an L-tripeptide unit associates preferentially with the corresponding D-unit[54,115]. Further, when glycyl-DL-alanyl-DL-phenylalanine methyl ester (LXXXIV) was treated in methanol solution with ammonia, the pure *meso*-compound, *cyclo*-(glycyl-D-alanyl-L-phenylalanyl-glycyl-L-alanyl-D-phenylalanyl) (LXXXV) crystallised[116] out. Various combinations of the stereoisomeric tripeptide monomers

$$\text{H·Gly·Ala·Phe·OCH}_3 \quad \longrightarrow \quad \begin{bmatrix} \rightarrow \text{Gly—Ala—Phe} \\ \text{Phe} \leftarrow \text{Ala} \leftarrow \text{Gly} \leftarrow \end{bmatrix}$$

(LXXXIV)

(LXXXV)

were subjected to these cyclisation conditions but in three cases only was cyclic hexapeptide produced. The compound (LXXXIV; DL) and its enantiomer (LXXXIV; LD) separately gave rise to the enantiomeric *cyclo*-hexapeptides (LXXXV) DLDL and LDLD respectively, whereas the racemic compound gave the *meso*-macrocycle as already mentioned. The other combinations of stereoisomeric tripeptides apparently cyclised so slowly that amidation became the predominant reaction. Stereospecific cyclodimerisation can be attributed to a lessening of steric interference of the side-chains on aggregation when these side-chains are part of residues possessing opposite configurations.

Several examples can be found in the *cyclo*-hexapeptide series of higher yields being obtained on cyclisation of sterically inhomogeneous peptides. The yield of *cyclo*-hexapeptide (LXXXII) from the linear hexapeptide azide (LXXXIII) is lower in the L–L and D–D series than for the L–D compound[115]. In the cyclisation of (LXXXVI), when both asymmetric centres had the L-configuration, the yield of cyclic hexapeptide was 31%, whereas it was 58% when the tyrosyl-residue of (LXXXVI) was racemic[106].

Enniatins

Three *cyclo*-hexadepsipeptides isolated from species of *Fusarium* have been extensively studied[1]. Two of these compounds, enniatin A, (LXXXVII; R = CHMeCH$_2$Me) and enniatin B (LXXXVII; R = CHMe$_2$), possess antimicrobial activity, but enniatin C (LXXXVII; R = CH$_2$CHMe$_2$) is inactive. The enniatins were originally formulated as *cyclo*-tetradepsipeptides, but synthetic studies established that there are six components in the ring[1]. A recently characterised toxin from *Beauveria bassiana*, beauvericin (LXXXVII; $R = CH_2Ph$), is an aromatic analogue of the enniatins[117] and has been synthesised[118].

Examination of synthetic analogues of the enniatins has established that changes in the ring size, configuration of the amino-acids, or modification of

H·Gly·Tyr·Gly·Gly·His·Gly·OH
|
BzL

(LXXXVI)

the hydroxy-acid side-chains cause marked loss of activity, although variation of the amino-acid side-chains has little effect[119]. *Enantio*-enniatin B[120] (LXXXVIII; $R = CHMe_2$; all the ring components have the opposite configuration to those in enniatin B), however, closely resembles enniatin B in both its antimicrobial activity and its physical properties (except that its o.r.d. curve is antipodal). If the formula (LXXXVIII; $R = CHMe_2$) is rotated through 60° in the plane of the paper, subsequent superimposition on (LXXXVII; $R = CHMe_2$) shows all superimposed centres to possess the same chirality, although each ester group is replaced by an *N*-methyl-amide group and vice versa. Thus, with respect to a receptor *enantio*-enniatin B possesses an identical spatial distribution of side-chains. This is the first known example of the enantiomer of a naturally occurring substance possessing the same biological activity. An analogue of enniatin A (LXXXVIII; $R = CHMeCH_2Me$) in which the amide and ester bonds have been interchanged is indistinguishable from the natural material in its activity and o.r.d. behaviour[121]. The variation in the o.r.d. spectra of enniatins A and B with solvent polarity is rather less than that of enniatin C. This implies that while β-branched side-chains favour a reasonably rigid conformation, the γ-branching of enniatin C leads to a more flexible ring which is conformationally less suitable for receptor interaction[122].

Depsipeptides such as the enniatins are able to increase selectively the permeability of certain natural and synthetic membranes for alkali metal cations[123-126]. In general, a parallel exists between the antimicrobial activity of these depsipeptides and their ability to complex with alkali metal cations in lipophilic media[121]. Their enhancement of K^+ transport compared to that of Na^+ is connected with the highly specific conformation-dependent interaction of the *cyclo*-depsipeptide with the cation. The ion is included in the internal sphere of the molecule through ion–dipole interactions with the amide and ester groups (LXXXIX), and the complexes can be isolated in the crystalline state[127] or their formation inferred from physicochemical data such as conductometric measurements[122]. The variations in the o.r.d. spectrum of enniatin B in solvents of varying polarity has been interpreted in terms of an equilibrium between two conformers (N and P). One form (N) predominates in non-polar solvents, but as the polarity of the solvent medium is increased so the proportion of (P) increases. The o.r.d. spectrum of the K+ complex of enniatin B resembles that of form P, indicating that their conformations are similar[121,128].

In $CS_2:CD_3C_6D_5$ (2:1) at 24°C the n.m.r. spectrum of enniatin B shows a sharp N—CH_3 singlet, but at −127°C there are three singlets of equal intensity. Similarly the doublet due to the α-hydrogens of the methylated amino-acid residues becomes three equal doublets. The conformation of enniatin B is therefore not symmetrical under these conditions. Theoretical calculations and studies with models led to a proposal for the conformation[128] of N (XC), which is compact and does not possess a central cavity; three

Conformation of enniatin B in non-polar solvents

adjacent isopropyl groups are *quasi*-axial and the other three *quasi*-equatorial. Support for this conformation comes from the agreement of measured and calculated dipole moment (3·35 ± 0·1D in CCl_4 and 3·5 ± 0·5D) and also the large spin–spin coupling constants of the α- and β-hydrogens, which are consistent with their *trans*-orientation.

The conformation of the K+ complex of enniatin B (similar to the P form) was elucidated by studying the n.m.r. behaviour of the tri-*N*-desmethyl analogue (XCI) (the o.r.d. curves of the two *cyclo*-depsipeptides indicated that they both possess similar conformations in polar solvents and on complexation). Two models were proposed; a choice between them was made by studying the variation of $^3J_{\text{NH-CH}}$ of the Li$^+$, Na$^+$, K$^+$ and Cs$^+$ complexes of (XCI). This increased from 5·1 to 8·4 Hz along the series and corresponds to the carbonyl groups opening out to accommodate the larger

Conformation of the K$^+$ complex of enniatin B

cations (likened to the opening of a flower bud). Furthermore, the vicinal spin–spin coupling constants for the α- and β-hydrogens of the K$^+$ complex of enniatin B indicate that in the amino-acid residues they are *trans* whereas in the α-hydroxy-acid residues they are essentially *gauche*. The potassium complex of enniatin B (XCII) thus possesses a compact arrangement of functional groups, the six carbonyl groups bending towards the central cation with *quasi*-equatorial orientation of the isopropyl groups[128] (the 'bracelet' conformation).

An X-ray analysis of the crystalline complex of enniatin B with potassium iodide[127,129] shows that the eighteen-membered ring adopts a disc shape with the backbone chain folded so as to provide the central potassium ion with an almost octahedral co-ordination from the six carbonyl oxygen atoms[129]. The chirality of the hydroxy- and amino-acid components is such that all the isopropyl groups point outwards from the centre of the disc, thus giving the molecule a lipophilic exterior; this arrangement in the crystal closely resembles the solution conformation mentioned above. In the crystal the discs are stacked in columns with their faces parallel. If a similar association were to occur in membranes, the Van der Waals attraction between peripheral isopropyl groups of adjacent molecules more than compensating for the electrostatic repulsion between the charged discs, there would be a kind of tunnel across the membrane which could facilitate ion transport[129].

Ferrichromes

The siderochromes are a group of fungal microbial growth factors which contain chelated ferric iron[130]. With the exception of ferrichrome A, which is inactive, the compounds exert a potent growth promoting activity on a number of unicellular organisms. These compounds possess the common structural feature of three bidentate hydroxamate ligands co-ordinating octahedrally with the iron. The sideramines exist as two sub-groups: the ferrioxamine type are linear heteromeric peptides[130] and will not be considered further, and the ferrichrome type are heteromeric *cyclo*-homodetic peptides.

The structures of the commonest ferrichromes are given in *Scheme 8*. The

$$\begin{array}{c} CH_2NHOH \\ | \\ CH_2 \\ | \\ CH_2 \\ | \\ H_2N \cdot CH \cdot CO_2H \end{array}$$

(XCIX)

determination of the structure of these compounds was complicated by degradation of δ-N-hydroxy-L-ornithine (XCIX) under the conditions of acid hydrolysis in the presence of iron[131,134] to ornithine. Accordingly, quantitative amino-acid analyses were obtained either by hydrochloric acid hydrolysis of the desferri-compounds, followed by catalytic hydrogenation, or alternatively by hydrolysis with hydriodic acid[132]. Raney nickel hydrogenation of ferrichrome (XCIII) yields *cyclo*-(triglycyl-tri-δ-N-acetyl-L-ornithyl)[133] (C; R = Ac, and R^1 = H) whose structure was proved by synthesis. Degradation of ferrichrome A (XCVIII) with a bacterial protease from *Pseudomonas* gave enough peptides such as seryl-glycyl-δ-N-acyl-N-hydroxy-ornithine to enable an amino-acid sequence to be proposed for the *cyclo*-hexapeptide portion[136] (*see Scheme 8*, (XCVIII)); this was subsequently confirmed by an X-ray crystallographic study[137] (*see* later).

The structure of the prosthetic moiety of the ferrichromes was deduced from the fact that the desferri-compounds contain three new ionisable groups which, when examined spectrophotometrically, exhibited the typical behaviour of an enol–enolate ionisation[131]. Cleavage with periodate[138] proved to be an invaluable tool in this work, enabling the acyl moiety to be split from the hydroxamic acid derivative. In the case of ferrichrome (XCIII), ferricrocin (XCIV), and ferrichrysin (XCV) the acyl moiety was shown to be acetyl. With desferri-ferrichrome A periodate cleavage liberated three moles of a dicarboxylic acid subsequently shown to be *trans*-β-methyl glutaconic acid[131] (*see* XCVIII), while with desferri-ferrirhodin[135] a lactone

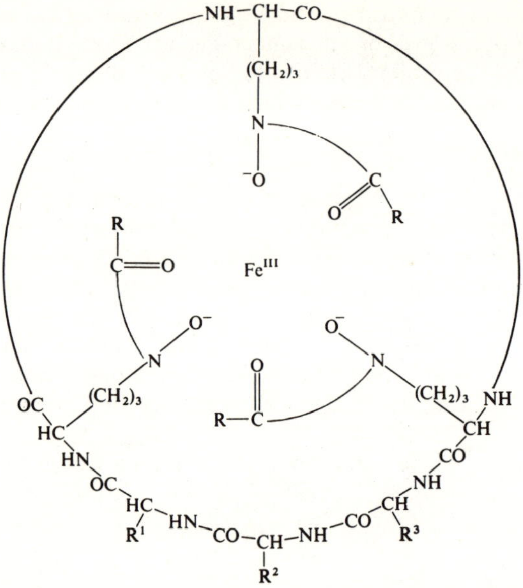

	R	R^1	R^2	R^3
(XCIII) Ferrichrome[131-133]	CH_3	H	H	H
(XCIV) Ferricrocin[134]	CH_3	H	H or $HOCH_2$	$HOCH_2$ or H
(XCV) Ferrichrysin[134]	CH_3	H	$HOCH_2$	$HOCH_2$
(XCVI) Ferrirhodin[135]	$\underset{H}{\overset{H}{>}}C=C\underset{CH_2-CH_2OH}{\overset{CH_3}{<}}$	H	$HOCH_2$	$HOCH_2$
(XCVII) Ferrirubin[135]	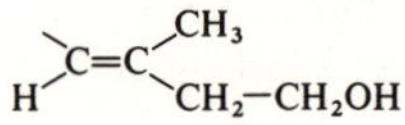 $\underset{H}{>}C=C\underset{CH_2-CH_2OH}{\overset{CH_3}{<}}$	H	$HOCH_2$	$HOCH_2$
(XCVIII) Ferrichrome A[131-133]	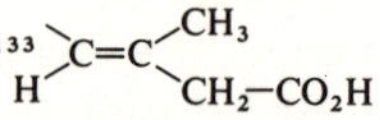 $\underset{H}{>}C=C\underset{CH_2-CO_2H}{\overset{CH_3}{<}}$	H	$HOCH_2$	$HOCH_2$
(CVII) Albomycin δ_2[151]	CH_3	$HOCH_2$	$HOCH_2$	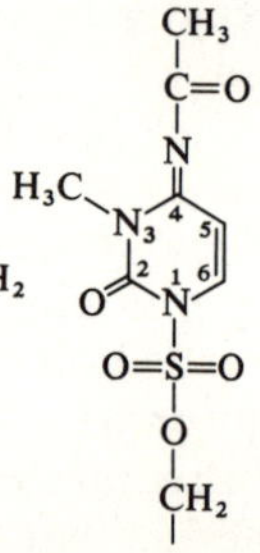 (see structure)
(CVIII) Albomycin δ_1[151]	CH_3	$HOCH_2$	$HOCH_2$	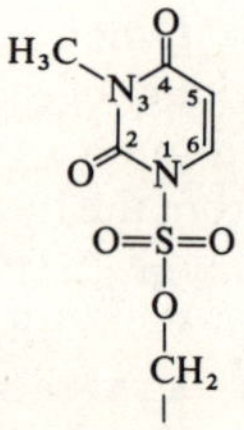 (see structure)

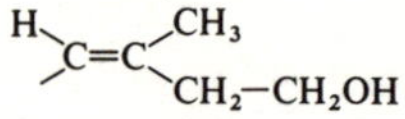

Scheme 8

(CI) was produced, indicating that the acyl group was *cis*-5-hydroxy-3-methyl-2-pentenoyl- (*see* XCVI). In the case of desferri-ferrirubin the corresponding *trans*-acid was isolated[135] (*see* XCVII).

Ferrichrome A (XCVIII) and ferrirubin (XCVII) have both been converted to ferrichrysin (XCV), thus showing their structural relationship. Ferrichrome A (XCVIII) in the desferri-state was subjected to mild hydrolysis and then acetylated with acetic anhydride (*see* Scheme 9). The *O*-acetyl

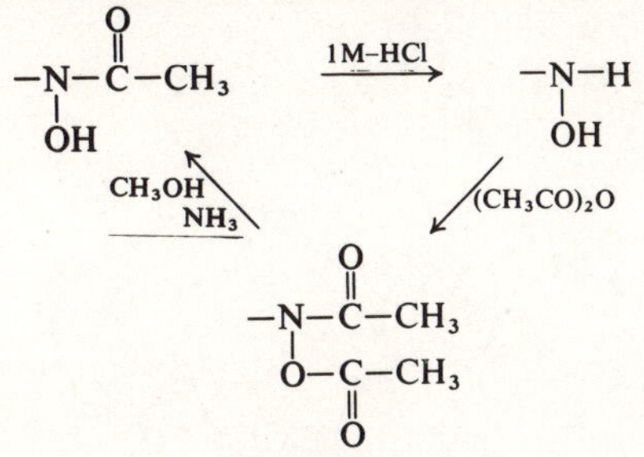

Scheme 9

groups were removed from the *O*,*N*-bis-acetyl compound by treatment with ammonia in methanol; the product on re-introduction of the iron was identical to ferrichrysin[135] (XCV). Thus the biologically inactive ferrichrome A was converted to the highly active ferrichrysin. Ferrirubin (XCVII) was similarly converted to ferrichrysin via the corresponding hexahydro derivative[135]. The hexahydro derivative obtained by the hydrogenation of ferrirhodin was shown to be identical with hexahydro-ferrirubin, thus establishing their common structural features[135].

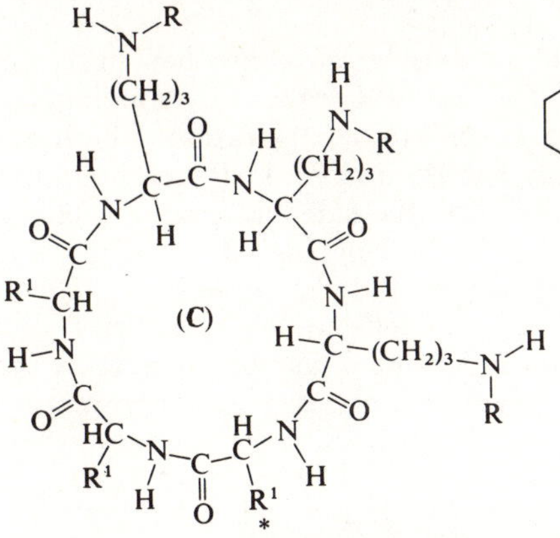

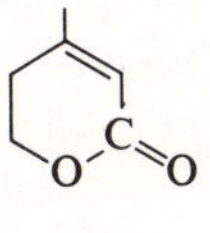

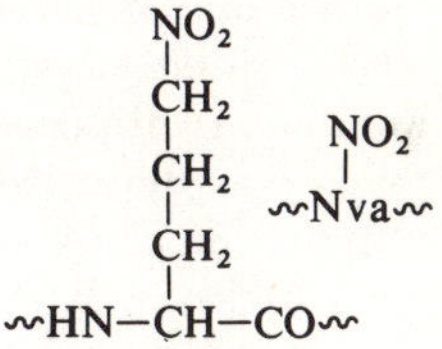

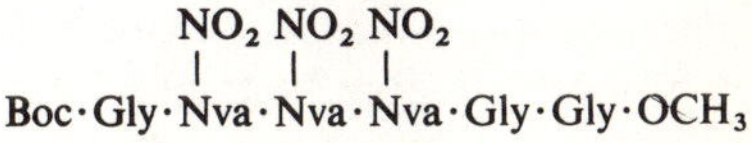

A total synthesis of ferrichrome (XCIII) has been reported[139]. The synthetic route was chosen to avoid use of the sensitive *N*-δ-hydroxy-ornithine. Accordingly a linear *N*-t-butyloxycarbonyl-hexapeptide methyl ester (CIII) containing three adjacent residues of δ-nitronorvaline (CII) was prepared. The methyl ester was hydrolysed with sodium carbonate, the acid converted to the corresponding *p*-nitrophenyl ester, the *N*-t-butyloxycarbonyl-group removed, and the product cyclised. The yield of *cyclo*-hexapeptide obtained was remarkably high (70 ~ 80%). The δ-nitro groups were reduced to the corresponding hydroxylamines and the crude product acetylated to give the tris-(N,O-diacetyl)cyclo-hexapeptide. Treatment with ammonia (*Scheme 9*) split off only the O-acetyl groups, and the product was purified as its Fe(III) complex to yield material with full biological activity[139].

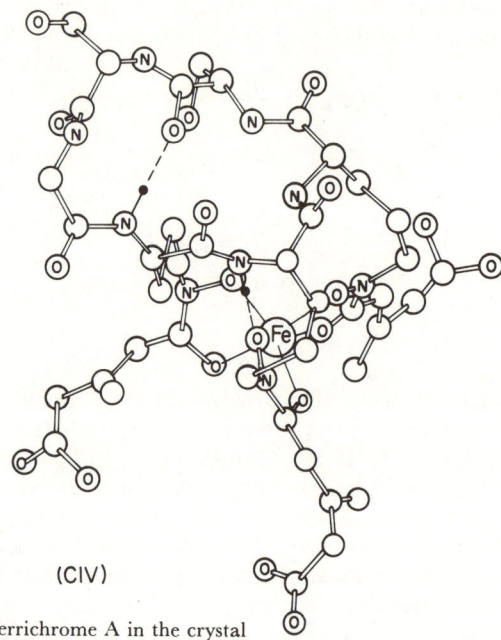

(CIV)

Conformation of ferrichrome A in the crystal

The molecular structure of ferrichrome-A tetrahydrate has been determined by single crystal X-ray diffraction techniques[137], including the use of anomalous dispersion to establish absolute configuration. The hexapeptide ring was shown to contain only one intra-chain hydrogen bond but another hydrogen bond was assumed to link the side-chain oxygen of the middle N-δ-hydroxy-ornithine residue with the ring (*see* CIV). All peptide links were found to be *trans*. The iron atom and the three hydroxamate ligands are bound in the configuration of a left-handed propeller (CV). Ferrichrome is unsuitable for n.m.r. work because of line broadening by the paramagnetic ion, but the solution conformation of desferri-ferrichrome and

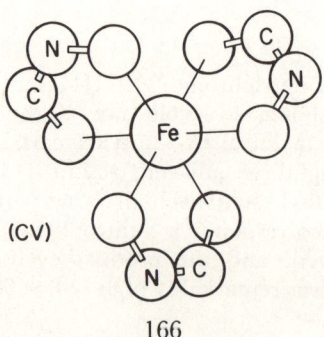

(CV)

its aluminium (Al^{3+}) complex (alumichrome) have been studied at 220 MHz in d^6-dimethyl sulphoxide[140]. The spectrum of desferri-ferrichrome showed six groups of amide N—H signals (spanning 7·7–8·3 ppm), whose multiplicity enabled them to be assigned to glycyl- or ornithyl-residues. However, in alumichrome these resonances showed a dramatic shift (spanning 6·4–10 ppm), thus showing that the mean chemical environment of the amide N–H groups is quite different in these two compounds. Double resonance experiments enabled the regions of the spectrum related by spin–spin coupling to be identified, but it was not possible to assign them to individual amino-acid residues in the sequence. The temperature dependence of the chemical shifts of the N—H resonances was used, as in the studies of Kopple mentioned earlier, to identify amide N—Hs which are solvent shielded. In desferri-ferrichrome the N—Hs of one glycyl- and one ornithyl-residue (the signals occurring at highest field in each set) show a reduced temperature dependence, whereas in alumichrome all the ornithyl N—Hs and one glycyl N—H show a reduced temperature coefficient. Hydrogen–tritium exchange experiments[141] indicate that ferrichrome and ferrichrome A possess, depending on pH, from two to four slowly exchanging hydrogens[140]. N.m.r. spectroscopy shows that two of the glycyl N—Hs of alumichrome in D_2O solution exchange instantaneously, while the other four N—Hs exchange more slowly. An analysis of the coupling constants ($^3J_{NH-CH}$) suggests that chelation with the metal ion does not greatly affect the conformation of the cyclic backbone. The evidence outlined is consistent with a conformation for alumichrome in d^6-dimethyl sulphoxide which resembles that of ferrichrome A in the crystal, but in addition the amide hydrogen of glycyl residue 3 (*see* CVI) is hydrogen bonded to the carbonyl group of ornithyl residue 3, and the amide N—H of ornithine residue 1 is shielded from solvent by the other side-chains.

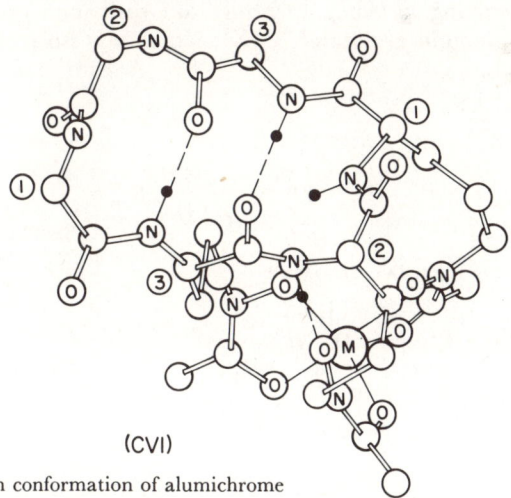

(CVI)

The solution conformation of alumichrome

H atoms are not shown with the exception of the four amide NHs that show reduced interaction with the solvent; of these the one belonging to Orn^1 is buried between the peptide backbone ring and the chelated ornithyl side-chains and the other three are intramolecularly H-bonded.

An antibiotic albomycin, isolated from *Actinomyces subtropicus*[142], is structurally related to the ferrichromes. Albomycin consists of a family of closely related substances[143], the compound designated δ_2 being the most active[144] and also extremely labile[144]. Albomycin contains ferric iron[145] chelated octahedrally by a tris-*N*-δ-acetyl-L-ornithyl ligand[146]. In addition the antibiotic contains sulphur[146] and a pyrimidine[147,148] nucleus. The amino-acid residues are part of a cyclic hexapeptide, a derivative of *cyclo*-(L-seryl-L-seryl-L-seryl-L-ornithyl-L-ornithyl-L-ornithyl)[149] (C; R = H, and R¹ = CH_2OH), which has been synthesised[150]. Partial acid hydrolysis of δ_2-albomycin (*Scheme 8*, CVII) yielded a fragment whose u.v. spectrum was very similar to that of 1,3-dimethyluracil[147]. Prolonged acid hydrolysis of this fragment gave 3-methyluracil together with equimolar amounts of serine and ornithine[147]. Alkaline hydrolysis of δ_2-albomycin forms 3-methylcytosine[151,152], five moles of acetate, and one mole of sulphate, and since albomycin could not be desulphurised with Raney nickel[151] it was concluded that the pyrimidine is linked through its *N*-1 to a sulphonyl group which is joined to a serine side-chain[151] (residue * in the above cyclic sequence (*see* C)). Since some albomycin fractions liberate 3-methylcytosine on hydrolysis it is presumed that δ_2-albomycin (CVII) contains a 4-acetylimino-3-methylcytosine moiety (which on acid hydrolysis liberates 3-methyluracil); this was substantiated by comparison of the u.v. spectrum of δ_2-albomycin with that of N^4-acetyl-1,3-dimethylcytosine[151]. Compounds such as δ_1-albomycin (CVIII) are regarded as degradation products of the highly labile δ_2-compound (*Scheme 8*, CVII). The similarity of the degradation products of the antibiotic grisein to those of albomycin make it likely that these two materials are identical[153].

CYCLO-HEPTAPEPTIDES

Naturally occurring *cyclo*-heptapeptides are rare, one example being the crystalline compound evolidine[1] (CIX), which was isolated from the leaves

Ser-Phe-Leu-Pro-Val-Asn-Leu
(CXI)

Val-Asn-Leu-Ser-Phe-Leu-Pro
(CXII)

Leu-Pro-Val-Asn-Leu-Ser-Phe
(CX)

of *Evodia xanthoxyloides* (a small tree found in north Queensland). Nesvadba and Young[154], and Studer and Lergier[104] independently prepared the same linear sequence (CX) of evolidine, but no material corresponding to evolidine could be isolated after treatment with dicyclohexylcarbodi-imide. Another linear sequence (CXI), corresponding to the product of partial acid hydrolysis of evolidine[155], was synthesised and cyclisation in dilute solution by the *p*-nitrophenyl active ester method gave a 20% yield of material identical with natural evolidine[104]. Recently another synthesis of evolidine from the linear sequence (CXII) has been reported[156]. In this case the hydroxyl side-chain of serine was masked as the O-acetyl derivative and cyclisation of the heptapeptide was brought about by the active ester method. O-Acetyl evolidine was obtained in only 1% yield. It is interesting to note that linear penta- or deca-peptides corresponding to the sequence of gramicidin S but with the same terminal groups as the linear evolidine heptapeptide (CXII) gave much higher yields of cyclic decapeptide, ~20%[54] and ~40%[157] respectively. This situation may be connected with the all L-configuration of the evolidine sequence.

CYCLO-DECAPEPTIDES

Gramicidin S

The antibiotic gramicidin S[1] (CXIII) is a cyclic decapeptide containing a repeated pentapeptide sequence which contains D-phenylalanine; it was isolated from *Bacillus brevis*. The antibiotic shows *in vitro* activity against Gram-positive bacteria but is inactive against Gram-negative bacteria. It possesses haemolytic and toxic properties and has been the subject of intensive investigation.

```
→ L-Val → L-Orn → L-Leu → D-Phe → L-Pro
   1        2       3       4       5
   5'       4'      3'      2'      1'
→ L-Pro ← D-Phe ← L-Leu ← L-Orn ← L-Val
```

(CXIII)

Gramicidin S was the first naturally occurring cyclic peptide to be synthesised[157]. The original synthesis involved cyclising the corresponding decapeptide *p*-nitrophenyl ester (CXIV; $n = 2$, and W = tosyl) in dilute solution in pyridine, but a more economical method of synthesis has been developed which involves cyclodimerisation of a pentapeptide active ester[54]

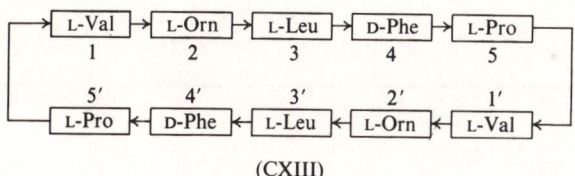

H-(L-Val-L-Orn-L-Leu-D-Phe-L-Pro)$_n$-O-⟨⟩-NO$_2$

(CXIV)

(CXIV; $n = 1$, and W = tosyl). As indicated in a previous section on cyclo-dimerisation, Schwyzer[109] attributed this phenomenon to the antiparallel aggregation of two pentapeptide units (*see* CXV) under the cyclisation conditions. This led to a postulated conformation for the *cyclo*-decapeptide (CXVI) which is in surprisingly good agreement with the conformation

(CXV)

deduced from spectroscopic studies (*see* later). Waki and Izumiya[158] found that the high dilution cyclisation of the pentapeptide *p*-nitrophenyl ester (CXIV; $n = 1$, and W = benzyloxycarbonyl) in pyridine gave rise to a mixture of protected gramicidin S and the corresponding *cyclo*-pentapeptide (semigramicidin S, CXVII). Cyclisation of linear pentapeptide analogues of gramicidin S has shown that the ratio of cyclic dimer to monomer is very sensitive to the nature of the terminal amino-acids (*see* Table 5.2). N-

Table 5.2. RELATIVE YIELD OF *cyclo*-PENTAPEPTIDE (M) AND *cyclo*-DECAPEPTIDE (D) ON CYCLISING THE PENTAPEPTIDE ACTIVE ESTER MONOMERS SHOWN AT A CONCENTRATION OF $\sim 3 \times 10^{-3}$ MOLAR. ANTIBACTERIAL ACTIVITY TOWARDS GRAM-POSITIVE BACTERIA, A_M = MORE ACTIVE THAN GRAMICIDIN S, A_{GS} = AS ACTIVE AS GRAMICIDIN S, A_L = LESS ACTIVE, AND I = INACTIVE.

L-Val	L-Orn	L-Leu	D-Phe	L-Pro	M	D	Activity of cyclo-dimer
Gly	L-Orn	L-Leu	D-Phe	Gly	100	0	I
Gly	L-Orn	L-Leu	D-Phe	L-Pro	100	0	—
Gly	L-Orn	L-Lys	L-Leu	D-Phe	L-Pro	100	0	—
L-Ala	L-Orn	L-Leu	D-Phe	L-Pro	91	9	A_{GS}
L-Val	L-Orn	L-Leu	D-Phe	Sar	90	10	A_{GS}
L-Val	L-Orn	L-Leu	D-Phe	β-Ala	89	11	I
L-Val	L-Orn	L-Leu	D-Phe	Gly	79	21	A_M
L-Leu	L-Orn	L-Leu	D-Phe	L-Pro	78	22	A_{GS}
L-Val	L-Orn	Gly	D-Phe	L-Pro	59	41	I
D-Phe	L-Leu	L-Orn	L-Val	L-Pro	53	47	A_L
L-Val	L-Orn	L-Ala	D-Phe	L-Pro	43	57	A_L
L-Val	L-Orn	L-Leu	D-Phe	L-Pro	32	68	A_{GS}
L-Val	L-Dab	L-Leu	D-Phe	L-Pro	32	68	A_{GS}
L-Val	L-Lys	L-Leu	D-Phe	L-Pro	29	71	A_{GS}
L-Val	L-Orn	L-Leu	D-Ala	L-Pro	23	77	A_L
L-Val	L-Orn	L-Leu	D-Val	L-Pro	17	83	A_{GS}
L-Val	L-Orn	L-Leu	D-Leu	L-Pro	15	85	A_{GS}
L-Val	L-Orn	L-Leu	Gly	L-Pro	0	100	A_L
*L-Val	L-Orn	L-Leu	L-Phe	L-Pro	40	60	A_L
L-Phe	D-Leu	D-Orn	D-Val	Gly	—	—	A_{GS}

*Cyclisation using *o*-phenylene chlorophosphite in diethyl phosphite at 100°C, concentration 5×10^{-3} molar.

Terminal glycine causes exclusive formation of cyclic monomer[159]. As the bulk of the N-terminal amino-acid is increased through alanine[159] to leucine[160] and valine[158] so the yield of cyclic monomer falls to 32%. However, C-terminal glycine also favours the production of cyclic monomer[161,162], so it is not surprising that cyclisation of the pentapeptide with both termini glycine yields the *cyclo*-pentapeptide[162] exclusively. The D-phenylalanine

(CXVI)

residue has a profound effect on the course of the cyclisation; if it is replaced by glycine no cyclic monomer is formed[163]. This is consistent with the earlier discussion of *cyclo*-pentapeptides. The proportion of semigramicidin S present in the cyclisation product increases as the initial concentration of linear pentapeptide active ester in pyridine decreases[158] (*see Table* 5.3). This

Table 5.3. RELATIVE PROPORTION (%) OF *cyclic* MONOMER (M) OR *cyclic* DIMER (D) AS A FUNCTION OF THE CONCENTRATION OF (CXIV; $n = 1$, AND W = BENZYLOXYCARBONYL-)

Initial concentration mole dm^{-3}	M	D
0.3×10^{-3}	45	55
3×10^{-3}	32	68
30×10^{-3}	29	71

behaviour is typical of a pair of competing reactions, one of which is intramolecular and therefore proceeds at a rate independent of concentration, while the other involves a bimolecular dimerisation and hence proceeds at a rate dependent on concentration.

The relationship of structure to antibiotic activity in gramicidin S has been studied by preparing numerous analogues. Ring size seems to be a critical factor, for *cyclo*-penta-[158] (CXVII), -hexa-[164] (CXVIII), and -hepta-[165] (CXIX) peptides containing the sequence of gramicidin S are quite inactive, as indeed is [5,5'-β-alanine]-gramicidin S[166] (CXX), which contains 32 atoms in the ring rather than the 30 of the natural product. The L-valyl residues at positions 1,1' in gramicidin S are not critical for activity and can be replaced by L-alanyl[159] or the γ-branched L-leucyl residues[160] without impairment of activity. The basic residues at positions 2,2' can be

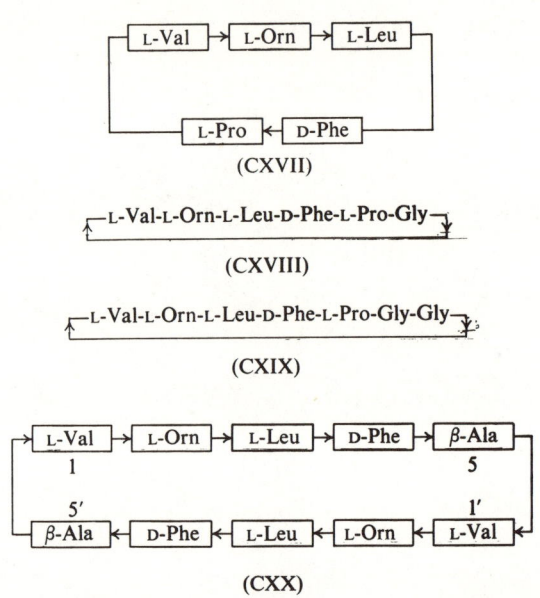

(CXVII)

(CXVIII)

(CXIX)

(CXX)

replaced by higher[167,168] (i.e., L-lysine) or lower homologues[168] (i.e., L-α,γ-diaminobutyric acid) without effect on the activity. Lessening the bulk of the leucyl side-chains at positions 3,3' leads to a lowering of the antibiotic activity with the L-alanyl analogue[169], and its complete loss in the glycyl analogue[169]. A change of configuration of the phenylalanyl residue at positions 4,4' leads to a marked weakening of the activity[170], comparable to that of [4,4'-glycine]-gramicidin S[163]. The D-phenylalanyl residues of the natural antibiotic can be replaced by D-leucyl[171] or D-valyl[171] without affecting the antibacterial activity, but it is lessened in the case of the D-alanyl analogue[172]. Finally, sarcosine can replace the 5,5' proline residues of gramicidin S without affecting the activity[173]; however, [5,5'-glycine]-gramicidin S is remarkable in that it possesses a higher activity[161,174] than the natural product against *Bacillus subtilis*. The high activity of this material gave impetus to the study of *retro*-compounds, which are now discussed.

The indistinguishable biological activities of enniatin B and *enantio-enniatin* B have already been mentioned. If the structures of [5,5'-glycine]-gramicidin S (CXXI) and its *retro-enantio*-analogue (CXXII; reversed direction of acylation) are compared it can be seen that the disposition of the side-chains is topologically similar but the direction of the peptide bonds is reversed. *Retro-enantio*-[5,5'-glycine]-gramicidin S does in fact exhibit the high antibacterial activity of the *cyclo*-diastereomer[175]. In gramicidin S itself the *retro-enantio*-analogue does not possess the same topological correspondence because of the presence of proline[176]. Based on the β-pleated sheet model of gramicidin S, the *retro* analogue (same configurations as gramicidin S but reversed sequence) would possess the hydrophobic proline residues on the same side of the ring as the hydrophilic ornithine residues. It was therefor not unexpected that it showed a weaker activity[177] than gramicidin S.

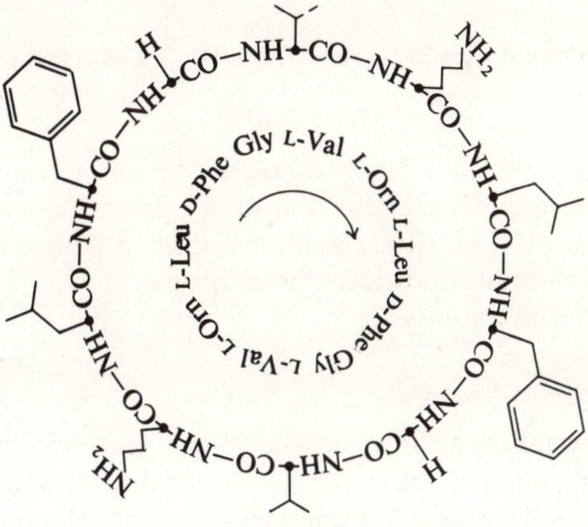

(CXXI)

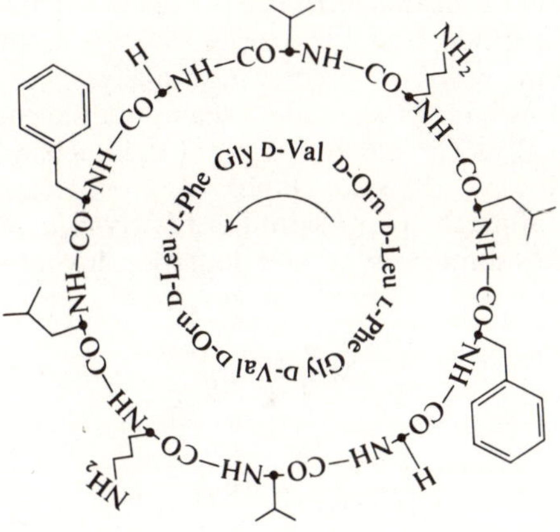

(CXXII)

Kato and co-workers[178] have observed a correlation between the o.r.d. spectra and antibacterial activity of gramicidin S analogues. Active analogues and gramicidin S itself[179] show a curve with a negative extremum at 232 nm; this is even true of analogues such as [4,4'-D-alanine]-gramicidin S where there are no aromatic chromophores, suggesting that this curve reflects the spatial arrangement of the amide chromophores only. Inactive analogues, on the other hand, show simple dispersion curves which decrease or increase steadily without any extrema. Thus it can be inferred that the active compounds possess similar conformations, which differ from those of the inactive ones[178].

The linear decapeptide derivative (CXXIII), isolated from *Bacillus brevis*,

$$\overset{H}{\underset{O}{\diagdown}}C\cdot\text{D-Phe-L-Pro-L-Val-L-Orn-L-Leu-D-Phe-L-Pro-L-Val-L-Orn-L-Leu}\cdot NH\cdot CH_2CH_2OH$$

(CXXIII)

$$\overset{H}{\underset{O}{\diagdown}}C\cdot\text{L-Val-L-Orn-L-Leu-D-Phe-L-Pro-L-Val-L-Orn-L-Leu-D-Phe-L-Pro}\cdot NH\cdot CH_2CH_2OH$$

(CXXIV)

has been suggested[180] as a biosynthetic precursor of gramicidin S. A synthetic sample of this compound was found to possess no antibiotic activity[181], although its o.r.d. spectrum resembled that of gramicidin S. A number of related linear decapeptides have been prepared (e.g. CXXIV) and exhibit weak antibacterial activity[182].

The Conformation of Gramicidin S

An X-ray crystallographic analysis[183] of gramicidin S and some of its derivatives in the dry and solvated crystal indicated that the molecule possesses a two-fold axis of symmetry and within the molecule there are two layers of atoms approximately 4·8 Å apart; the overall molecular dimensions are 11 × 19 × 9·5 Å. Hodgkin and Oughton[184] considered a number of molecular models for gramicidin S and concluded that a model based on two extended β-peptide chains, separated by 4·8 Å and linked together at the ends by proline residues, was the most likely.

Several attempts have been made to compute the conformation of gramicidin S theoretically. The conformation of the backbone of a peptide chain is determined by the dihedral angles about the single bonds $N-C_i^\alpha$ (ϕ_i) and $C_i^\alpha-C'$ (ψ_i) since the peptide unit is a fairly rigid planar group of atoms (*see* CXXV). The simplest type of calculation involves the computation of

(CXXV)

the values of ϕ_i and ψ_i which are sterically allowable, the atoms being regarded as hard spheres (contact criteria). A more sophisticated treatment considers the relative stabilities of the allowed conformers. An expression for the potential energy of the system as a function of the atomic co-ordinates of that system is subjected to an energy minimisation procedure to produce the most acceptable conformation. The factors taken into account are the energy of non-bonded interactions, electrostatic interactions, hydrogen bonding, torsion about the peptide bond, distortion of the peptide bond, solvent interactions, and of course ring closure in the case of cyclic peptides[185].

Scheraga et al.[186] produced a crude conformational model of gramicidin S using essentially contact criteria. A different conformation was calculated using energy minimisation techniques[187]; the structure possesses a two-fold symmetry axis and an antiparallel alignment of β-pleated sheet chains together with two intra-chain hydrogen bonds (between the amide NH groups of ornithyl residues and the amide carbonyl groups of the prolyl residues) (CXXVI). Liquori[188] and co-workers computed five low energy conformations for the two peptide unit system (called the stereochemical code) and used these data in computations on gramicidin S. The conformation

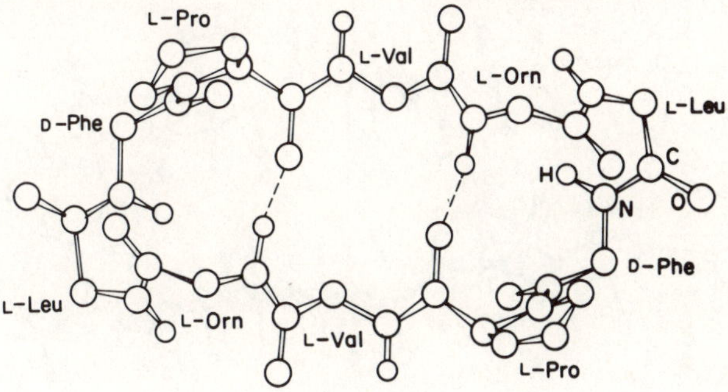

(CXXVI)

arrived at involved two distorted right handed α-helical tripeptide segments connected by two non-helical dipeptide segments, each α-helical segment containing one hydrogen bond (CXXVII). Later Scott and co-workers[189] re-evaluated these two conformations and found that at the energy minimum (CXXVI) was not precisely symmetrical, and conformation (CXXVII) was found to reside at a similar energy minimum providing that the hydrogen bonds mentioned earlier were absent. Momany and colleagues[190] then used the Hodgkin–Oughton–Schwyzer β-pleated sheet structure (CXVI) as the starting point for the energy minimisation calculations and arrived at a

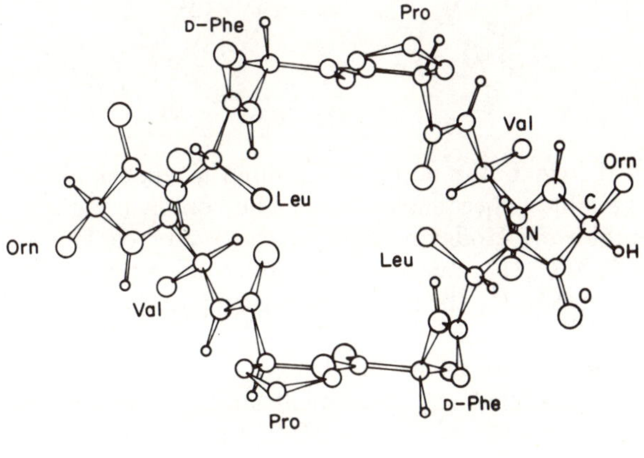

(CXXVII)

conformation of lower energy content than the previous conformations. It was devoid of formal hydrogen bonds, although some of the amide NHs are buried internally; its low energy arises from favourable nonbonded and electrostatic energy contributions. It cannot be overemphasised that the conclusions of these theoretical calculations are tentative due to the uncertainties involved. The diverse nature of the computed conformations may be related to these uncertainties or may result from energy minimisation leading to 'local' rather than 'global' minima. Recently, de Santis and Liquori[191] have recalculated the conformation of gramicidin S using their

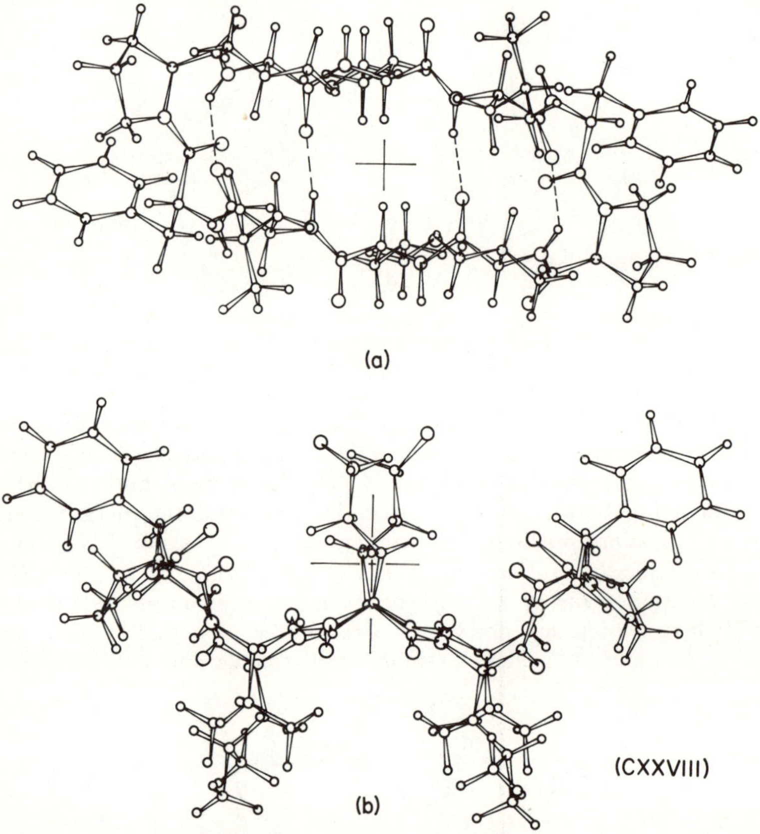

(a)

(b)

(CXXVIII)

stereochemical code, but in addition limiting the conformational possibilities using the available experimental data. The conformation deduced, which closely resembles the Hodgkin–Oughton–Schwyzer conformation (CXVI), is shown in (CXXVIII; (a) the projection is on a plane perpendicular to the two-fold axis, and (b) the projection is on a plane containing the two-fold axis).

Spectroscopic investigations of the conformation of gramicidin S in solution have paralleled the theoretical studies. Liquori and Conti[192,193] carried out a brief study of the 100 MHz n.m.r. spectrum of gramicidin S in d^6-dimethyl sulphoxide. Stern, Gibbons and Craig[194] carried out a rigorous analysis of the 100 and 220 MHz n.m.r. spectrum of gramicidin S in CD_3OD and d^6-dimethyl sulphoxide. The strong singlet appearing at 7·3 ppm in the CD_3OD spectrum was assumed to result from ten phenyl hydrogens of the two D-phenylalanine residues, and this was used as a basis for integrating the other signals. Four separate NH peaks were identified, each integrating for two hydrogens. Decoupling experiments and the invariability of the spin–spin coupling constants $^3J_{NH-CH}$ with field strength showed that the splitting of the amide peaks was due to spin–spin coupling. The relationship between the amide, α-, and β-hydrogens was determined by

(CXXIX)

Assignment of the chemical shifts of gramicidin S in CD_3OD in ppm

double resonance experiments; the assignment of the hydrogens of gramicidin S in CD_3OD is shown in (CXXIX). At 220 MHz the amide region of gramicidin S in d^6-dimethyl sulphoxide shows 3 doublets, 1 singlet and a broad peak (6H) due to the ornithine ammonium groups. The four amide and five α-hydrogen resonances each correspond to two hydrogens, indicating that each pair of hydrogens is in a similar electronic environment, which is consistent with a two-fold symmetry axis. Information on the stereochemistry of the gramicidin backbone was obtained from the magnitude of $^3J_{NH-CH}$. The three upfield amide doublets possess similar values of $^3J_{NH-CH}$ (~8·5–9·0 Hz) which suggests that the N—C dihedral angle for the leucyl, ornithyl, and valyl residues is small (i.e. NH—CH is nearly *trans*), whereas the N—C dihedral angle of the phenylalanyl residues is large (i.e. NH—CH is nearly *cis*). Hydrogen deuterium exchange experiments revealed that the ornithyl and phenylalanyl amide hydrogens exchange rapidly with deuterium, whereas the leucyl amide hydrogens exchange only slowly and the valyl ones even more slowly (CD$_3$OD solution, $t_{\frac{1}{2}}$ N—H of Phe, Orn, Leu, and Val ≈ 0.5 h, 24 h, 1 week, and >2 weeks respectively).

A conformational model of gramicidin S[194] which takes these facts into account (*see* CXVI) contains four intra-annular hydrogen bonds involving leucyl N—H and valyl C=O, and valyl N—H and leucyl C=O groups; the valine hydrogen bonds are perforce shorter than the leucine ones, which ties in with their exchange behaviour. The phenylalanyl and ornithyl N—Hs are not solvent shielded and so would be expected to exchange rapidly. Using a rapid dialysis technique, Laiken, Printz and Craig[195] studied the tritium hydrogen exchange of gramicidin S in aqueous solution and obtained data consistent with the above model. Eight peptide hydrogens were found to be exchangeable. Above pH 3·0 there are four rapidly exchanging hydrogens, and two groups each of two slowly exchanging hydrogens. Below pH

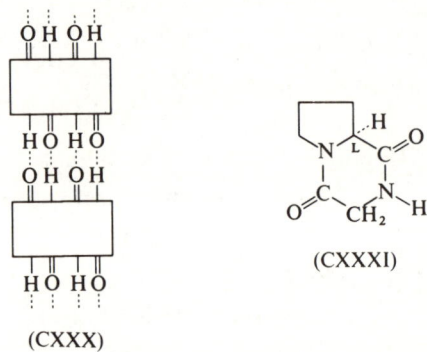

(CXXX)   (CXXXI)

3·0 there are four slowly exchanging hydrogens. The model is also consistent with i.r. dichroic studies[196], the molecular dimensions from crystallography[183], and from surface film measurements[197]. In the latter case Few[197] found that gramicidin S will form a condensed solid monolayer on water which results from extensive intermolecular association (CXXX); at zero compression the molecular area was found to be ~ 170 Å^2. It was assumed from the agreement of theoretical and experimental dipole moment data that the molecule in the film is flat with hydrophobic side-chains pointing into air and hydrophilic side-chains pointing into the aqueous phase.

Ohnishi and Urry[198] also studied the 100 and 220 MHz n.m.r. spectrum of gramicidin S and obtained essentially similar results to Stern and co-workers. In addition they studied the temperature dependence of the amide hydrogen chemical shifts in methanol. The amide hydrogens of the leucyl and valyl residues exhibit a low temperature coefficient, whereas the amide hydrogens of the ornithyl and phenylalanyl residues exhibit a high temperature coefficient similar to that of *N*-methylacetamide. Schwyzer and Ludescher[88] carried out similar studies, but they concentrated on the solvent dependence of the amide hydrogen chemical shifts. Using the amide hydrogen assignment of Stern and co-workers[194] it was found that the signals due to leucyl and valyl hydrogens moved downfield on passing from the amide hydrogen acceptor solvent, CD$_3$OD, to CF$_3$CO$_2$H, while the ornithyl and phenylalanyl amide hydrogens moved upfield. On passing from an amide acceptor solvent to trifluoroacetic acid, solvent-exposed amide hydrogens became more shielded (e.g. the single amide hydrogen of *cyclo*-(glycyl-L-prolyl) (CXXXI)), while solvent-protected hydrogens (i.e. those involved in intramolecular hydrogen bridges) became less shielded. The downfield shift of exposed peptide hydrogens on going from CF_3CO_2H to d^6-dimethyl sulphoxide or deutero-methanol can be attributed to stronger hydrogen bonding with the more basic solvents, whereas the upfield shift of internal peptide hydrogens can be attributed to shielding resulting from the greater electron density of the adjacent amide, since this bond is no longer protonated.

Schwyzer and Ludescher[199] also used a completely different approach for studying the conformation of gramicidin S. They showed that the change in chemical shift of the aromatic hydrogens of the phthalimido group could

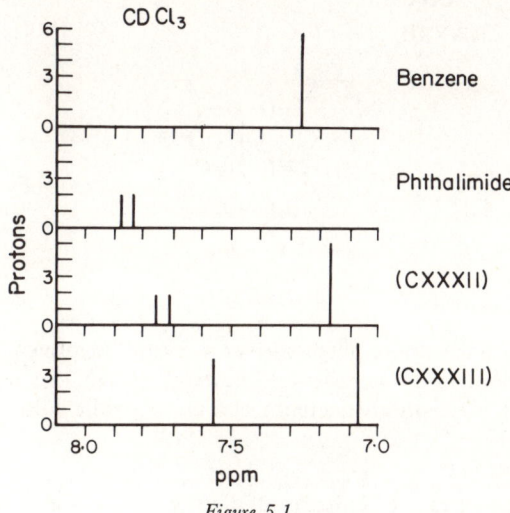

Figure 5.1

be used as a sensitive indicator for signalling the proximity of a phenyl side-chain[200]. Model compounds such as *N*-phthaloyl-L-phenylalanine methyl ester (CXXXII), and *N*-phthaloyl-*cis*-2-phenylcyclopentylamine (CXXXIII) enabled the effect of mutual ring current shielding on the chemical shifts to be evaluated (the relevant data are summarised in *Figure 5.1*). With compound (CXXXIII), relative to the standards shown, an upfield shift of 0·30 and 0·26 ppm for the phthaloyl 3,6-hydrogens and 4,5-hydrogens respectively occurred, and on average 0·20 ppm for the phenyl hydrogens. Assuming a face-to-face orientation of the two aromatic rings, the theoretical ring current shielding was found to agree closely with the observed values[200]. The 100 MHz n.m.r. spectrum of diphthaloyl-gramicidin S in CD_3OD was compared with that of phthaloylglycine methyl ester[199]. Upfield shifts of the hydrogens of the phthaloyl groups on the ornithine side-chains (0·10 ppm for the 3,6-hydrogens and 0·2 ppm for the 4,5-hydrogens) indicate that diphthaloyl-gramicidin S exists in a conformation where the phenyl and phthaloyl groups are in close proximity[199]. There is good agreement between the calculated and observed shielding for the pleated sheet model. Gramicidin S thus exists in a conformation or conformational equilibrium wherein the ornithyl and phenylalanyl side-chains can be brought into close proximity.

Another investigation demonstrates, from a chemical rather than spectroscopic point of view, that it is possible to bring the two ornithine side-chains of gramicidin S into close proximity[201]. Treatment of the *cyclo*-peptide with

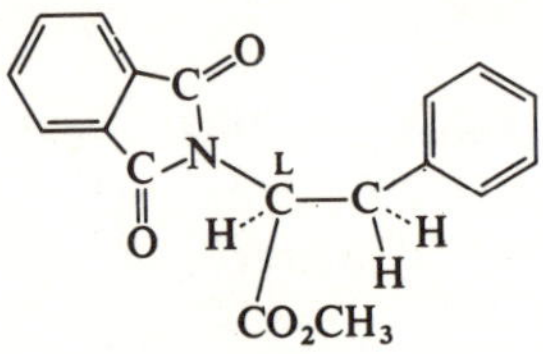

(CXXXII) 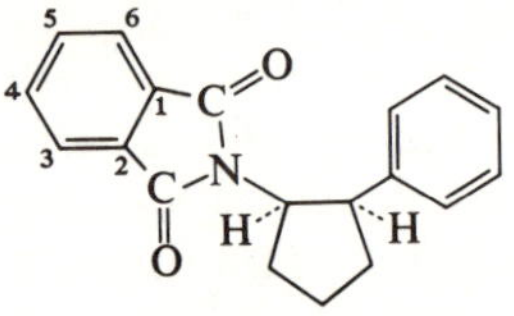   (CXXXIII)

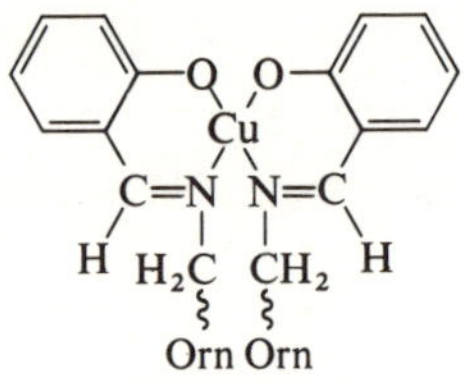

(CXXXIV)

salicylaldehyde and copper acetate gives a crystalline olive-green compound, copper(II) gramicidyl bis(salicylaldiminate) (CXXXIV), which exhibits the typical u.v. absorption spectrum of a copper salicylaldimine chelate.

Tyrocidines

Tyrothricin, an extract of cultures of *Bacillus brevis*, yields a family of antibiotics known as tyrocidines[1] which are structurally related to gramicidin S. Tyrocidine A (CXXXV) is a cyclic decapeptide containing two D-amino-acid residues and the same pentapeptide sequence that is repeated in gramicidin S; it is active against some Gram-positive bacteria but less active against Gram-negative ones. In the determination of the structure[202] of tyrocidine C (CXXXV) use was made of a specific cleavage reaction of acylproline derivatives with lithium aluminium hydride[203]. The initial linear decapeptide formed, terminating in 2-amino-3-phenylpropionaldehyde (CXXXVI), was reduced to the corresponding D-phenylalaninol peptide (CXXXVII) with sodium borohydride and cleaved with pepsin. The structures of the two (tetra- and hexa-peptide) fragments were determined by standard methods. Tyrocidine D[204] (CXXXV) is produced when large amounts of L-tryptophan are added to the culture medium of the bacterium. Comparison with other tyrocidines showed that it contains an L-tryptophan residue in place of L-tyrosine of tyrocidine A. Tyrocidine E[205] (CXXXV) (in which the L-tyrosine of tyrocidine A is replaced by L-phenylalanine) was biosynthesised using the partially purified tyrocidine enzyme synthesising system of *Bacillus brevis* in a cell-free system in the absence of both tyrosine and tryptophan. Its structure was indicated by the content of L- and D-phenylalanine in the hydrolysate.

A number of these antibiotics have been synthesised[206,207] from linear decapeptides with C-terminal proline residues via active ester cyclisations. This enabled the hitherto unknown quantitative features of the antibacterial

```
→ L-Val → L-Orn → L-Leu → D-Phe → L-Pro ┐
                                          │
└── X ← L-Gln ← L-Asn ← Y ← Z ←──────────┘
         (CXXXV)
```

	X	Y	Z
Tyrocidine A	L-Tyr	D-Phe	L-Phe
B	L-Tyr	D-Phe	L-Trp
C	L-Tyr	D-Trp	L-Trp
D	L-Trp	D-Phe	L-Phe
E	L-Phe	D-Phe	L-Phe

$$\text{(CXXXVI)} \xrightarrow{NaBH_4} \text{(CXXXVII)}$$

activity of tyrocidine B[208] to be defined. Synthetic tyrocidine B, C[209], and E[210] possess a similar level of activity towards Gram-positive microorganisms as tyrocidine A. A mixture of tyrocidine (80%) and gramicidin (20%) is still used therapeutically for the treatment of oral infections.

The tyrocidines are characterised by a pronounced tendency to aggregate in solution[202,211], a tendency absent in gramicidin S. An attempt has been made to find the structural features responsible for aggregation by studying the thin-film dialysis[211] of a number of tyrocidine derivatives. Aggregation when present is apparent since the rate of dialysis increases as the concentration of the peptide inside the membrane decreases. Such modifications as hydrogenation of the aromatic rings of tyrocidine A, succinylation of the ornithine δ-amino-group in tyrocidine B, and methylation of the phenolic group of tyrosine in succinyl-tyrocidine B failed to eliminate the association. Thus aromatic groups, ornithine side-chains, or the phenolic group of tyrosine do not seem responsible for the association. The open chain sequence of tyrocidine B does not associate, and only those derivatives which associate exhibit antibacterial activity[211]. Since the association is greatest in solvents of high dielectric constant, hydrophobic interactions may be involved.

The o.r.d. curve[211] of tyrocidine B is similar to that of gramicidin S despite the association of the former. A study has been made of the effect of aggregation on the c.d. spectra[212] and the n.m.r. spectra[213] of the tyrocidines. In the latter case a preliminary study revealed that a 6% solution of tyrocidine B in D_2O at room temperature gives a featureless spectrum, no signals being observable. However, on raising the temperature, reducing the concentration, or adding methanol, peaks develop. The featureless nature of the spectrum is ascribed to dipolar line-broadening caused by the high molecular weight of the aggregate. Similarities in the i.r. spectra of gramicidin S and tyrocidine B suggest that their conformations may be alike[213].

HIGHER CYCLO-PEPTIDES

Valinomycin

Valinomycin was isolated from the mycelium of an actinomyces strain of the species *Streptomyces fulvissimus*[1]. It is highly active against certain bacteria and, although originally thought to be a *cyclo*-octadepsipeptide, synthesis[1] later showed it to be a *cyclo*-dodecadepsipeptide. The macrocycle is thus

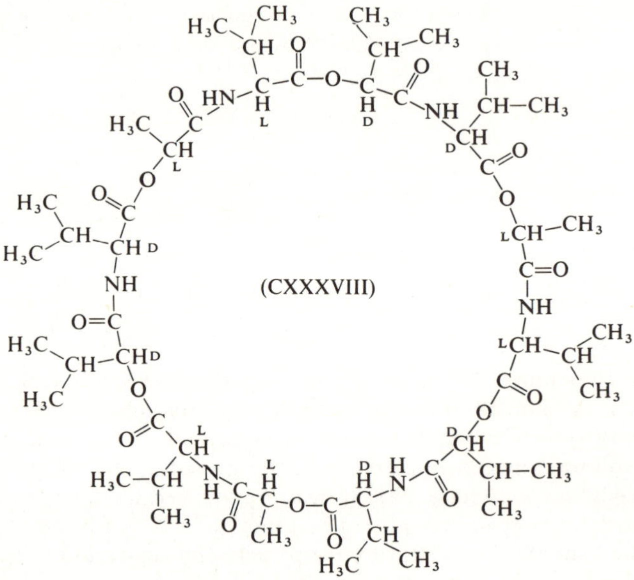

(CXXXVIII)

twice the size of the enniatins (i.e. 36 atoms in the ring) with alternating D-valyl, L-lactyl, L-valyl; and D-α-hydroxy-valeryl residues (CXXXVIII). The Merrifield solid-phase technique has recently been used to prepare the linear dodecadepsipeptide of valinomycin[214]. The didepsipeptides *N*-t-butyloxycarbonyl-L-valyl-D-α-hydroxyisovaleric acid (CXXXIX) and *N*-t-butyloxycarbonyl-D-valyl-L-lactic acid (CXL) were coupled alternately to resin bound D-valyl-L-lactate (CXLI) using dicyclohexylcarbodi-imide. The presence of an ester linkage in these carboxyl components prevents the possibility of oxazolinone formation, and hence racemisation during the coupling steps. The resultant linear dodecadepsipeptide was cyclised by the acid chloride method.

The dependence of the antibacterial activity of valinomycin on structure is rather similar to that of enniatin derivatives. Optimum activity is achieved with the 36-membered ring, and the structure and configuration of a few amino-acid residues can be changed without greatly affecting the activity. Change in the structure or configuration of the hydroxy residues leads in general to complete inactivation[215], but ester groups can be replaced by amide groups without appreciably affecting the activity[215].

Valinomycin exerts a remarkably specific effect on the permeability of certain membranes to monovalent cations. This effect, which is related to the antimicrobial activity of the depsipeptide, is specific for K^+ but not for Na^+. Valinomycin is able to induce active transport of K^+ in animal mitochondria[123,124,216] (leading to enhanced respiration), in lipid bilayers[125], and under non-aqueous conditions it forms a complex[217] with K^+ much more readily than with Na^+, a property which is undoubtedly related to its effect on membranes. The nature of the o.r.d. curve of valinomycin depends on solvent polarity, a phenomenon ascribed to a conformational equilibrium. The nature of these conformers and the conformation of the potassium ion

$$H_3C-\underset{\underset{CH_3}{|}}{\overset{\overset{CH_3}{|}}{C}}-O-\overset{\overset{O}{\|}}{C}-NH-\underset{L}{\overset{\overset{CH_3\diagdown CH_3}{CH}}{\underset{|}{CH}}}-CO_2-\underset{D}{\overset{\overset{CH_3\diagdown CH_3}{CH}}{\underset{|}{CH}}}-CO_2H$$

(CXXXIX)

$$H_3C-\underset{\underset{CH_3}{|}}{\overset{\overset{CH_3}{|}}{C}}-O-\overset{\overset{O}{\|}}{C}-NH-\underset{D}{\overset{\overset{CH_3\diagdown CH_3}{CH}}{\underset{|}{CH}}}-CO_2-\underset{L}{\overset{\overset{CH_3}{|}}{CH}}-CO_2H$$

(CXL)

$$H_2N-\underset{D}{\overset{\overset{CH_3\diagdown CH_3}{CH}}{\underset{|}{CH}}}-CO_2-\underset{L}{\overset{\overset{CH_3}{|}}{CH}}-CO_2-CH_2-RESIN$$

(CXLI)

complex was determined by detailed spectroscopic studies[121,198,217]. The n.m.r. spectrum of valinomycin in carbon tetrachloride shows two doublets in the amide hydrogen region; one of these shifts downfield when the hydrogen bond acceptor solvent d^6-dimethyl sulphoxide is used[217]. This has been interpreted in terms of the internally hydrogen-bonded conformation being present in non-polar solvent, but in more polar solvents the amide hydrogens of the L-valine residues are thought to hydrogen bond to solvent[217]. I.r. studies and a study of the temperature dependence[198] of the amide hydrogen chemical shifts confirm this view. The temperature coefficients of the amide resonances in methanol are similar to those in N-methylacetamide, but in d^6-dimethyl sulphoxide the high field NH resonance shows a reduced temperature coefficient. A detailed analysis of the spin–spin coupling constants[66] $^3J_{NH-CH}$ and $^3J_{C_\alpha H-C_\beta H}$ enabled the 'bracelet' conformation (CXLII) of valinomycin in non-polar solvents to be deduced[217]. The molecule exists as a rigid puckered ring approximately 8 Å in diameter and 4 Å high. The puckering of the backbone chain is achieved by a regular hydrogen bonding between all the amide bonds (CXLIII; the ring has been opened out in this diagram). The dipole moment calculated for this conformation is in agreement with the experimental value. The spectroscopic data for valinomycin in polar solvents have been interpreted in terms of the existence of two nearly equivalent 'disc' structures (CXLIV) containing only three intramolecular hydrogen bonds[198,217].

The addition of sodium chloride to a solution of valinomycin in methanol

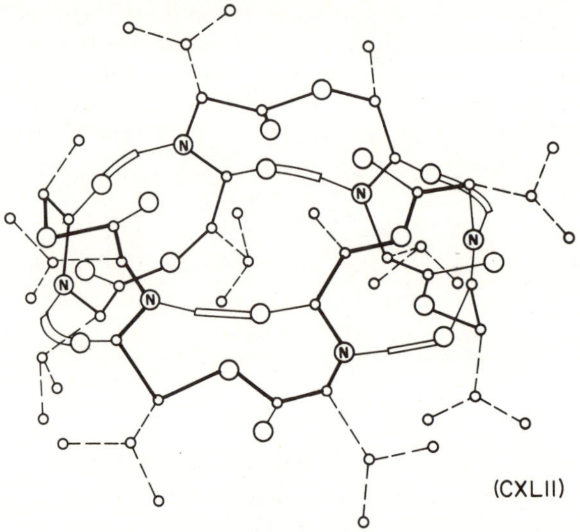

(CXLII)

∘C ◯O Ⓝ N ⊏⊐ H−bond

Conformation of valinomycin in non-polar solvents

has little effect on the temperature coefficient of the amide hydrogen chemical shifts. However, addition of potassium bromide dramatically reduces the temperature coefficients[198], indicating that all six amide hydrogens are internally hydrogen bonded in the K+ complex. I.r. studies confirm this view, and a detailed analysis of coupling constants enabled the conformation (CXLV) to be deduced[198,217]. A similar conformation was deduced from the X-ray crystallographic analysis[218] of the crystalline valinomycin–potassium aurichloride complex. The potassium ion is in fact co-ordinated to the carbonyl oxygen atoms of each of the six valine residues. The puckering

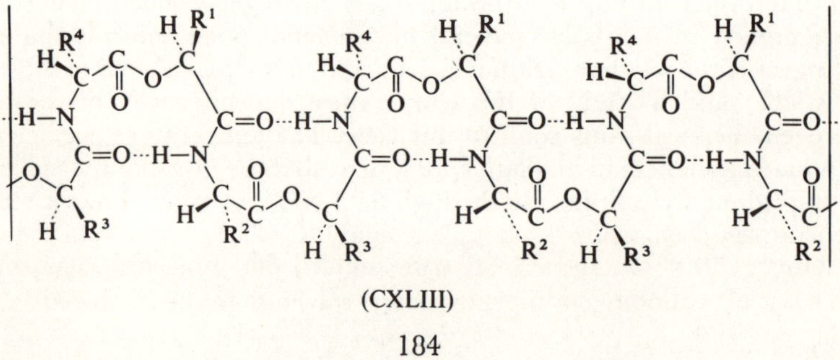

(CXLIII)

(CXLIV)

of the peptide backbone forms a 'bracelet' or 'pore' structure, the array of lipophilic side-chains on the periphery of this cylindrical peptide presumably serving to hinder access of solvent to the chelated potassium ion.

Three models have been proposed to explain the manner in which a cyclic antibiotic can bring about cation transport[219]. Several active molecules could stack together forming an ion navigable channel which could breach

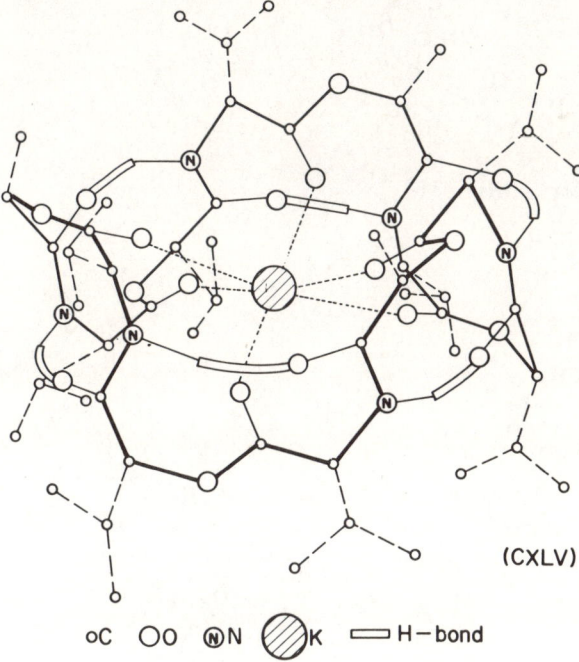

(CXLV)

oC oO ⓝN ▨K ▭ H−bond

Conformation of the K⁺ complex of valinomycin

the membrane, or an active molecule could attach itself to the membrane surface providing a 'pore' through which cations could penetrate. Alternatively, the antibiotic could act as a mobile carrier, i.e. form a lipid soluble complex with the cation which is then able to diffuse across the membrane and release the cation on the other side. Exploratory experiments on the application of n.m.r. spectroscopy to the determination of the rate of complexation have utilised the line broadening of the lactyl methyl signal which occurs on complexation[219]. A measurable rate of transfer of the cation between valinomycin molecules in the medium 80% methanol–20% $CDCl_3$ was observed; however, no measurable exchange was observed in $CDCl_3$ which argues against the tunnelling of ions being involved in their transport[219]. Experiments with a U-tube containing a gently stirred solution of valinomycin in chloroform in the centre section and aqueous solutions on either side in the two limbs have been used to illustrate in a simple way the transport of anion–cation pairs by the antibiotic[218]. Clearly, in this case the antibiotic is acting as a mobile carrier. Transport of all potassium salts does not occur. Picrate salts are more easily transported than aurichlorides, and simple chloride is not carried through the lipophilic barrier in this system. Obviously, the anion must have a reasonable affinity for the organic phase in order to be transported through it[218].

Mycobacillin

Mycobacillin is a macrocyclic peptide antibiotic with antifungal properties. It was obtained from the culture filtrate of *Bacillus subtilis* B_3[220], and contains

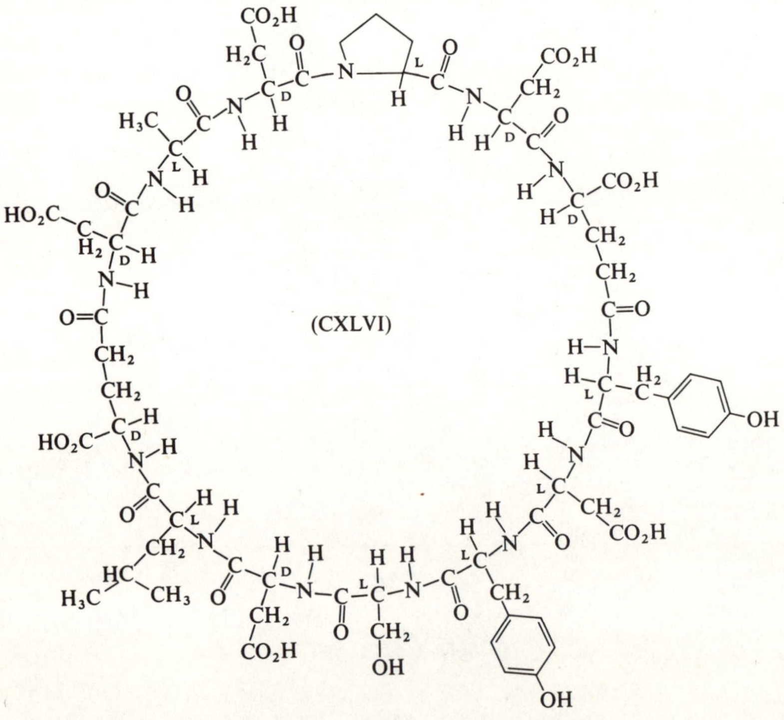

(CXLVI)

seven different amino-acids[221]. Although originally formulated as completely α-linked[222], recent work has shown that mycobacillin (CXLVI) has a 43-membered ring containing two γ-glutamyl linkages[223]. *N*-Bromosuccinimide decarboxylation of mycobacillin shows that two free α-carboxyl groups are present. Oxidation with sodium hypobromite and subsequent hydrolysis led to the detection of succinic acid, which indicates that at least one γ-linked glutamic acid unit is involved. Hydrazinolysis of the natural product yields the α-hydrazide of aspartic acid and the γ-hydrazide of glutamic acid. The involvement of both glutamic acid residues in γ-linkages was confirmed by

$$\begin{array}{cc} & CH_2-OH \\ CH_2-OH & CH-NH_2 \\ CH_2 & CH_2 \\ H_2N-CH-CO_2H & CH_2-CO_2H \\ (CXLVII) & (CXLVIII) \end{array}$$

total acid hydrolysis of the lithium borohydride reduction product of the heptamethyl ester of mycobacillin. Quantitative analysis showed that the molar ratio of α-amino-γ-hydroxybutyric acid (CXLVII) to γ-amino-δ-hydroxyvaleric acid (CXLVIII) was ∼5:2. Titration confirmed the presence of seven free carboxyl groups and two tyrosine phenolic groups. The location of the D- and L-amino-acid residues was determined after partial acid hydrolysis. The fifteen peptides produced were subjected separately to total acid hydrolysis, and each amino-acid mixture was re-examined after incubation with L- or D-amino-acid oxidase. The peptides which contained aspartic or glutamic acid (which are resistant to amino-acid oxidases) were incubated with L-glutamate decarboxylase (which had L-aspartate decarboxylase activity) and the amino-acids which remained after hydrolysis were identified. In two peptides the aspartyl residue was converted to alanyl, indicating the L-configuration, while the other aspartate peptide yielded aspartic acid, which has therefore the D-configuration. The composition of the two L-aspartate peptides enabled their positions in the macrocycle to be determined.

CYCLIC PEPTIDES CONTAINING UNUSUAL AMINO-ACIDS

Alamethicin

Alamethicin is a cyclic peptide antibiotic from the micro-organism *Trichoderma viride*[224], and like valinomycin it can transport cations through synthetic membranes. It consists of 18 amino-acid residues which include seven α-aminoisobutyric acid residues, two glutamine residues (one of which is pendant to the ring), and there is one free carboxyl group and a γ-glutamyl linkage in the ring (CXLIX); in this respect it resembles mycobacillin. The determination of its amino-acid composition was complicated by non-integral analyses of alanine and α-aminoisobutyric acid, which were ascribed to microheterogeneity of the sample. The predominant form possesses the composition Aib_7, Ala_2, Gln_2, Glu, Gly, Leu, Pro_2, and Val_2, all optically active amino-acids possessing the L-configuration[224,225]. Enzymic degradation

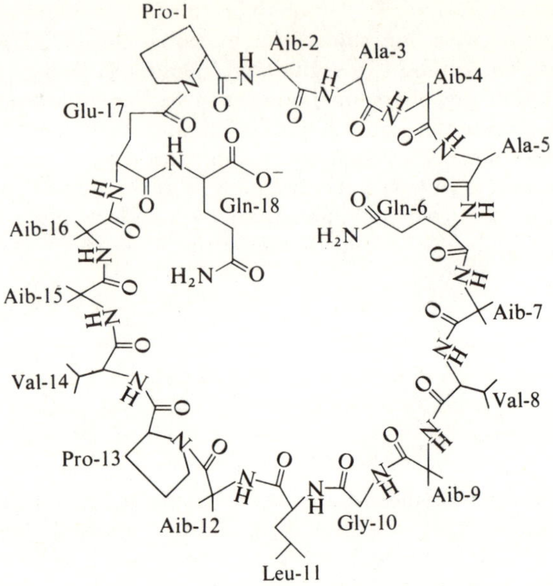

(CXLIX)

of alamethicin or linear sequences of alamethicin proved ineffective, probably due to the stereochemistry of the unusual α-aminoisobutyric acid residues[225]. Partial acid hydrolysis under mild conditions yielded three peptides, one of which corresponded to the full linear sequence of alamethicin, while the other two together corresponded to this sequence. All three peptides were shown to possess N-terminal proline residues[225]. More severe hydrolysis conditions released a whole family of smaller peptides which were sequenced by the Edman method, thus establishing the total sequence. The analysis was complicated by the tendency of N-terminal α-aminoisobutyric acid residues to be cleaved incompletely. A single free carboxyl group was detected by titration but no free amino groups were observed[225]. Two primary amides were detected and it was shown that glutamic acid residues six and eighteen were in fact amidated[225]. The presence of a peptide linkage between the imino group of proline-1 and the γ-carboxyl group of glutamic acid-17 was inferred by reducing the free carboxyl group of glutamine-18 under mild conditions with diborane. Subsequent hydrolysis and

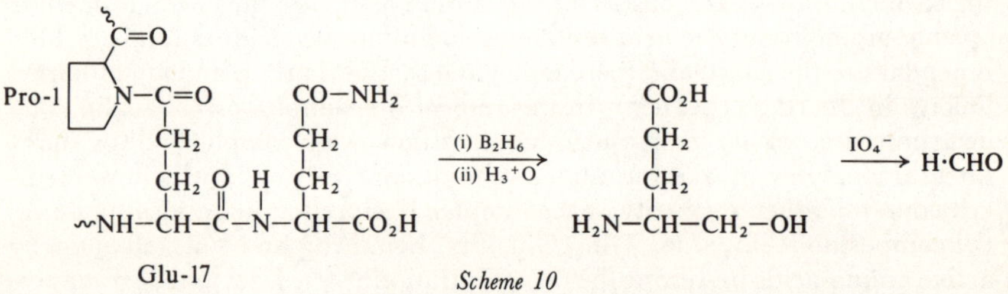

Scheme 10

treatment with periodate led to the detection of formaldehyde, a similar result being obtained with free glutamine[225] (see Scheme 10).

Alamethicin forms lipid soluble alkali-metal ion complexes over a wide pH range which can be either neutral or positively charged, but it shows poor discrimination between the different cations.

Monamycin

In 1959 an investigation of cultures of the soil micro-organism *Streptomyces jamaicensis* led to the isolation of small quantities of a crystalline chlorine-containing antibiotic, which was named monamycin[226]. This material only became available in reasonable quantity some years later following improvements in the technique of submerged culture of the micro-organism. Recent work has shown it to be a mixture of fifteen *cyclo*-hexadepsipeptides. All components are variations on one structure, (CL), which has a sequence of alternate D and L residues and contains single residues of *N*-methyl-D-leucine,

(CL)

one α-hydroxy-acid [either L-2-hydroxy-3-methylpentanoic acid (CL; R^1 = Me) or L-2-hydroxy-3-methylbutanoic acid (CL; R^1 = H)], and (3S, 5S)-5-hydroxypiperazic acid.* The remaining residues are L-proline (or *trans*-4-methyl-L-proline), D-isoleucine (or D-valine), and (3R)-piperazic acid [or (3R, 5S)-5-chloropiperazic acid; both of the D-configuration][227]. The sequence of the major component, monamycin D_1, (CL; R^1 = Me, R^2 = H, R^3 = Me, and R^4 = H) was the first to be determined by a combination of chemical degradation and mass spectrometry. The structures of the other components were related to this by mass spectrometry.

Cyclic amino-acids of the piperazic acid type have not been previously reported. The structure of piperazic acid itself was established by comparison with racemic synthetic material prepared by the successive catalytic hydrogenation and acid-catalysed hydrolysis of 1,2,3,6-tetrahydro-1,2-phthaloyl-pyridazine-3-carboxylic acid, the adduct of phthalazinedione and penta-2,4-dienoic acid (Scheme 11). Identification of the reduction product of the

* Piperazic acid is hexahydropyridazine-3-carboxylic acid.

natural amino-acid as D-ornithine established its configuration[227]. The stereochemistry of the 5-substituted position of the modified piperazic acids has been established (CLI; R = OH or Cl)[228]. In the crystalline mixture of monamycins there are traces of compounds in which Δ^4- and Δ^5-dehydropiperazic acids occur; products similar to these acids can be obtained from 5-chloropiperazic acid by dehydrohalogenation.

Two dioxopiperazines (CLII and CLIII) have been isolated from partial acid hydrolysates of monamycin D_1. Their formation is ascribed to the presence during hydrolysis of a significant proportion of a protonated form of the piperazic acid residue while still bound in the parent molecule which favours an intramolecular cyclisation (*Scheme 12*; the reaction leading to CLII is depicted)[229].

The occurrence of D-isoleucine in these peptides is interesting in view of the suggestion that all the D-residues which occur in peptides arise from the corresponding members of the L-series by C-2 epimerisation only after incorporation of the L-residue into the peptide chain[230]. On this basis the D-isoleucine residue would be derived from L-*allo*-isoleucine, an isomer which itself has not yet been found to occur naturally. L-Isoleucine and D-*allo*-isoleucine, on the other hand, are well-known constituents of natural peptides.

Cycloheptamycin

Cycloheptamycin (CLIV) is an antibiotic isolated from an unidentified *Streptomyces* species, and it contains some unusual residues[231]. When treated with diazomethane in the presence of boron trifluoride, the free hydroxyl group is methylated but the lactone remains intact. Subsequent remethylation with deuteromethyl iodide causes fission of the lactone, and a mass spectrum of the resulting linear product established the amino-acid sequence.

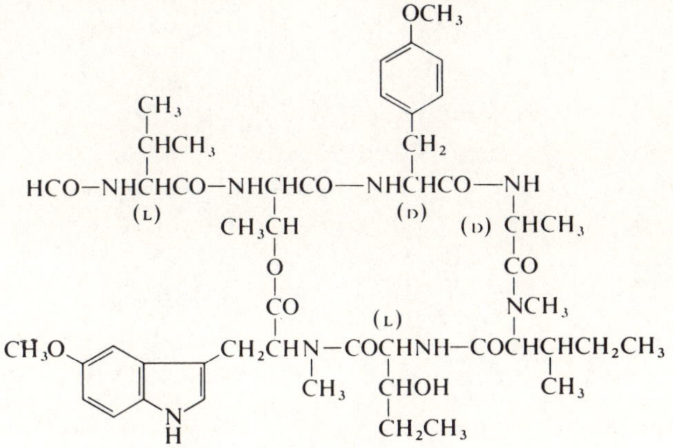

(CLIV)

An unlabelled methyl ether occurred in the sixth residue from the *N*-terminus, indicating that the hydroxyl group of the β-hydroxynorvaline is free in the parent compound. The methyl ether of threonine, on the other hand, showed up as fully labelled and must be involved in the lactone of the native compound. The methyl groups of the *O*-methyl-D-tyrosine and *N*-methyl-5-methoxytryptophan residues show up clearly in the n.m.r. spectrum of cycloheptamycin, as does the *N*-formyl group. Enzymic degradation of acid hydrolysates with D- and L-amino-acid oxidases established the configuration of four of the constituent amino-acids, but the threonine and the *N*-methyl residues are resistant to these enzymes. Neither β-hydroxynorvaline nor *N*-methyl-5-methoxytryptophan have been previously detected in natural products, and only the L-antipode of *O*-methyltyrosine has been hitherto recorded[231].

Viomycin

Elucidation of the structure of the tuberculostatic antibiotic viomycin has proved to be a difficult task. Acid hydrolysis liberates L-serine, L-α,β-diaminopropionic acid, L-β-lysine, and a new basic amino-acid viomycidine, together with carbon dioxide, ammonia, urea, and a trace of glycine. The absence of a free carboxyl group shows viomycin to be a cyclic peptide, but although most of the sequence has been established the nature of the chromophore and the position of a guanidine residue are still subjects of controversy. A tentative structure (CLV; $R^1 = R^2 = OH$, $R^3 = H$) has been proposed for this antibiotic, but it is possible that the positions of the β-lysyl and one of the seryl units may be reversed[232,233]. The structure of viomycidine[234] (CLVI) has been confirmed by X-ray crystallography[235]; it is probably an artefact of the hydrolysis since after hydrogenation, acidic

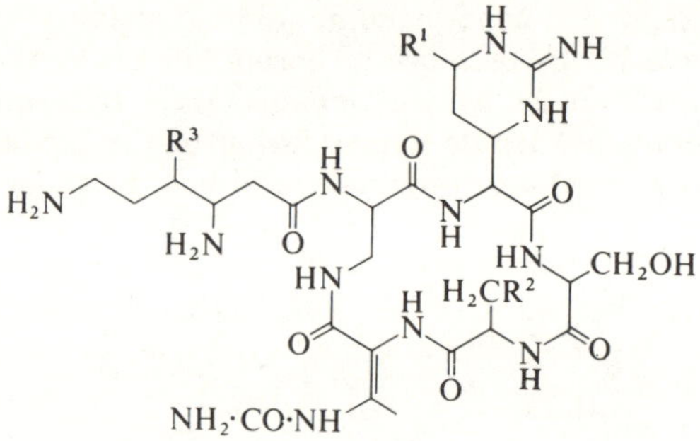

(CLV)

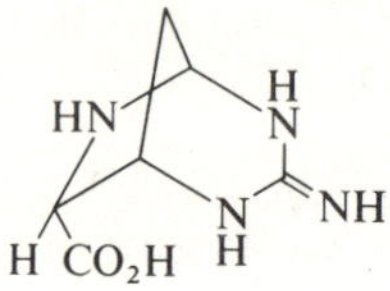

(CLVI) 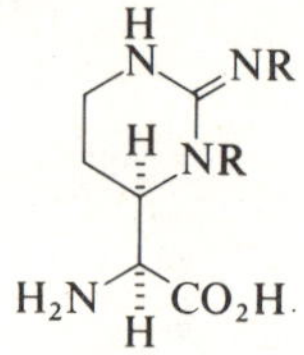   (CLVII)

degradation of viomycin gives another amino-acid (CLVII; R = H) instead. This latter amino-acid (capreomycidine) has also been isolated from closely related antibiotics, e.g. capreomycin 1A (CLV; $R^1 = H$, $R^2 = NH_2$, $R^3 = H$)[236,238] and tuberactinomycin (CLV; $R^1 = R^2 = R^3 = OH$)[236,237]. Another degradation product of viomycin is viocidic acid

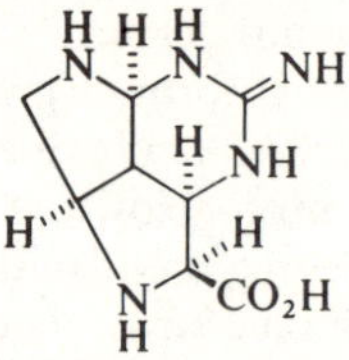

(CLVIII) 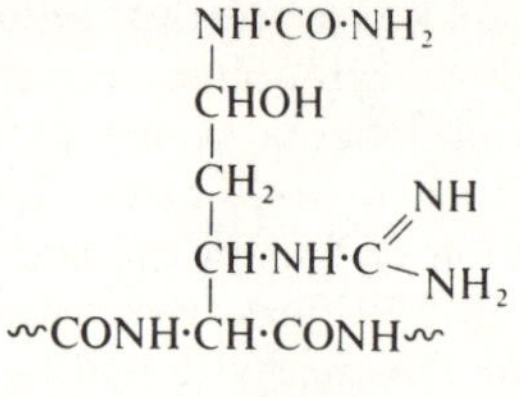   (CLIX)

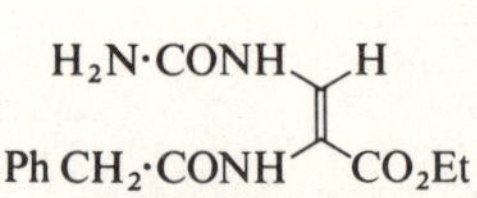

(CLX)

(CLVIII). It has been suggested on the basis of a positive Sakaguchi reaction and the enzymatic removal of an amidine group from viomycin that the antibiotic contains the partial structure (CLIX)[239]. However, support for the dehydroamino-acid chromophore proposed in (CLV) comes from a study of the properties of the synthetic model compound (CLX)[240].

Stendomycin

The name stendomycin has been given to a mixture of closely related compounds which are antifungal antibiotics. The mixture has so far proved impossible to separate, but a structure (CLXI) has been proposed for the dominant member of this family[241]. In other members isomyristic acid is replaced by its lower homologues and *allo*-isoleucine by valine or leucine. There are two unusual residues in this type of antibiotic. The only basic amino-acid liberated on acid hydrolysis, stendomycidine (CLVII; R = CH_3) has a cyclic arginine structure which is an *N*-methyl derivative of the compound isolated from reduced viomycin. Since there are no α-amino-dicarboxylic acids in the molecule and the C-terminal carboxyl group is bound in a lactone, the presence of dehydrobutyrine was initially inferred from the detection of rather more than one molecule of ammonia in the acid hydrolysate. The u.v. absorption spectrum of stendomycin supports the presence of a dehydroamino-acid residue, and the isolation of the 2,4-dinitrophenylhydrazone of α-ketobutyric acid from the normal acid hydrolysate, and α-aminobutyric acid from the acid hydrolysate after hydrogenation of the intact peptide, further confirms this. The peptide chain contains amino-acids with bulky side-chains in adjacent positions, and the steric hindrance to acid hydrolysis of the amide bonds is such that 90 h reaction times with 6M hydrochloric acid at 110°C are required to obtain the correct ratios of amino-acids[241].

Telomycin

The complete structure (CLXII) of this cyclic peptide was established in 1968[242]. Three novel amino-acids, *erythro*-β-hydroxyleucine and *cis*- and *trans*-3-hydroxyprolines, were characterised from the acid hydrolysate, and the presence of β-methyltryptophan was indicated by mass spectrometry. The principal chromophore of telomycin is thought to be dehydrotryptophan; the u.v. spectra of models of this system are in accord with that of the antibiotic. Further evidence comes from a study of the products of alkaline hydrolysis. Indole-3-aldehyde is formed (possibly by a retro-aldol reaction) together with a yellow compound whose properties are in agreement with those of an unsaturated dioxopiperazine (CLXIII). Indolyl-3-acetic acid and 0·46 of a mole of DL-tryptophan per mole of telomycin are also present in the basic hydrolysate. It has been suggested that these can arise from indolylpyruvic acid, the expected degradation product of dehydrotryptophan in alkali, by an initial formation of a ketone–ammonia adduct followed by dehydration and decarboxylation to give (CLXIV). Hydrolysis of (CLXIV) would yield indole-3-acetic acid and a 50% maximum recovery of DL-tryptophan[242].

Peptide Antibiotics Containing 3-Hydroxypicolinic Acid

Stendomycin and telomycin are members of a group of peptide antibiotics which contain as part of the peptide ring a lactone formed from the carboxyl group of an amino-acid residue and the hydroxyl group of a hydroxyamino-acid. Compounds of this type have come to be called peptide lactones, and in many of them the amino group of the hydroxyamino-acid is acylated by a

(CLXI) and (CLXII)

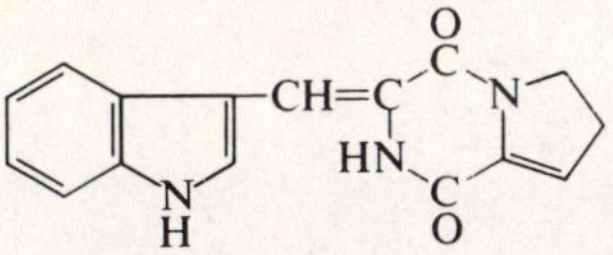

(CLXIII)

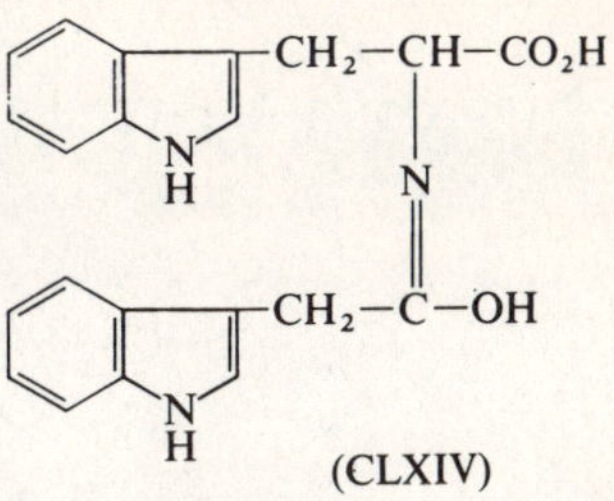

(CLXIV)

heterocyclic acid. One such compound is pyridomycin, an antibiotic produced by *Streptomyces albidofuscus* Okami et Umezawa. The total structure of this molecule (CLXV) required X-ray analysis for its solution[243], but confirmatory chemical degradation has been carried out[244]. Biosynthetic studies have shown that both the 3-hydroxypicolinic acid and the γ-amino-acid, 4-amino-3-hydroxy-2-methyl-5-(3-pyridyl)pentanoic acid, are derived from L-aspartic acid and glycerol[245]. 3-Hydroxypicolinic acid is quite a common constituent of the peptide lactones; it occurs in etamycin (CLXVI), staphylomycin (CLXVII; R = H), and the pristinamycins (CLXVII is pristinamycin I_A when R = NMe_2). The pristinamycins are also known as the osteogrycins and vernamycins[246]. Recent mass spectral studies of the principal components of etamycin and staphylomycin show the most important characteristic fragmentations to be the loss of the elements of the ester

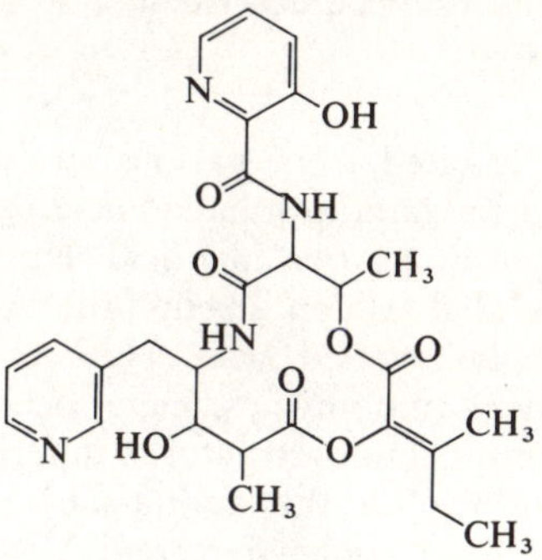

(CLXV)

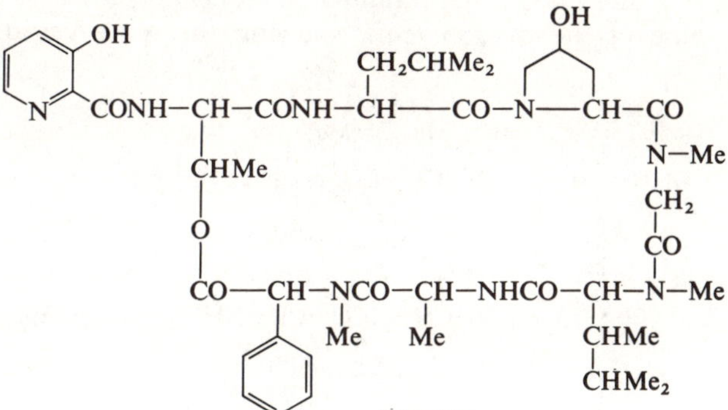

(CXLVI)

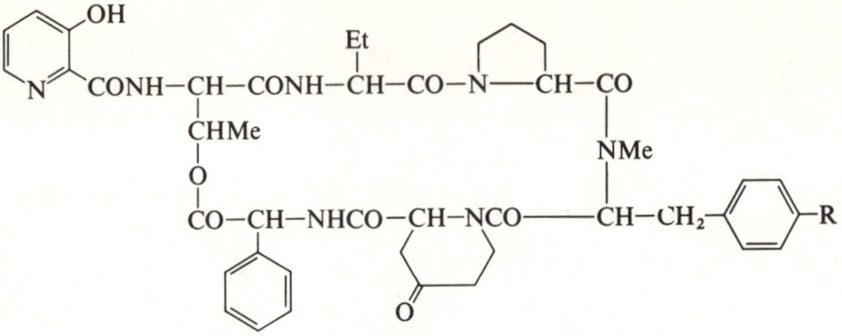

(CLXVII)

link, followed by a sequential loss of amino-acid residues in such a way that the positive charge always remains on the 3-hydroxypicolinic acid residue[247].

Staphylomycin and the pristinamycins are accompanied in their fermentation liquors by a second group of antibiotics of unrelated structure. The biological potency of this second group is considerably enhanced by addition of the peptide lactone components. Although the synthetic work on peptide lactones has so far been concentrated on the actinomycins, this synergistic activity has stimulated a few other synthetic studies. An analogue of pristinamycin I$_A$ in which proline replaces pipecolic acid and phenylalanine replaces *p*-dimethylaminophenylalanine has been synthesised and found to possess the synergistic activity of the natural antibiotic[248]. The synthesis of two other analogues has also been described[249,250].

Quinoxaline antibiotics

Ten peptide lactones isolated from various strains related to *Streptomyces aureus* have been found to contain quinoxaline-2-carboxylic acid. These are all *cyclo*-octapeptides possessing two identical tetrapeptide sequences which are joined in an antiparallel fashion through the hydroxyl group of D-serine. The first member to be studied was echinomycin[251] (quinomycin A); subsequently five other quinomycins have been characterised. All six antibiotics have the same basic structure, differing only in the type of aliphatic side-chain of two of the amino-acids (CLXVIII; X = N-Me valine, N-Me or N-diMe *allo*isoleucine, and Y = N-Me or N-diMe *allo*-isoleucine). These peptides are in fact bicyclic as the two cysteine residues are linked as a disulphide. The amino-acid sequences of the triostins closely resemble those of the quinomycins, showing the same variations in the

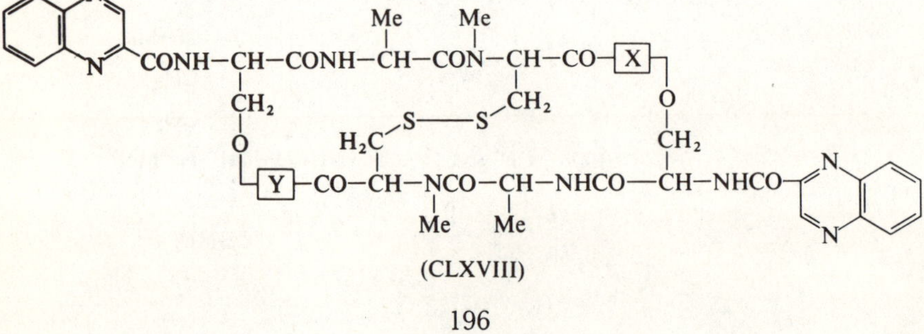

(CLXVIII)

residues esterified by the serine hydroxyls, except that the α-hydrogen atoms of the cysteine residues seem to be replaced by direct bonding to sulphur to give a dithian ring system (CLXIX)[252]. Radiotracer work has shown that tryptophan is an efficient precursor of the quinoxaline moiety of the triostins[253].

Actinomycin

The actinomycins are a series of yellowish-red antibiotics produced by a variety of *Streptomyces*. They are highly toxic compounds, and much of the interest in their chemistry has been due to their cytotoxic action and growth inhibitory effects on tumours. Their structures were elucidated by Brockmann and his co-workers in the early 1960s; the structures of the three most widely studied components are shown in (CLXX). Another half dozen members of the series are known; in these both X and Y are D-valine, but the L-proline residues are wholly or partially replaced by γ-oxoproline or hydroxyproline.

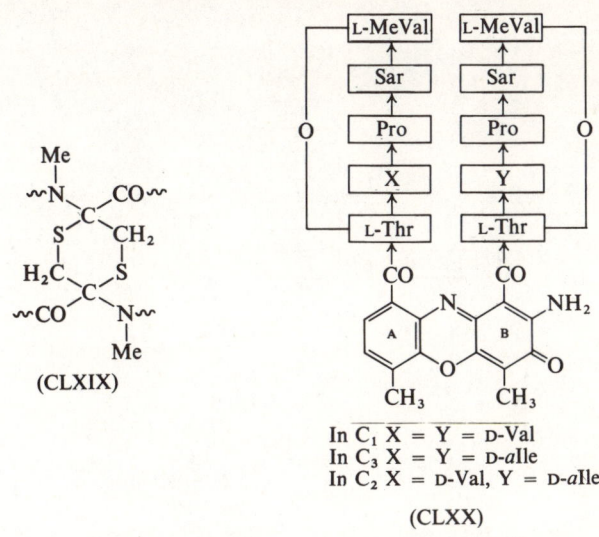

(CLXIX)

In C_1 X = Y = D-Val
In C_3 X = Y = D-*a*Ile
In C_2 X = D-Val, Y = D-*a*Ile

(CLXX)

The two pentapeptide lactone rings of the actinomycins are linked through the chromophore 3-amino-4,5-dicarboxyl-1,8-dimethyl-2-phenoxazone (actinocin). The actinomycins can be divided into two groups. In the *iso*-series there are two identical pentapeptide lactone rings, and in the *aniso*-series the ring sequences differ.

The total synthesis of the *iso*-antibiotic actinomycin C_3 was first fully reported in 1967. In this synthesis a linear tetrapeptide chain was built up on 2-nitro-3-benzyloxy-4-methylbenzoic acid, and the hydroxyl group of the threonine residue was esterified with carbobenzoxy-L-*N*-methyl valine to give (CLXXI). Hydrogenation of (CLXXI) reduced the nitro group to an amino group and concomitantly cleaved the benzyl ether, benzyl ester, and *N*-carbobenzoxy protecting groups. On subsequent oxidation with potassium ferricyanide dimerisation occurred (*Scheme 13*) to give the required phenoxazine derivative (CLXXII), bis-*seco*-actinomycin C_3, in high yield. The synthesis was completed by ring closure using the mixed anhydride method of peptide bond formation, the product being identical with the natural antibiotic[254]. This synthesis is not an unequivocal one as ring closure could have occurred to give a decapeptide di-lactone ring. However, hydrogen peroxide oxidation of both 2-desamino-2-hydroxy actinomycins C_3 and C_2 (derivatives formed by the action of acid on the native antibiotics) gives rise to an N-oxalyl-*cyclo*-pentapeptide whose structure has been established as (CLXXIII)[255]. A subsequent alternative but similar synthesis of actinomycin C_3 involved a final ring closure between the N-methylvaline and

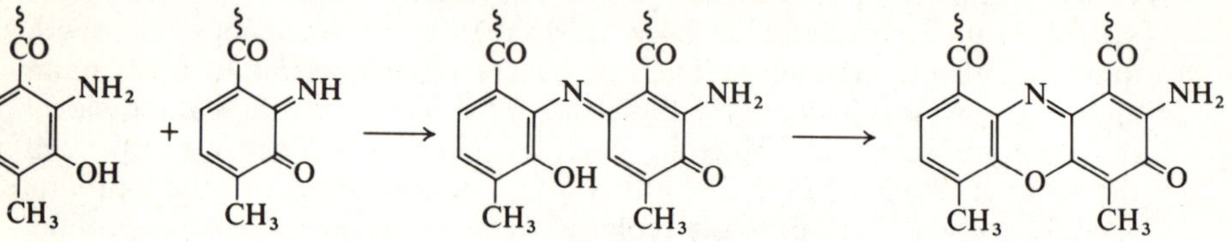

Scheme 13

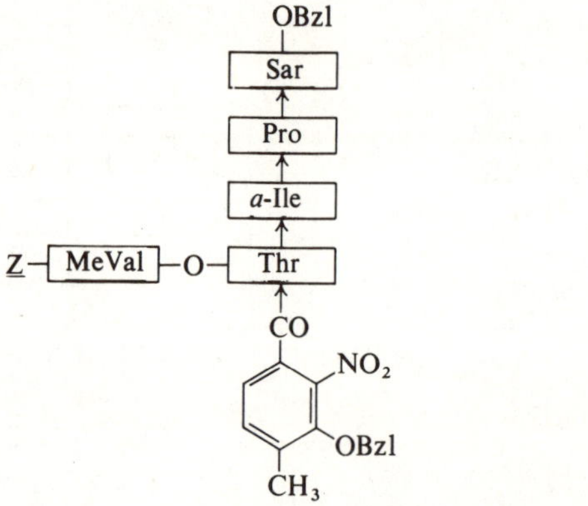

Z = carbobenzoxy

(CLXXI)

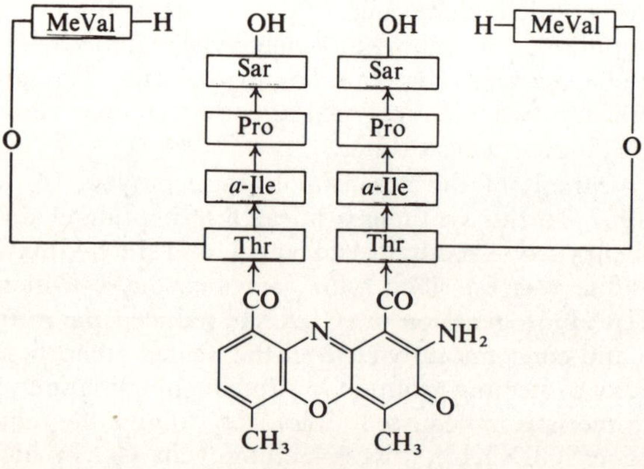

(CLXXII)

threonine residues[256]. In a third synthesis the pentapeptide lactone ring has been closed before the oxidative generation of the chromophore. The identification of the products of such syntheses as actinomycins C_3 and C_1 conclusively rules out a decapeptide di-lactone system[257,258].

Although the oxidation of actinocinyl-bis-L-threonine and similar derivatives with pendant peptide chains gives almost exclusively one product on oxidation (*Scheme 13*)[254,259], this is not the case if the hydroxyl and amino groups are interchanged in position. Oxidation of O-acetyl-N-(3-amino-2-

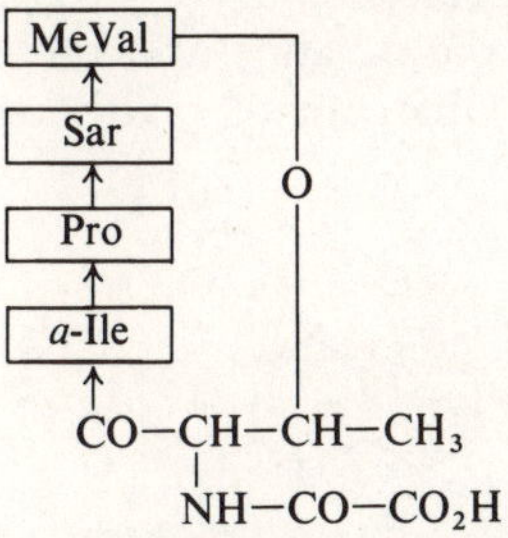

```
         MeVal
           ↑
          Sar
           ↑       O
          Pro
           ↑
         a-Ile
           ↑
       CO—CH—CH—CH₃
             |
          NH—CO—CO₂H
```
(CLXXIII)

hydroxy-4-methylbenzoyl)-L-threonine methyl ester (CLXXIV) gives a mixture of pseudo-actinocinyl-bis-L-threonine (CLXXV) and a product (CLXXVI) in which the threonine groups are on opposite sides of the

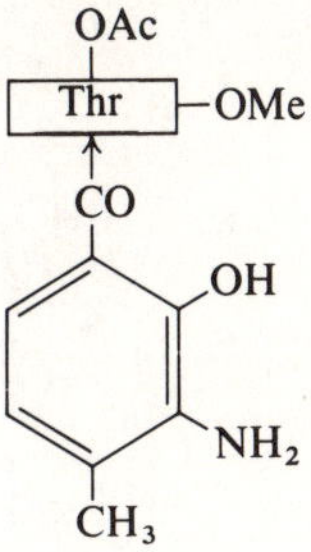

(CLXXIV)

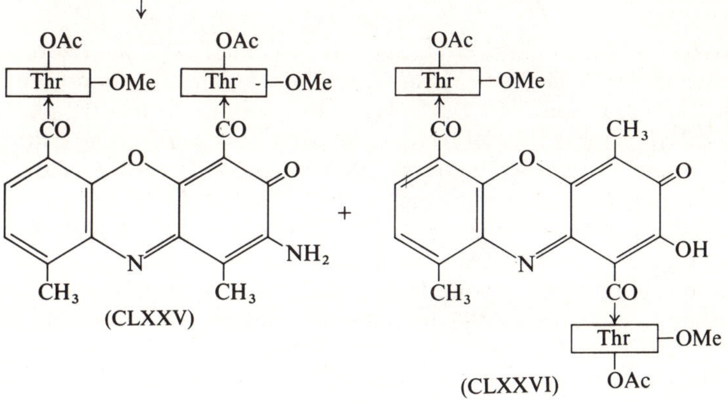

↓

(CLXXV) + (CLXXVI)

molecule. Analogues of the actinomycins based on such systems are being currently investigated[260].

The synthesis of the *aniso*-actinomycins presents difficulties which are not encountered in the *iso*-series. Present approaches utilise the oxidation of mixtures of N-(2-amino-3-hydroxy-4-methylbenzoyl)pentapeptide lactones rather than the separate addition or building up of the two different peptide chains on the actinocinyl nucleus. Oxidation of equimolar amounts of (CLXXVII) and (CLXXVIII) gives a mixture of the four possible products, actinomycins C_1 (CLXXIX), C_3 (CLXXX), C_2 (CLXXXII), and an isomer of C_2 which has the D-valine and D-*allo*isoleucine residues reversed in position (CLXXXI). Unfortunately, although C_1 and C_3 can be readily

R^1 = L-Thr·D-*a*Ile·L-Pro·Sar·L-MeVal (lactone via O)

R^2 = L-Thr·D-Val·L-Pro·Sar·L-MeVal (lactone via O)

R^3 = H

removed from the mixture, C_2 and *iso*-C_2 cannot be separated chromatographically[258,261]. However, it has been found that oxidation of a mixture of (CLXXVII) and (CLXXVIII), where in the former R^1 = L-Thr-D-aIle–L-Pro–Sar–L-MeVal–OH, gave C_2 and *iso*-C_2 precursors which could be separated by chromatography. Subsequent ring closure of the isolated components afforded pure actinomycins C_2 and *iso*-C_2[262].

A decision between structures (CLXXXI) and (CLXXXII) for actinomycin C_2 was originally made on the basis of degradative evidence. Not only is the D-*allo*isoleucine residue found in the oxalyl derivative isolated after oxidation of 2-desamino-2-hydroxy C_2, indicating that this lactone ring was attached to ring B of actinocin, but the quinone (CLXXXIII) has been identified as a product of the alkaline hydrolysis of a degradation product

(CLXXXIV) of C_2, and such a compound is unlikely to have been formed under these conditions from ring A^{263}. This assignment has been confirmed recently by carrying out the synthesis of actinomycin C_2 as described above using a precursor labelled with deuterium in the 6 position (CLXXVII; $R^3 = D$). The actinomycin C_2 isolated contained no deuterium (CLXXXII), but the *iso*-C_2 was labelled in position 8 of the phenoxazine system (CLXXXI; $R^3 = D$). If the lactone ring containing D-*allo*-isoleucine had been incorporated attached to what became ring A of the chromophore as in (CLXXXI), then the deuterium should not be lost on oxidation. Actinomycin C_2 therefore must be $(CLXXXII)^{262}$.

It was discovered early in the studies on the actinomycins that replacement

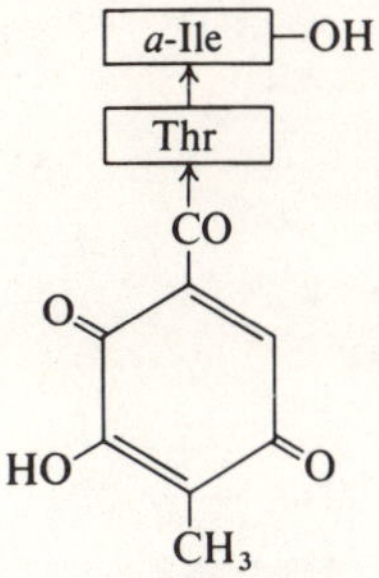

(CLXXXIII) 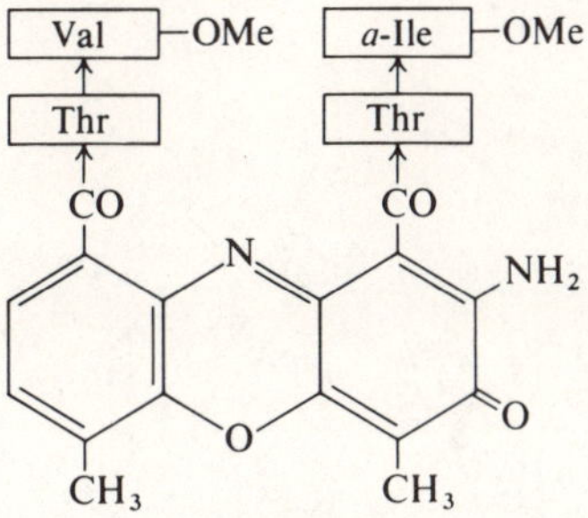   (CLXXXIV)

of the 2-amino group with a hydroxyl group destroyed the biological activity of the molecule. More recently, 4,6-didesmethyl-4,6-dimethoxy-actinomycin C_1 has been synthesised and its antibiotic activity found to be much lower than native $C_1{}^{264}$. The synthesis of an interesting hybrid of gramicidin S and actinomycin (CLXXXV) has been described, but this molecule has no antibiotic activity. The final stage of this synthesis involved a transannular oxidative coupling to generate the actinocin bridge^{265}.

Thiostrepton

Thiostrepton was first isolated from *Streptomyces azureus* in 1954. Its solubility in aqueous systems is low, and it induces resistance to itself in bacteria before

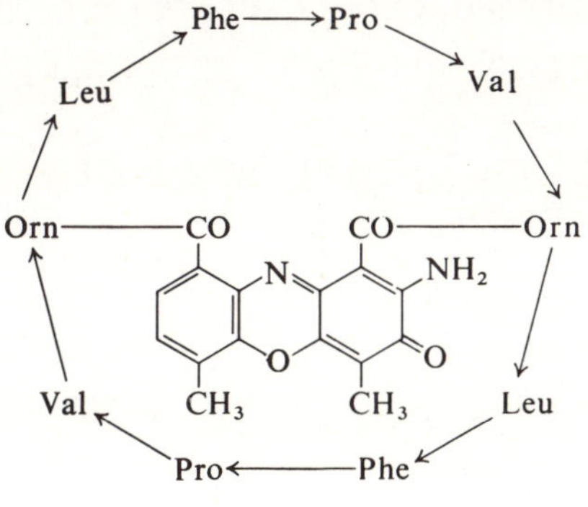

(CLXXXV)

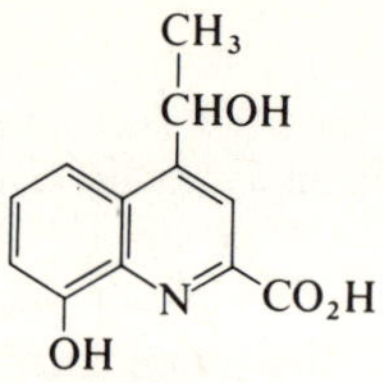

(CLXXXVI)

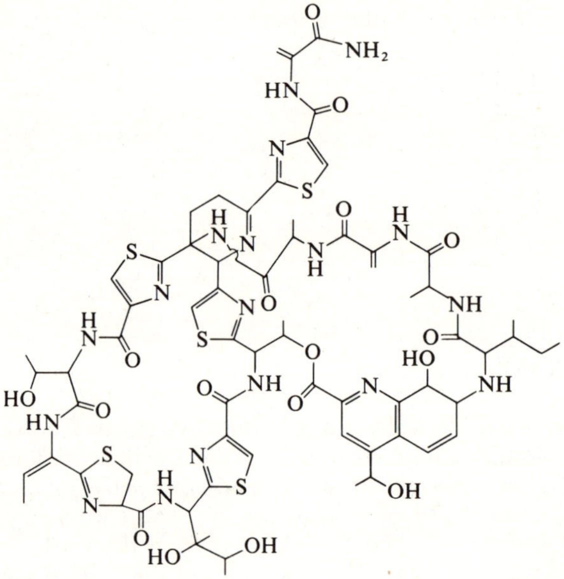

(CLXXXVII)

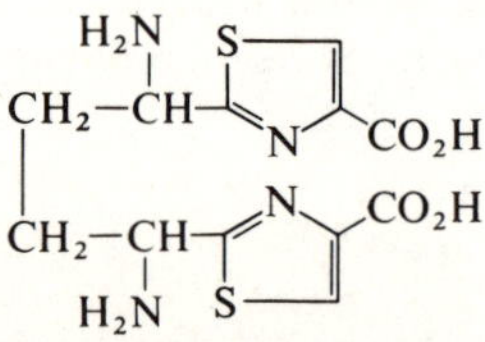

(CLXXXVIII)

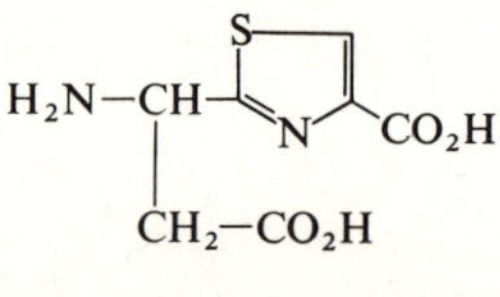

(CLXXXIX)

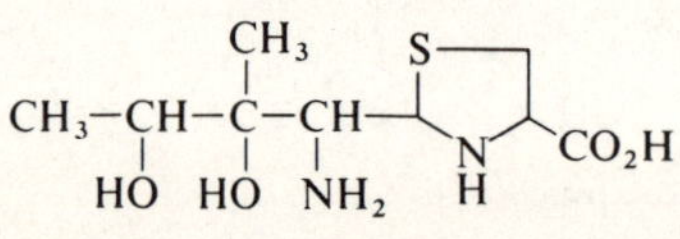

(CXC)

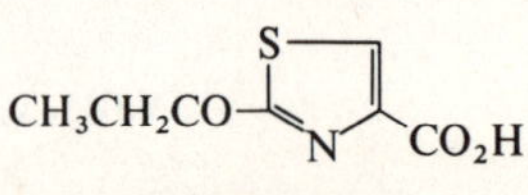

(CXCI)

its concentration becomes therapeutically sufficient. Consequently, although its spectrum of activity against Gram-positive bacteria is comparable to that of penicillin, it has not been developed for medicinal purposes. Thiostrepton has a complex chemical structure. Degradation established that the sulphur content was bound up in thiazole rings, and acid hydrolysis was found to liberate 4-(α-hydroxyethyl)-8-hydroxyquinaldic acid (CLXXXVI)[266,267], but the large size of the molecule and the variety of products produced on hydrolysis made its structural elucidation difficult. However, a recent X-ray crystallographic study[268] has established many features of the molecule, and a basic structure (CLXXXVII) has been put forward. Thiostreptoic acid (CLXXXVIII) is obtained in 72% yield on acid hydrolysis of thiostrepton, but if performic acid oxidation precedes the hydrolysis 2-(1-amino-2-carboxyethyl)thiazole-4-carboxylic acid (CLXXXIX) is formed instead[269]. This suggests that the double bond must shift to allow the reduced pyridine ring to be degraded in these two different ways. On acid hydrolysis the thiazole parts of the molecule are also degraded to thiostreptine (CXC) and 2-propionylthiazole-4-carboxylic acid (CXCI)[266,267].

Some doubt remains as to the nature of the side-chain of thiostrepton. This part of the molecule is not stabilised conformationally by intramolecular forces as it extends into a pool of disordered solvent, and bond lengths, angles, and even atom types are difficult to determine by X-ray crystallography. Hydrolysis of thiostrepton yields two molecules of pyruvic acid; one originates from the dehydroalanine residue in the cyclic part of the molecule, and since thiostrepton contains no free carboxyl groups the other has been tentatively assigned as terminating the side-chain as an α-aminoacrylamide group[268]. Other amino-acids in thiostrepton besides cysteine have been modified from their normal forms. Dehydrogenation has given rise to two

(CXCII)

α-aminoacryl groups and one α-aminodehydrobutyryl residue, and hydroxylation has given rise to dihydroxyleucine. As far as the multiple thiazole rings in the part of thiostrepton giving rise to thiostreptoic acid are concerned, the same biosynthetic pattern can be observed in micrococcinic acid (CXCII), a degradation product of the antibiotic micrococcin P[270].

Bacitracin

A modified cysteine residue also occurs in the antibiotic bacitracin A, which was originally assigned a *cyclo*-hexapeptide structure (CXCIII). Bacitracin A has no free sulphydryl group, but acid hydrolysis liberates cysteine. The antibiotic is unstable outside the pH region 4·5–6·5, and an oxidative

$$\begin{array}{c}
\text{Et} \quad \text{Me} \\
\diagdown \diagup \\
\text{CH} \quad \text{S—CH}_2 \\
| \quad \diagup \\
\text{H}_2\text{N—CH—C} \\
\diagdown \\
\text{N—CH—CO—L-Leu} \\
\quad \text{NH}_2 \quad \downarrow \\
\quad \nearrow \text{D-Asp·OH} \quad \text{D-Glu} \\
\text{L-His} \rightarrow \text{L-Asp} \nwarrow \quad \downarrow \\
\text{D-Phe} \quad \text{D-Lys} \leftarrow \text{L-Ile} \\
\nwarrow \quad \nearrow \\
\text{L-Ile} \leftarrow \text{D-Orn} \\
\\
\text{(CXCIII)}
\end{array}
\qquad
\begin{array}{c}
\text{Et} \quad \text{Me} \\
\diagdown \diagup \\
\text{CH} \quad \text{S} \\
| \quad \diagdown \\
\text{CO} \quad \quad \\
\diagdown \quad \diagup \\
\text{N} \quad \quad \text{CO}_2\text{H} \\
\\
\text{(CXCIV)}
\end{array}$$

deamination of the amino-terminal leucine and conversion of the thiazoline to a thiazole seems to occur since acid hydrolysis of the oxidised product yields (CXCIV)[1]. In 1966 a new method of identifying *endo*-asparaginyl residues was applied to bacitracin. Treatment with ethylene dichlorophosphite gave a β-cyanoalanine derivative which on Birch reduction followed by acid hydrolysis yielded 2,4-diaminobutyric acid *(Scheme 14)*[271]. This amino-acid could only have arisen if the parent molecule contained an asparagine residue linked as part of a peptide chain, and bacitracin is now thought to contain a *cyclo*-heptapeptide ring with the aspartic acid residue bridging the asparagine and lysine residues.

$$\begin{array}{c}
\text{CONH}_2 \\
| \\
\text{CH}_2 \\
| \\
\sim\sim\text{NH—CH—CO}\sim\sim
\end{array}
\longrightarrow
\begin{array}{c}
\text{CN} \\
| \\
\text{CH}_2 \\
| \\
\sim\sim\text{NH—CH—CO}\sim\sim
\end{array}
\longrightarrow
\begin{array}{c}
\text{CH}_2\text{NH}_2 \\
| \\
\text{CH}_2 \\
| \\
\sim\sim\text{NH—CH—CO}\sim\sim
\end{array}$$

Scheme 14

The peptide chain containing the *cyclo*-hexapeptide ring originally thought to be present has been synthesised, but attempts to cyclise the cysteine residue to the thiazoline ring by treatment with acid were not successful. This is perhaps not surprising since it is known that peptides containing a cysteine residue adjacent to the *N*-terminal residue will only cyclise to the thiazoline ring when protected by an acetyl or benzyloxycarbonyl group[272]. However, on the basis of chemical evidence it is not clear whether the *N*-terminal isoleucine residue in bacitracin has a completely free amino group. The isolation of some phenylalanylisoleucine after partial acid hydrolysis has led to the idea of some sort of link between the D-phenylalanyl residue and the *N*-terminal isoleucine[1]. Zinc ions have been found to bind by coordinate bonds to four positions in bacitracin; since two of these bonds involve the histidine and thiazoline residues, the close proximity of these residues receives some confirmation[273]. The subtleties of the bacitracin A structure, however, as yet remain to be solved.

Cyclic Peptides from Amanita Fungi

More than 95% of the fatal cases of mushroom poisoning are due to the green mushroom *Amanita phalloides* and its white relative *Amanita verna*. Wieland and his colleagues have carried out extensive studies of the poisonous principles of these fungi, and by early 1971 had published 43 papers on this topic[274]. The mushrooms contain the same two series of toxic *cyclo*-hepta- and octa-peptides, known respectively as phallotoxins (CXCV) and amatoxins (CXCVI). The two types of toxin differ in the time of onset of the symptoms after ingestion. In large doses, the phallotoxins cause death within two hours, but the amatoxins do not exert a lethal effect until about 15 hours have elapsed after their consumption. However, on a weight basis the amatoxins are 10–20 times more toxic than the more rapidly acting phallotoxins. Both types of toxin primarily affect the liver.

An unusual feature of these toxins is their bicyclic nature. A bridge is formed across the α-linked peptide ring through a cysteine sulphydryl group

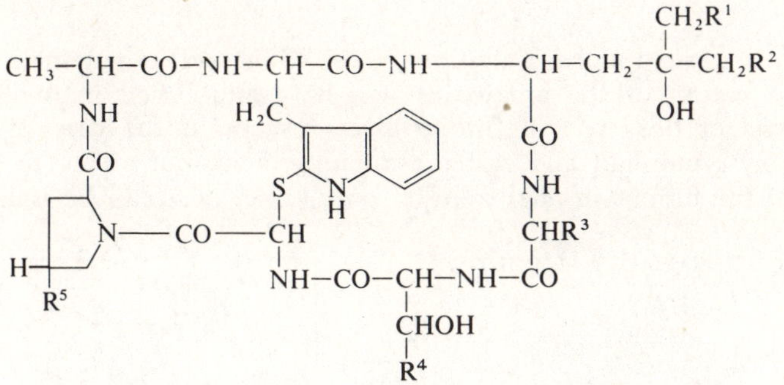

	R^1	R^2	R^3	R^4	R^5	Refs.
Phalloidin	OH	H	CH_3	CH_3	OH	278
Phalloin	H	H	CH_3	CH_3	OH	274
Phallisin	OH	OH	CH_3	CH_3	OH	277
Phallicidin	OH	H	$CH(CH_3)_2$	COOH	OH	274
Phallin B (tentatively)	H	H	CH_2Ph	CH_3	H	276

(CXCV)

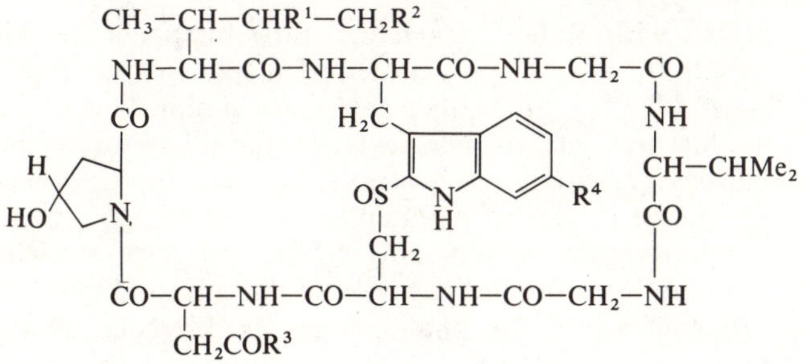

	R^1	R^2	R^3	R^4	Refs.
α-Amanitin	OH	OH	NH_2	OH	275
β-Amanitin	OH	OH	OH	OH	275
γ-Amanitin	H	OH	NH_2	OH	275
Amanin	OH	OH	OH	H	277
Amanullin	H	H	NH_2	OH	278
ε-Amanitin	H	OH	OH	OH	278

(CXCVI)

and the indole nucleus of a tryptophan residue. This double amino-acid residue has been called tryptathionine. Each ring also contains a γ-, $\gamma\delta$- or $\gamma\delta\delta$-hydroxylated amino-acid residue. Such γ-hydroxyacids were first characterised from these toxins[279], but recently γ-hydroxyleucine, which occurs in phalloin, has been identified in a hydrolysate of gelatin[280]. Hydrochloric acid hydrolysis of the phallotoxins splits the thioether bridge,

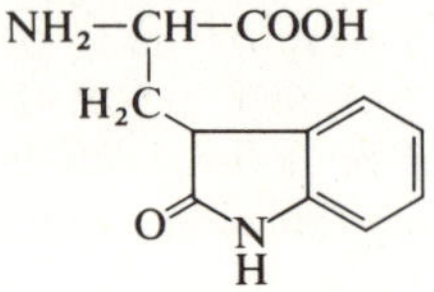

(CXCVII)

liberating L-β-oxindolyl-3-alanine (CXCVII) and L-cysteine. The sulphoxide nature of the amatoxins was only established in 1968[281]. Acid hydrolysis of this type of bridge gives (except in the case of amanin) 6-hydroxytryptophan and cysteine-sulphinic acid, but owing to the instability of the former in acid solution only traces of it can be isolated. The

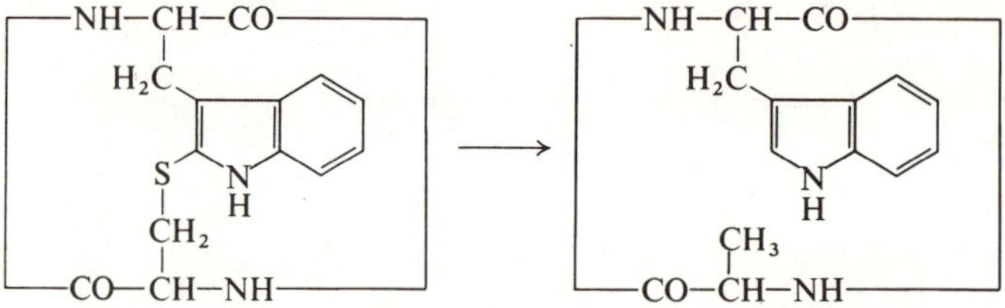

Scheme 15

sulphoxide assignment is based largely upon comparisons of the u.v. spectra of the amatoxins with those of the sulphoxide derived from phalloidin by peroxide oxidation and of other model sulphoxides such as 3-methyl-2-indolyl ethyl sulphoxide[281].

When treated with Raney nickel in boiling methanol the phallotoxins suffer loss of sulphur and the tryptathionine is split to give tryptophan and alanine (*Scheme 15*). The amatoxins undergo a similar fission, and in both cases the desthio-products are non-toxic. In the phallotoxins the presence of the γ-hydroxyl group is not essential for activity. Conversion of the $\gamma\delta$-dihydroxyleucine residue of phalloidin, for instance, to a methyl ketone or an n-propyl residue (norphalloin; *Scheme 16*) does not diminish the toxicity[282]. Amanullin is the one member of this group of peptides which is non-toxic. In contrast to the phallotoxins, the presence of a γ-hydroxy residue is apparently essential for the toxic action of the amatoxins[274].

The peptide ring of both types of toxin can be opened selectively. Treatment with 50–80% trifluoroacetic acid at room temperature for two hours splits the peptide bond adjacent to the γ-hydroxy residue, leading to the formation of a lactone (*Scheme 17*); the resulting *seco*-compounds are totally non-toxic. The sequence of amino-acids in the *seco*-phallotoxins prepared from the desthio compounds was readily established by applying six cycles of the Edman degradation. Preparation of the desthio-derivatives of the

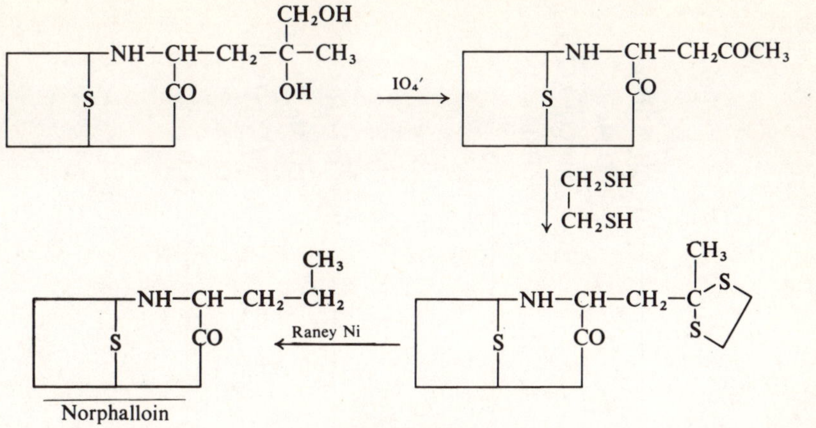

Scheme 16

amatoxins prior to sequence studies is unnecessary as the acid conditions required to form the *seco*-derivatives suffices to cause loss of the sulphur concomitantly.

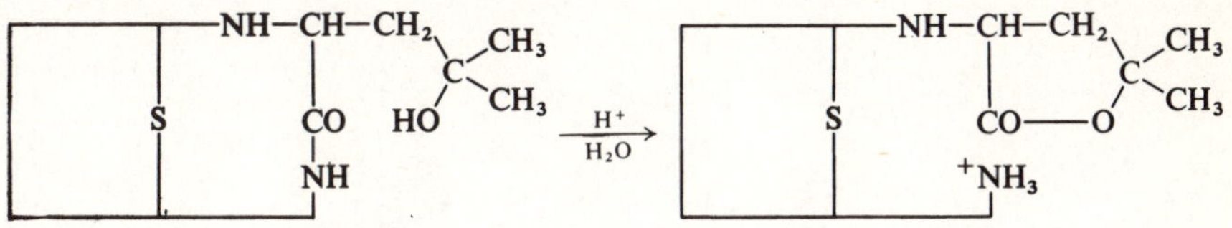

Scheme 17

A synthesis of norphalloin has recently been reported[283], and is outlined in *Scheme 18*. The thioether bridge was formed by treating a cysteine-containing tripeptide in acetic acid with *N*-chlorosuccinimide to form the sulphenyl chloride, and then adding a tryptophan-containing tetrapeptide at the moment of maximum absorption at 380 nm. Optimum conditions for such couplings were previously established with model compounds[284]. The product was cyclised by the mixed anhydride method, protecting groups were removed from two amino-acid residues, and a second cyclisation carried out under similar conditions to the first. The toxicity of the product was comparable to the material derived from phalloidin (*Scheme 16*). Cyclisation of *seco*-ketophalloidin (CXCVIII) by the mixed anhydride method gives a mixture of two neutral substances, one of which is identical to keto-phalloidin. Cyclisation using thionyl bis-imidazole gives only the material isomeric with keto-phalloidin. It has been suggested that this may be an atropisomer as models show that bridged cyclic compounds of the phalloin type may exist in two atropisomeric forms because the trypthathionine bridge is too bulky to pass through the *cyclo*-peptide ring[274].

During the isolation of toxins from *amanita phalloides* one partially purified fraction, despite its known content of the poisonous phallin B, was found to be non-toxic. This observation led to the isolation from this fraction of antamanide, a crystalline *cyclo*-decapeptide (CXCIX) which, if administered

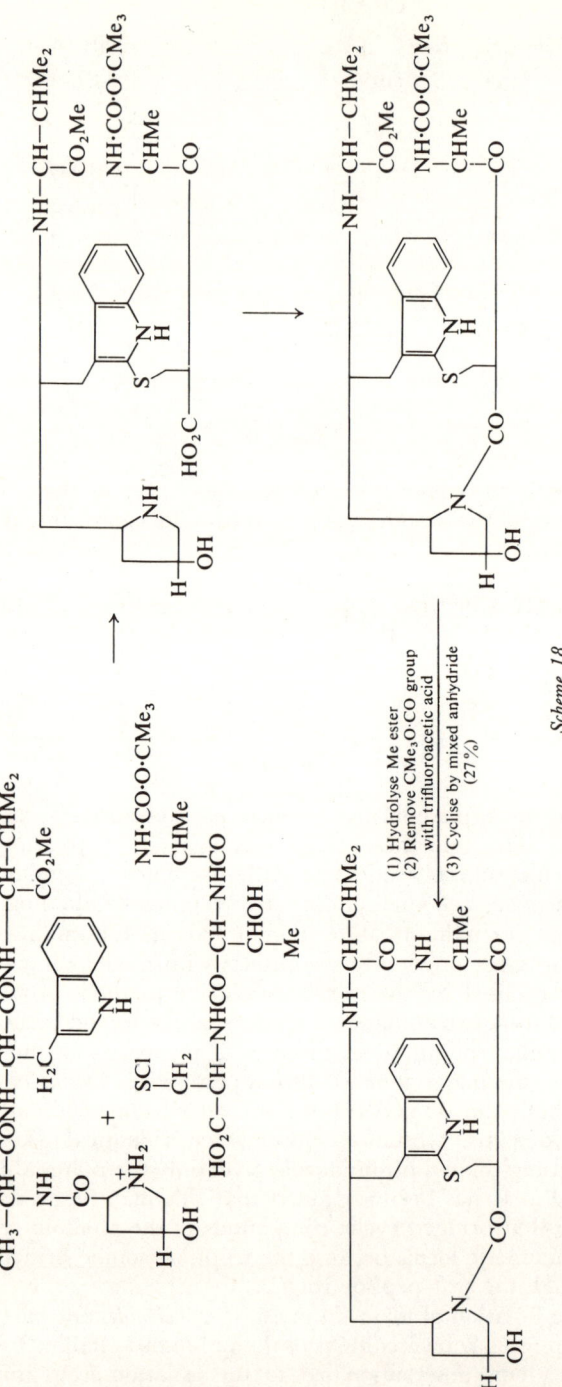

Scheme 18

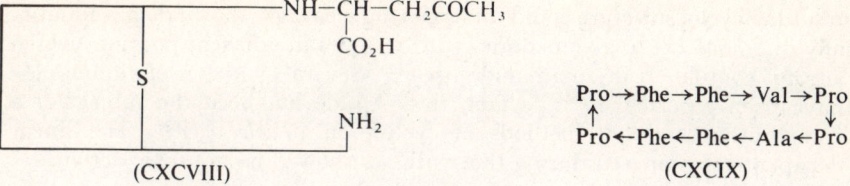

(CXCVIII)   (CXCIX)

before or simultaneously with amanita toxins, counteracts their lethal action[285]. Its natural concentration in the fungus is, of course, so low that the toxic action predominates. The amino-acid sequence of this lipophilic antitoxic principle proved difficult to determine by classical methods, but mass spectrometry was successfully applied. In the mass spectrometer antamanide rapidly breaks down into two linear peptides, each with an N-terminal

Scheme 19

prolylproline sequence, which then decompose stepwise. A combination of gas chromatography and mass spectrometry applied to partially methanolysed products confirmed this sequence, although some of the peptides characterised had sequences which were not consistent with the proposed structure. Their occurrence has been attributed to the occurrence of a transannular interaction during acid-catalysed methanolysis. The

Table 5.4. YIELDS OF ANTAMANIDE ON CYCLISATION OF LINEAR DECAPEPTIDES

Amino-acids linked by cyclisation	Method of Cyclisation				
	p-Nitrophenyl ester	Thiophenyl ester	Mixed anhydride	Dicyclohexyl-carbodi-imide*	Ethoxy-acetylene
Ala—Phe	—	—	—	36·5%	—
Phe—Phe	7%	40%	25%	27%	none isolated
Pro—Phe	29%	—	10·5%	—	—

* In the presence of N-hydroxysuccinimide.

resulting cyclol structure could undergo ring cleavage (shown diagrammatically in *Scheme 19*) to give peptides with residues in adjacent positions which are not contiguous in antamanide itself[285]. Several syntheses of antamanide have been reported[285,286]; in fact, this peptide has been the subject of a comparative study of methods of cyclisation (*Table 5.4*)[286]. The linear decapeptides prepared during the syntheses showed no antitoxic activity.

Peptide Alkaloids

Since 1967 about twenty alkaloids from a variety of sources (principally the *Rhamnaceae*) have been found to contain a *p*-alkoxystyrylamino residue as part of a 14-membered ring system. These alkaloids can be represented by a general formula (CC). The structures of most of these compounds have been determined by mass spectrometry (see *Table 5.5*); the fragmentation pattern is largely independent of the nature of the amino-acid residues and highly characteristic for this type of peptide alkaloid. The ring consists of two α-linked amino-acids and a dehydrodecarboxytyrosine residue whose hydroxyl group forms an ether link with the hydroxyl group of β-phenylserine or β-hydroxyleucine (except in the case of Ceanothine D). Acid

Table 5.5

	R^2	R^1	R^3	Refs
Adouetine Y′	—CHMe$_2$	NMe$_2$Phe—	—CHMeEt	287, 288
Adouetine X	—CHMe$_2$	NMe$_2$Leu—	—CHMeEt	288, 289
Adouetine Y	—Ph	NMe$_2$Phe—	—CHMeEt	288, 289
Adouetine Z	—Ph	NMe$_2$Phe·Pro—	—CH$_2$Ph	288
Americine	—CHMe$_2$	NMeVal—	—CH$_2$—(indole)	289, 290
Aralionin A	—Ph	NMe$_2$Ile—	—COPh	291
Aralionin B	—Ph	NMePhe—	—CHMeEt	292
Ceanothine B	—CHMe$_2$	NMePro—	—CH$_2$Ph	293
Ceanothine C	—CHMe$_2$	NMePro—	—CH$_2$CHMe$_2$ or —CHMeEt	289
Ceanothine D	R^2 is part of β-hydroxyvaline	NMePro—	—CH$_2$CHMe$_2$	289
Ceanothine E	—Ph	NMe$_2$Phe—	—CH$_2$CHMe$_2$	289
Franganine	—CHMe$_2$	NMe$_2$Leu—	—CH$_2$CHMe$_2$	289, 294
Frangufoline	—CHMe$_2$	NMe$_2$Phe—	—CH$_2$CHMe$_2$	289, 294
Frangulanine	—CHMe$_2$	NMe$_2$Ile—	—CH$_2$CHMe$_2$	295
Integerresine	—Ph	NMe$_2$Val—	—CH$_2$Ph	296
Integerrenine	+–Ph	NMe$_2$Ile—	—CH$_2$CHMe$_2$	296
Integerrine	—Ph	NMe$_2$Val—	—CH$_2$—(indole)	297
Lasiodine B	—CHMe$_2$	NMePhe·Pro—	—CH$_2$CHMe$_2$	299
Scutianine	—CHMe$_2$	NMe$_2$Phe·Pro—	—CH$_2$Ph	298

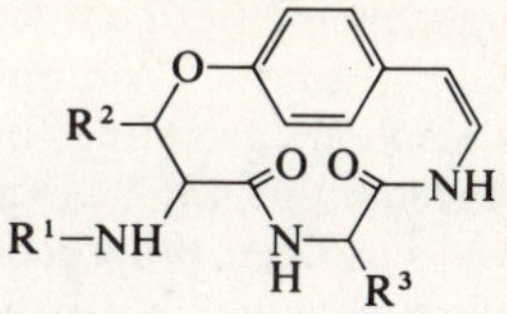

(CC)

hydrolysis of the dihydroalkaloids liberates *p*-tyramine, but the styrylamino linkage of the native alkaloids gives no characteristic fragments on hydrolysis. Quite extensive decomposition of the β-hydroxy residue involved in the ether linkage occurs on acid hydrolysis. Besides some free β-phenylserine, those peptides containing this amino-acid also liberate phenylpyruvic acid and a product of its dimerisation, β-phenylnaphthalene. In the case of β-hydroxyleucine, the free amino-acid liberated has (in several compounds) been established as *threo*-β-hydroxyleucine by chromatographic comparison

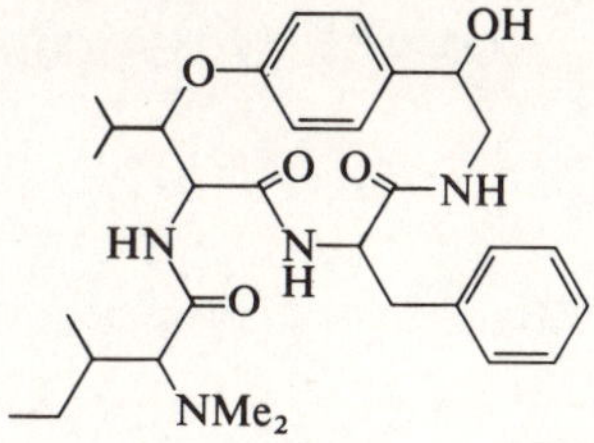

(CCI)

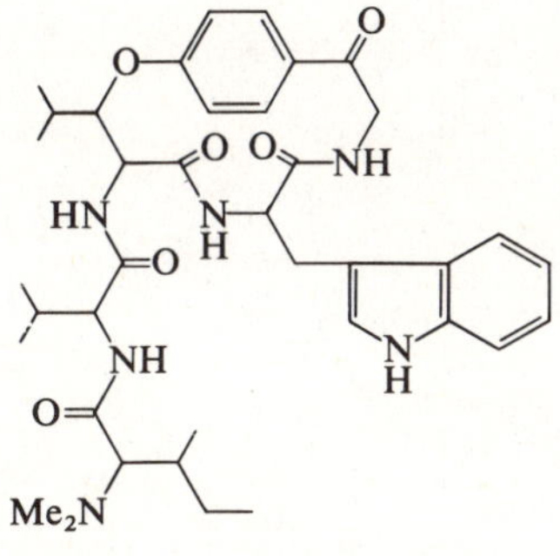

(CCII) 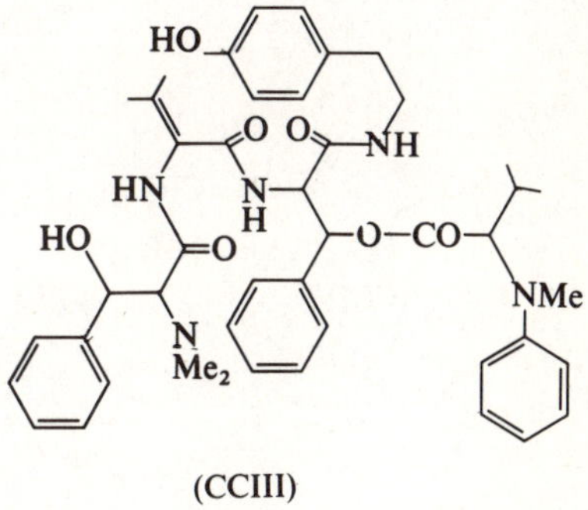 (CCIII)

with authentic material[287,294,295,298]. According to the conditions used, varying amounts of 2-keto-4-methylenepentanoic acid, leucine, and glycine are formed from the bound β-hydroxyleucine[287,294,302].

Related peptide alkaloids are known which contain hydrated (pandamine; CCI)[300] and subsequently oxidised (hymenocardine; CCII)[301] forms of the styryl double bond. Lasiodine A has an acyclic structure (CCIII) in which dehydrovaline occurs instead of the usual β-hydroxyacid derivative[299]. The chemical degradation of pandamine has been studied more extensively than most of the peptide alkaloids. Hydrolysis with methanolic alkali liberates phenylalanine, some leucine but no α-ketoacid, 2-*p*-hydroxyphenyl-2-methoxyethylamine (CCIV), and *N,N*-dimethylisoleucine. Pandamine itself contains a secondary alcoholic grouping, being readily oxidised to the corresponding ketone. A mechanism which explains the formation of the hydrolysis products observed has been proposed (*Scheme 20*)[300]. The formation of leucine from β-hydroxyleucine in alkali is quite well known[302]. A peptide alkaloid of a rather different type is zizyphine (CCV) which contains a prolyl-hydroxyprolyl dioxopiperazine with the hydroxyl group of the hydroxyproline forming an ether link[303].

CYCLIC PEPTIDES CONTAINING CYSTINE

Of the types of bond which occur in the heterodetic cyclic peptides, the disulphide bridge between two cysteine residues is the least stable. It can be ruptured in most cases without chemically degrading the rest of the molecule, and often reformed under mild conditions. It is unstable to extremes of pH and, if more than one such bridge is present in a molecule, disulphide interchange may occur. The most important members of this class of cyclic peptide contain only those amino-acids of the L-configuration normally found in proteins.

The discussion of disulphide bridged cyclic peptides will be limited in this review to compounds whose synthesis has been described and which are of current interest, and to those larger molecules whose study has made some contribution to molecular biology. Only those aspects of the chemistry which are relevant to their cyclic nature are outlined, and the amino-acid sequences of the larger peptides are not detailed; these molecules are pictured diagrammatically in order to emphasise the important features relating to their disulphide links.

Malformin

An interesting bicyclic peptide in which a disulphide bridge spans a cyclic pentapeptide ring occurs in various strains of *Aspergillus niger*. This peptide, malformin A (CCVI), was first isolated in 1962[304]; it is a plant toxin, and assayed by its malformation effect on bean seedlings. Two other malformins have been characterised by mass spectrometry. In these the isoleucine residue is replaced by *allo*-isoleucine (B_1) or valine (B_2)[305]. It was not until 1969 that a successful synthesis of malformin A was reported. In the final stages of this synthesis the protected pentapeptide (CCVII) was cyclised in a dilute solution using a mixture of dicyclohexylcarbodi-imide and N-hydroxy-succinimide, the S-benzyl protecting groups were cleaved with sodium in liquid ammonia, and the di-thiol oxidised with 1,2-di-iodoethane to give material with 90% of the biological activity of natural malformin[306]. A variety of sulphydryl compounds inhibit the promotion of plant curvatures by malformin. This is due to thiol–disulphide interchange (*Scheme 21*). With β-mercaptoethanol and some other thiols the mixed disulphide is insoluble and precipitates out of solution[307].

Neurohypophyseal Hormones

The study of cyclic peptides in which a disulphide group forms part of the ring was initially stimulated in 1953 when the structure of the pituitary hormone oxytocin (CCVIII) was established. Oxytocin and other pituitary hormones of similar size have been the subject of intense synthetic activity

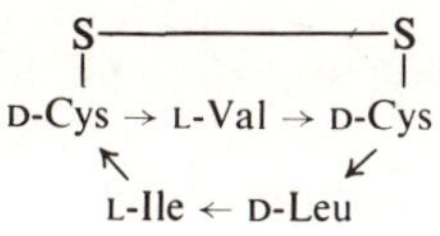

```
   S————————S
   |          |
D-Cys → L-Val → D-Cys              S·Bzl           S·Bzl
   ↖          ↙                     |               |
   L-Ile ← D-Leu            H·D-Leu·L-Ile·D-Cys·L-Val·D-Cys·OH
                                         (CCVII)
      (CCVI)
```

```
   ┌———S                    ┌———SH
   |   |   + RSH  →         |
   └———S                    └———S—SR
```

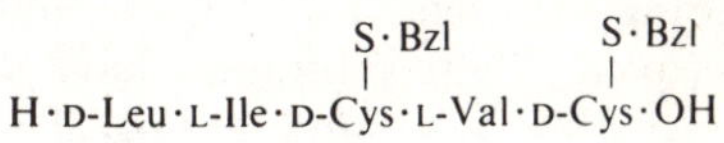

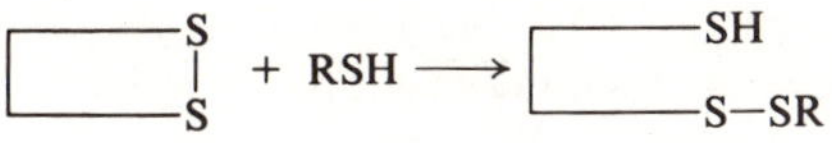

Scheme 21

and over 150 structural analogues have been prepared to probe the structure–activity relationships of compounds of this type. This aspect of the chemistry of oxytocin has been well reviewed[308], and only a few of the more interesting analogues will be mentioned here. [Threonine-4]-oxytocin possesses about

$$\text{H·Cys·Tyr·Ile·Glu·Asp·Cys·Pro·Leu·Gly·NH}_2$$
with NH₂ NH₂ side groups and an S—S bridge between the two Cys residues.

(CCVIII)

twice the oxytocic effect of the natural hormone, and as its vasopressor potency and antidiuretic effect are one tenth and one third respectively of those of oxytocin it may have clinical application as a highly selective oxytocic agent. This seems to be a genuine case of a synthetic hormone showing an improvement on the natural material. Desamino-oxytocin has about the same oxytocic activity as the [threonine-4]-analogue, but in this instance the higher activity is thought to be a reflection of its longer lifetime *in vivo* as the molecule is no longer susceptible to attack by the specific aminopeptidase which normally breaks it down[309].

Syntheses in which the sulphur atoms of oxytocin have been replaced by methylene groups (carba-oxytocins) show that the disulphide linkage is functionally insignificant in the neurohypophyseal hormones. Both desamino-[carba-1]- and desamino-[carba-6]-oxytocins (the synthesis of desamino oxytocins is rather simpler) are about twice as active as native oxytocin, although the spectrum of their activity differs markedly. The antidiuretic activity of the [carba-6]-analogue exceeds by an order of ten that of any other analogue yet prepared. Desaminodicarba-oxytocin is biologically active, but much less potent than oxytocin[310]. The cyclic nature of oxytocin is essential to its activity; if the disulphide bridge is broken all activity is lost[1]. Two isomeric dimers of oxytocin have been isolated as by-products

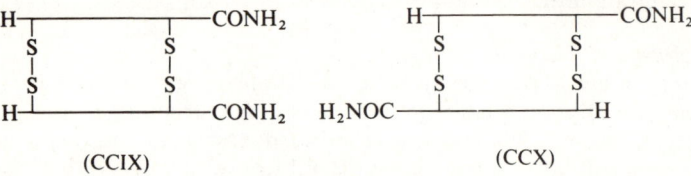

(CCIX)   (CCX)

during the oxidation of the linear disulphydryl peptide oxytoceine to oxytocin. These substances have low oxytocic activity, and can only be the parallel (CCIX) and antiparallel (CCX) dimers since both compounds, after reduction with sodium in liquid ammonia, give a good yield of oxytocin on re-oxidation. Oxytocin undergoes disulphide interchange in the presence of bases such as triethylamine to give a mixture of these dimers[311].

Hypophysectomised rats have been found unable to acquire conditioned avoidance responses, e.g. learning to jump a barrier in response to a buzzer signal to avoid an electric shock, but when treated with porcine pituitary extracts their responses resemble those of normal rats. This observation has led to the isolation and identification of the active factor as desglycinamide-[lysine-8]-vasopressin (CCXI). This octapeptide has no hormonal activity, but can readily be prepared from [lysine-8]-vasopressin by incubation with trypsin; it probably arises naturally by this route[312].

$$\begin{array}{c} \quad\quad\quad\quad H_2N \quad NH_2 \\ \quad\quad\quad\quad\ \ | \quad\quad\ | \\ H\cdot Cys\cdot Tyr\cdot Phe\cdot Glu\cdot Asp\cdot Cys\cdot Pro\cdot Lys\cdot OH \\ \quad\ \ | \quad\quad\quad\quad\quad\quad\quad | \\ \quad\ \ S\rule{3cm}{0.4pt}S \end{array}$$

(CCXI)

C.d. studies of oxytocin and some of its analogues show a band at 250 nm which is attributed to the disulphide group and corresponds to a sulphur dihedral angle close to 90 degrees[313]. Laser-excited Raman spectroscopy shows promise as an alternative method of studying the conformation of disulphide bridges. Good spectra have also been obtained from aqueous solutions of proteins, and the relative intensities of the C—S and S—S bands in lysozyme and ribonuclease differ greatly from those in free cystine. A comparison with model compounds suggests a correlation between the relative intensities of these bands and the average dihedral angle of the disulphide bridges[314].

Calcitonin

In the last few years there has been a tremendous interest in the hormone porcine calcitonin (often called thyrocalcitonin) which inhibits the action of osteoclasts, the cells which resorb bone. Its therapeutic use in the treatment of diseases causing bone loss is currently under investigation. The amino-acid sequence of this dotriacontapeptide (CCXII) has been deter-

S————————————S
| NH$_2$ NH$_2$ | |

H·Cys·Ser·Asp·Leu·Ser·Thr·Cys·Val·Leu·Ser·Ala·Tyr·Trp·Arg·Asp·Leu·
  1   2   3   4   5   6   7   8   9   10   11   12   13   14   15   16

NH$_2$
|
Asp·Asp·Phe·His·Arg·Phe·Ser·Gly·Met·Gly·Phe·Gly·Pro ·Glu·Thr·Pro·NH$_2$
 17   18   19   20   21   22   23   24   25   26   27   28   29   30   31   32

(CCXII)

mined by three groups working independently[315], and two syntheses[316] have been recorded. Both synthetic schemes involve the addition of sequence 1–9 containing the preformed disulphide bridge as the last step in construction of the peptide chain. Synthesis of this cyclic nonapeptide has been used to demonstrate the utility of a method of forming a disulphide bridge in high yield by direct treatment of a linear peptide with iodine in methanol. Prior removal of the *S*-trityl groups is thereby unnecessary[317]. In contrast to the situation with oxytocin, preliminary experiments indicate that the intra-chain disulphide bridge of calcitonin is not essential to biological activity; β-mercaptoethanol reduced material exhibits undiminished potency[318].

O.r.d. and c.d. studies suggest that there is little ordered structure in aqueous solutions of calcitonin; 10% α-helix may be present[319]. The structures of salmon calcitonin[320] and human calcitonin [321] have also been determined. They differ extensively in amino-acid sequence from the porcine variety; in fact, although there is only one change in the sequence of the cyclic peptide portion of the molecule, in the tail sequence 8–32 only three amino-acid residues (9, 28 and 32) are common to all three varieties. Salmon calcitonin is notable in that it is more than twenty times as active as the porcine variety. Human calcitonin was isolated from thyroid tumour tissue, where it occurs together with the antiparallel cyclic dimer. This dimer reverts to the monomer on treatment with ammonium hydroxide[321].

Oxidative Studies on Model Cysteine Peptides

In forming disulphide bridges, as has been illustrated in the case of oxytocin, combination to give products other than cyclic monomer can occur. The extent to which the number of amino-acid residues separating two cysteine residues affects the nature of the products formed on oxidation has been investigated in twelve model peptides of the L-cysteinyl-polyglycyl-L-cysteine series. The lower members of the series give mixtures containing varying proportions of cyclic monomer and antiparallel dimer. The hexapeptide and higher members, in contrast, give predominantly the monomeric cyclic disulphides. The amount of these monomers formed agrees quite well with those predicted by a simple statistical theory, indicating that the nature of the oxidation products is controlled mainly by the probability of encounter of the two thiol groups when the two sulphur atoms are separated by sixteen or more atoms[322]. Preferential formation of dimers owing to the occurrence of disulphide bond synthesis between two peptide chains held side by side through inter-chain hydrogen bonding can therefore be discounted as being important in the higher members of this series (up to fifteen glycine residues). The effect of varying the amino-acid side-chains on the spectrum of oxidation products has as yet received no systematic study.

Parallel dimers formed by the oxidation of peptides containing two cysteine residues have not been as well characterised as their antiparallel isomers; in fact, few cases have been authenticated. Hiskey and his co-workers have examined some parallel peptide bis-disulphides (CCXIII;

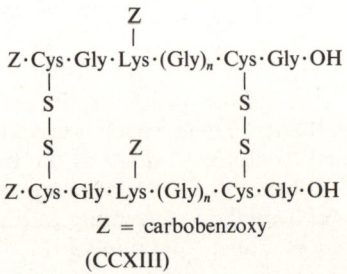

(CCXIII)

n = 1 or 2) synthesised unambiguously by forming the disulphide bridges in succession. In the presence of bases such as triethylamine, or even the dimethylamine occurring naturally in dimethylformamide, rearrangement to the corresponding cyclic monomer occurs. Conveniently, this conversion can be followed polarimetrically[323].

Ribonuclease

Ribonuclease is the most widely studied protein containing only one amino-acid chain but several disulphide bridges. This enzyme contains 124 amino-acid residues and four disulphide bridges. Despite the number of different cysteine pairings which are possible (105) conditions have been established which enable reduced ribonuclease to be oxidised to largely restore the original cysteine pairings. However, if the oxidation is carried out in solutions containing urea, the product is enzymically inactive as isomers containing incorrectly linked cysteine residues are formed. After removal of the urea, such 'scrambled' isomers undergo disulphide interchange in the presence of thiols to regenerate the native (by implication lowest energy) form. These observations have led to the hypothesis that the sequence of amino-acids contains all the information necessary to control the conformation of the peptide chain so that the cysteine residues will be in pairs in close proximity[324]. The synthesis of peptides with the sequence of ribonuclease[325,326] which possess enzymic activity, albeit of a low order, provided the first direct evidence for this hypothesis. The amino-acids at the C-terminal end of ribonuclease are essential for reformation of the native conformation. From a peptic digest of reduced ribonuclease a peptide lacking four amino-acids at this end of the molecule can be isolated. This compound is enzymically inactive since one of the amino-acid residues involved in the active site has been lost, but o.r.d. and c.d. studies show that it is not conformationally disordered. However, it will not refold correctly on oxidation; a vital element of information has been lost with the C-terminal tetrapeptide[327].

The function of intra-chain disulphide bridges in proteins would therefore seem to be one of stabilising the conformation which is formed naturally by the peptide chain rather than imposing on the chain a pattern of folding that it would not, left to itself, naturally adopt. This cross-linking will, of course, increase the conformational stability in less favourable environments which may interfere with intra-chain hydrogen bonding, and it also seems to confer resistance to enzymic degradation. In ribonuclease enzymic activity is not dependent on all the disulphide bridges being intact. Exposure of the enzyme to the sulphydryl reagent dithiothreitol causes reduction of disulphide bridges, but two of the four react more rapidly and the half reduced material can be isolated after stabilisation as its di-mercury derivative. This derivative is enzymically as active as the native protein[328]. It has also been found possible to insert mercury between the sulphur atoms in all four disulphide bridges. This elongation of the S—S distance by 3 Å gives a material retaining 25 per cent of the enzymic activity of ribonuclease towards cytidine 2′,3′-cyclic monophosphate[329].

Insulin

The insulin molecule (shown diagrammatically in CCXIV) is composed of two peptide chains and three disulphide bridges, one of them forming an

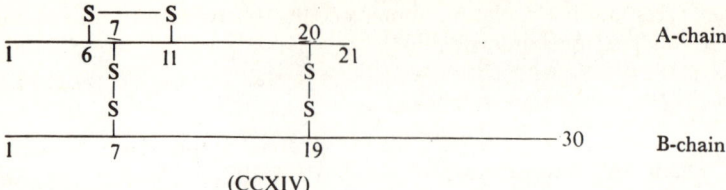

(CCXIV)

intra-chain loop. The presence of two peptide chains introduces a new complicating factor in the pairing of cysteine residues on oxidation. The sequence of the two chains seems able to exert only a moderate control over disulphide bridging in favour of the native form on oxidation of mixtures of their reduced forms derived from natural insulin. Early experiments gave only low activity on recombination, but improved experimental conditions (including the use of excess of A chain) have led to up to 50% recovery of activity[330].

A total synthesis of insulin has been described by three groups of workers[331], but the yield of active insulin when the synthetic chains are combined by oxidation has always been much lower (only 1–2%) than that obtained by recombination of the chains derived from natural insulin. This reflects the side reactions occurring during the removal of the S-protecting groups used during assembly of the peptide chains. Better results have been obtained using A or B chains prepared without using a conventional type of S-protecting group. In one such synthesis the symmetrical disulphides (CCXV) and (CCXVI) of sequences 1–16 and 17–30 of the B chain were coupled by the

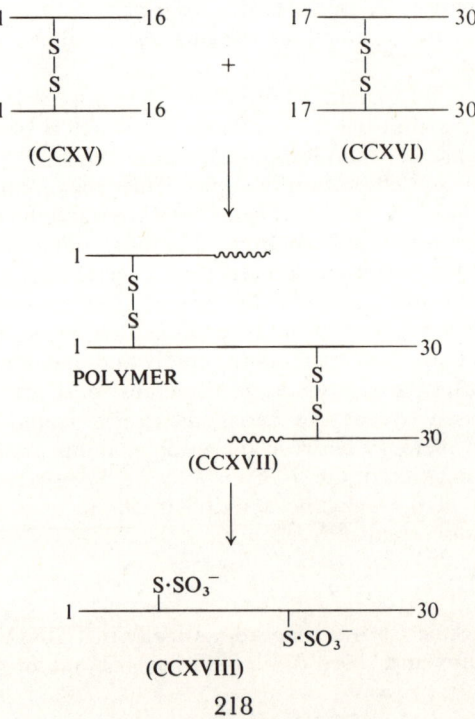

azide method. The resulting polymer (CCXVII) was stripped of its protecting groups by treatment with hydrogen bromide in trifluoroacetic acid, and oxidative sulphitolysis (see the following equation) freed the B-chain as its bis-S-sulphonate (CCXVIII).

$$R-S-S-R + 2SO_3^{2-} \rightarrow 2RS-SO_3^- + 2e^-$$

In contrast to material prepared using S-benzyl protection, this S-sulphonate formed insulin of the same potency as B-chain isolated from natural insulin by oxidative sulphitolysis[332]. The A chain of insulin has been prepared by incorporating the cysteine residues as their mixed disulphides with ethanethiol. After thiolysis of the S-ethylmercapto groups at the end of the synthesis and recombination with natural B chain, the insulin produced was again more potent than a sample prepared similarly using S-benzyl protection[333].

Much work has also been directed towards a completely unequivocal synthesis of insulin. This requires the use of three types of sulphur protecting group which can be removed in succession to enable the individual disulphide bridges to be constructed unambiguously one at a time. In view of the lability of disulphide bridges, once one has been formed there is a considerable limitation on reagents which can subsequently be employed. Four new S-protecting groups for cysteine have been developed in the last few years (*Table* 5.6), and better ways of removing S-trityl[317,340] and S-benzyl[341] groups have become available. Although several studies involving different

Table 5.6

S-Protecting group	Method of removal	Used in the synthesis of
Acetamidomethyl	Hg^{2+} at pH 4[334]	Ribonuclease[326]
2,2-Diethoxycarbonyl	Alkali[335]	Glutathione[336]
Ethylmercapto	Thiophenol[337]	Oxytocin, insulin A chain[338]
1-Phenylcyclohexyl	Hot trifluoroacetic acid[339]	H·Leu·Cys·Gly·OH[339]

combinations of S-protecting groups have been made (notably by Zervas and his group in Athens), a synthesis of insulin involving a stepwise building of the disulphide links has not yet been achieved. Perhaps the most notable progress in this direction has been the synthesis of the insulin model (CCXIX) by Hiskey[342]. Peptide (CCXX) was prepared from the corresponding t-butyl ester using boron trifluoride in acetic acid and coupled with a tripeptide to yield (CCXXI). The S-trityl group of this model A chain reacted with the sulphenyl thiocyanate of a model B chain to give the 9–12 disulphide bridge (CCXXII). (This is an example of a quite general method of disulphide bond formation.) Treatment of (CCXXII) with trifluoroacetic acid in the presence of thiocyanogen removed the benzhydryl groups and formed the third disulphide link at 2–16. Partial cleavage of the t-butyl ester occurred under these conditions, and fission of the ester was completed by treatment with boron trifluoride in acetic acid. The resultant

```
            S─────────S
            |         |
   Z─Cys─Cys─Gly─Phe─Gly─Cys─Phe─Gly─OH  +  H─Cys─Gly─Val─OBuᵗ
   1   |2   3   4   5   6   7   8              |
      SBzh       (CCXX)                       STri
              dicyclohexylcarbodi-imide  │(74% yield)
                                         ▼
            S─────────S
            |         |
   Z─Cys─Cys─Gly─Phe─Gly─Cys─Phe─Gly─Cys─Gly─Val─OBuᵗ   model A chain
   1   |2   3   4   5   6   7   8  |9   10  11
      SBzh                         STri
                   (CCXXI)

                           S·SCN            SBzH
                            |                |
      (76% yield) │   Z─Cys─Gly─Gly─Gly─Cys─Gly─OBuᵗ   model B chain
                  │       12  13  14  15  16  17
                  ▼
            S─────────S
            |         |
   Z─Cys─Cys─Gly─Phe─Gly─Cys─Phe─Gly─Cys─Gly────Val─OBuᵗ
   1   |2   3   4   5   6   7   8  |9   10    11
      SBzh                          S
                                    |
                  SBzh              S
                   |                |
         Buᵗ O─Gly─Cys─Gly─Gly─Gly─Cys─Z   (CCXXII)
              17   16  15  14  13  12
                       │ (i) trifluoroacetic acid–acetic acid, (SCN)₂
          (72% yield)  │ (ii) BF₃-acetic acid
                       ▼
            S─────────S
            |         |
   Z─Cys─Cys─Gly─Phe─Gly─Cys─Phe─Gly─Cys─Gly─Val─OH   model A chain
   1   |2   3   4   5   6   7   8  |9   10  11
       S╲                          S
         ╲S                        S
          |                        |
      HO─Gly─Cys─Gly─Gly─Gly─Cys─Z   model B chain
          17   16  15  14  13  12
                 (CCXIX)
```

peptide (CCXIX) contains the essential structural features of insulin, differing only in the size and sequence of its peptide chains[342].

Insulin containing cystathionine in place of the intra-chain disulphide bridge ('carba' insulin) has been synthesised. The hypoglycaemic activity of this analogue is of the same order as that of native insulin[343].

Pro-insulin

Evidence that insulin biosynthesis might occur through a single chain precursor protein was first obtained in 1967 by Steiner and Oyer during a study of the uptake of ³H-leucine into slices of human islet cell adenoma[344]. This precursor, pro-insulin, was subsequently isolated in a pure form and found to possess a loop of 33 amino-acid residues linking the A and B chains (CCXXIII)[345]. Trypsin will cleave the bond between the A chain and the loop in the correct place, but frees the B chain one amino-acid too far down,

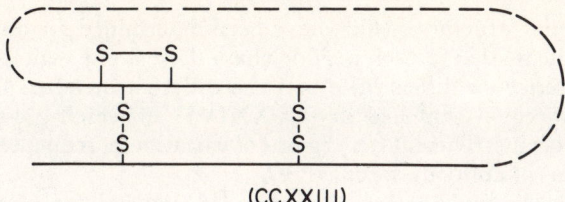

(CCXXIII)

liberating desalanine insulin. It is not known which enzymes are involved naturally in the pro-insulin to insulin conversion, but experiments have shown that bovine pro-insulin is cleaved in the secretion granule into insulin and 'C-peptide'; both remain stored together. C-peptide is the central connecting chain which has lost two arginine residues from its B-chain end and a lysine and an arginine residue from its A-chain end[346].

Fully reduced pro-insulin undergoes oxidation in air in dilute alkaline solution to give a high proportion of native pro-insulin (as evidenced by the regeneration of immunological activity; pro-insulin has only a low hypoglycaemic activity). Subsequent limited incubation with trypsin forms biologically active desalanine insulin[347]. Although c.d. evidence indicates that the connecting chain in pro-insulin is largely random coil in nature, the insulin part of pro-insulin having a configuration similar to that of insulin itself[348], it seems that in its absence the A and B chains do not align themselves in the best position for correct recombination on oxidation of their reduced forms.

Immunoglobulins

The immunoglobulins are a group of blood serum proteins which function as antibodies. Normal blood contains a complex mixture of these proteins, but certain plasma cell tumours produce large quantities of chemically homogeneous immunoglobulins which are more suitable for structural studies. The complete covalent structure of a human myeloma Gl molecule was announced in 1969[349]. It contains four peptide chains, two identical 'light' chains (214 residues) and two identical 'heavy' chains (446 residues), and these are held together by only four inter-chain disulphide bridges out of a total of 16 such links. There is also a small amount (2–3%) of carbohydrate attached to the heavy chains. The arrangement of the peptide chains is shown in (CCXXIV). The evidence available suggests that all immunoglobulins have the same general structural features. Comparison of Gl

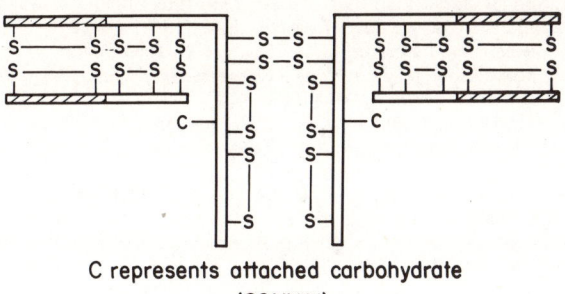

C represents attached carbohydrate
(CCXXIV)

with the partial structures which have been determined for other immunoglobulins indicate that in each peptide chain there is one region in which the amino-acid sequence differs little between different members of the group, and another region (hatched in (CCXXIV)) in which gross amino-acid substitution occurs. These latter regions of variation of sequence are thought to be the basis of antibody specificity[350].

Studies on the re-oxidation of a fully reduced poly-DL-alanyl immunoglobulin (the protein was modified by attachment of this polyamino-acid to increase the solubility of the reduced form) show that up to 50% of the antigenic activity can be restored, even though in this case twenty-three disulphide bridges are being re-connected[351]. Reoxidation of the reduced form of the heavy chain of a poly-DL-alanyl immunoglobulin in the absence of the light chain also gives a product with considerable antigenic activity. This suggests that the conformation of that site of the heavy chain which combines with other proteins is solely dependent upon its own primary amino-acid sequence[352].

Manuscript received August 1971.

REFERENCES

1. SCHRÖDER, E. and LÜBKE, K., *The Peptides*, Academic Press, New York and London (1965)
2. CAESER, F., JANSSON, K. and MUTSCHLER, E., *Pharm. Acta Helv.*, **44**, 676 (1969)
3. KHOKLOV, A. S. and LOKSHIN, G. B., *Tetrahedron Lett.*, 1881 (1963); RAO, K. U. and CULLEN, W. P., *J. Am. chem. Soc.*, **82**, 1127 (1960)
4. BROWN, R., KELLEY, C. and WIBERLEY, S. E., *J. org. Chem.*, **30**, 277 (1965); VONDRÁČEK, M. and VANĚK, Z., *Chemy. Ind.*, 1686 (1964)
5. SHIN, C., CHIGIRA, Y., MASAKI, M. and OHTA, M., *Tetrahedron Lett.*, 4601 (1967); see also SHIN, C., MASAKI, M. and OHTA, M., *J. org. Chem.*, **32**, 1860 (1967)
6. QUILICO, A., *Res. Prog. org. biol. med. Chem.*, **1**, 225 (1964); ROMANET, R., CHEMIZART, A., DUHOUX, S. and DAVID, S., *Bull. Soc. chim. Fr.*, 1043 (1963); BIRCH, A. J., BLANCE, G. E., DAVID, S. and SMITH, H., *J. chem. Soc.*, 3128 (1961)
7. QUILICO, A. and PANIZZI, L., *Ber. dt. chem. Ges.*, **76**, 348 (1943)
8. NAKASHIMA, R. and SLATER, G. P., *Can. J. Chem.*, **47**, 2069 (1969); see, however, *Tetrahedron Lett.*, 4433 (1967)
9. HOUGHTON, E. and SAXTON, J. E., *Tetrahedron Lett.*, 5475 (1968)
10. WESTLEY, J. W., CLOSE, V. A., NITECKI, D. N. and HALPERN, B., *Analyt. Chem.*, **40**, 1888 (1968)
11. WEINDLING, R., *Phytopathology*, **31**, 991 (1941); BELL, M. R., JOHNSON, J. R., WILDI, B. S. and WOODWARD, R. B., *J. Am. chem. Soc.*, **80**, 1001 (1958)
12. BEECHAM, A. F., FRIDRICHSONS, J. and MATHIESON, A. MCL, *Tetrahedron Lett.*, 3131 (1966)
13. FRIDRICHSONS, J. and MATHIESON, A. MCL., *Acta crystallogr.*, **23**, 429 (1967)
14. SYNGE, R. L. M. and WHITE, E. P., *Chemy. Ind.*, 1546 (1959); DONE, J., MORTIMER, P. H. and TAYLOR, A., *J. gen. Microbiol.*, **26**, 207 (1961)
15. HODGES, R. and SHANNON, J. S., *Aust. J. Chem.*, **19**, 1059 (1966)
16. RAHMAN, R. and TAYLOR, A., *Chem. Commun.*, 1032 (1967)
17. RAHMAN, R., SAFE, S. and TAYLOR, A., *J. chem. Soc. C*, 1665 (1969)
18. HODGES, R., RONALDSON, J. W., TAYLOR, A. and WHITE, E. P., *Chemy. Ind.*, 42 (1963); SHANNON, J. S., *Tetrahedron Lett.*, 801 (1963)
19. HERRMANN, H., HODGES, R. and TAYLOR, A., *J. chem. Soc.*, 4315 (1964)
20. FRIDRICHSONS, J. and MATHIESON, A. MCL., *Tetrahedron Lett.*, 1265 (1962); *Acta crystallogr.*, **18**, 1043 (1965)
21. RAHMAN, R., SAFE, S. and TAYLOR, A., *Q. Rev.*, **24**, 233 (1970); SAFE, S. and TAYLOR, A., *J. chem. Soc. C*, 1189 (1971)
22. NAGARAJAN, R., HUCKSTEP, L. L., LIVELY, D. H., DE LONG, D. C., MARSH, M. M. and NEUSS, N., *J. Am. chem. Soc.*, **90**, 2980 (1968)
23. NEUSS, N., NAGARAJAN, R., MOLLOY, B. B. and HUCKSTEP, L. L., *Tetrahedron Lett.*, 4467 (1968)

24. COSULICH, D. B., NELSON, N. R. and VAN DEN HENDE, J. H., *J. Am. chem. Soc.*, **90**, 6519 (1968)
25. MONCRIEF, J. W., *J. Am. chem. Soc.*, **90**, 6518 (1968)
26. NAGARAJAN, R., NEUSS, N. and MARSH, M. M., *J. Am. chem. Soc.*, **90**, 6518 (1968)
27. MINATO, H., MATSUMOTO, M. and KATAYAMA, T., *Chem. Commun.*, 44 (1971)
28. HAUSER, D., WEBER, H. P. and SIGG, H. P., *Helv. chim. Acta*, **53**, 1061 (1970)
29. FISCHER, E., *Ber.*, **39**, 2893 (1906)
30. BLÁHA, K., *Colln. Czech. Chem. Commun.*, **34**, 4000 (1970)
31. GRAHL-NIELSON, O., *Tetrahedron Lett.*, 2827 (1969)
32. KOPPLE, K. D. and GHORAZIAN, H. C., *J. org. Chem.*, **33**, 862 (1968)
33. LICHTENSTEIN, N., *J. Am. chem. Soc.*, **60**, 560 (1938)
34. NITECKI, D. E., HALPERN, B. and WESTLEY, J. W., *J. org. Chem.*, **33**, 864 (1968)
35. RAMACHANDRAN, G. N. and LAKSHMINARAYANAN, A. V., *Biopolymers*, **4**, 495 (1966)
36. CAILLET, J., PULLMANN, B. and MAIGRET, B., *Biopolymers*, **10**, 221 (1971)
37. COREY, R. B., *J. Am. chem. Soc.*, **60**, 1598 (1938)
38. DEGEILH, R. and MARSH, R. E., *Acta crystallogr.*, **12**, 1007 (1959)
39. BENEDETTI, C., CORRADINI, P. and PEDONE, C., *Biopolymers*, **7**, 751 (1969)
40. SLETTEN, E., *J. Am. chem. Soc.*, **92**, 172 (1970)
41. BENEDETTI, C., CORRADINI, P. and PEDONE, C., *J. phys. Chem.*, **73**, 2891 (1969)
42. BENEDETTI, C., CORRADINI, P., GOODMAN, M. and PEDONE, C., *Proc. natn. Acad. Sci. U.S.A.*, **62**, 650 (1969)
43. GROTH, P., *Acta. chem. Scand.*, **92**, 172 (1970)
44. KOPPLE, K. D. and MARR, D. H., *J. Am. chem. Soc.*, **89**, 6193 (1967)
45. GAWNE, G., KENNER, G. W., ROGERS, N. H., SHEPPARD, R. C. and TITLESTAD, K., *Peptides 1968*, ed, Bricas, E., North-Holland Publishing Co., Amsterdam, 28 (1968)
46. KOPPLE, K. D. and OHNISHI, M., *J. Am. chem. Soc.*, **91**, 962 (1969)
47. JOHNSON, C. E. and BOVEY, F. A., *J. Chem. Phys.*, **29**, 1012 (1958)
48. GREENFIELD, N. J. and FASMAN, G. D., *Biopolymers*, **7**, 595 (1969)
49. BALASUBRAMANIAN, D. and WETLAUFER, D. B., *J. Am. chem. Soc.*, **88**, 3449 (1966)
50. SCHELLMAN, J. A. and NIELSON, B. E., *Conformations of Biopolymers*, ed, Ramachandran, G. N., Academic Press, New York, Vol. I, 109 (1967)
51. EDELHOCH, H., LIPPOLDT, R. E. and WILCHEK, M., *J. biol. Chem.*, **243**, 4799 (1968)
52. SVEC, H. J. and JUNK, G. A., *J. Am. chem. Soc.*, **86**, 2278 (1964)
53. NAGARAJAN, R., OCCOLOWITZ, J. L., NEUSS, N. and NASH, S. M., *Chem. Commun.*, 359 (1969)
54. SCHWYZER, R. and SIEBER, P., *Helv. chim. Acta*, **41**, 2186 (1958)
55. BLÁHA, K., SMOLÍKOVÁ, J. and VÍTEK, A., *Colln. Czech. Chem. Commun.*, **31**, 4296 (1966)
56. MIYAZAWA, T., *J. Molec. Spectrosc.*, **4**, 155 (1960)
57. See for example, HARDY, P. M., in *Amino-acids, Peptides, and Proteins*, Vol. 1, Specialist Periodical Reports of the Chemical Society, p. 114
58. ROTHE, M., STEFFEN, K.-D. and ROTHE, I., *Angew. Chem. Int. Edn.*, **4**, 356 (1965)
59. DALE, J. and TITLESTAD, K., *Chem. Commun.*, 656 (1969)
60. VENKATACHALAM, C. M., *Biochim. biophys. Acta*, **168**, 397 (1968)
61. MYOKEI, R., SAKURAI, A., CHANG, C-F. and KODAIRA, Y., *Tetrahedron Lett.*, 695 (1969)
62. SCHWYZER, R., ISELIN, B., RITTEL, W. and SIEBER, P., *Helv. chim. Acta*, **34**, 872 (1956)
63. RAMAKRISHNAN, C. and SARATHY, K. P., *Biochim. biophys. Acta*, **168**, 400 (1968); (a) FRIDKIN, M., PATCHORNIK, A. and KATCHALSKI, E., *J. Am. chem. Soc.*, **87**, 4646 (1965)
64. GROTH, P., *Acta. chem. scand.*, **24**, 780 (1970)
65. DALE, J. and TITLESTAD, K., *Chem. Commun.*, 1403 (1970)
66. BYSTROV, V. H., PORTNOVA, S. L., TSETLIN, V. I., IVANOV, V. T. and OVCHINNIKOV, YU. A., *Tetrahedron*, **25**, 493 (1969)
67. KONNERT, J. and KARLE, I. L., *J. Am. chem. Soc.*, **91**, 4888 (1969)
68. OVCHINNIKOV, YU. A., IVANOV, V. T., KIRYSHKIN, A. A. and SHEMYAKIN, M. M., *Dokl. Akad. Nauk SSSR*, **153**, 122 (1963); OVCHINNIKOV, YU A., IVANOV, V. T., PECK, G. YU., SHEMYAKIN, M. M., *Acta Chim. Acad. Sci. Hung.*, **44**, 211 (1965)
69. HARDY, P. M., KENNER, G. W. and SHEPPARD, R. C., *Tetrahedron*, **19**, 95 (1963)
70. MIYAO, K., *Bull. Agric. Chem. Soc. Japan*, **24**, 23 (1960)
71. STUDER, R. O., *Experientia*, **25**, 899 (1969)
72. BOHMAN, G., *Tetrahedron Lett.*, 3065 (1970)
73. See HASSALL, C. H., SANGER, D. G. and THOMAS, J. O., in *Peptides 1968*, ed, Bricas, E., North-Holland Publishing Co., Amsterdam, 70 (1968)

74. SHEMYAKIN, M. M., OVCHINNIKOV, YU A., ANTONOV, V. K., KIRYUSHKIN, A. A., IVANOV, V. T., SHCHELOKOV, V. I. and SHKROB, A. M., *Tetrahedron Lett.*, 47 (1964)
75. HASSALL, C. H., MARTIN, T. G., SCHOFIELD, J. A. and THOMAS, J. O., *J. chem. Soc. C*, 997 (1967)
76. HASSALL, C. H. and THOMAS, J. O., *J. chem. Soc. C*, 1495 (1968)
77. DALE, J., *Angew. Chem. Int. Edn*, **5**, 1000 (1966)
78. BROWN, C. J., *J. chem. Soc. C*, 1108 (1966)
79. HASSALL, C. H. and THOMAS, W. A., *Chem. Br.*, 145 (1971)
80. KENNER, G. W., THOMSON, P. J. and TURNER, J. W., *J. chem. Soc.*, 4148 (1958)
81. RAMAKRISHNAN, C. and SARATHY, K. P., *Int. J. Protein Res.*, **1**, 63 (1969)
82. SCHERAGA, H. A., *Chem. Rev.*, **71**, 195 (1971)
83. SCHWYZER, R., *Experientia*, **26**, 577 (1970)
84. SCHWYZER, R., TUN-KYI, A., CAVIEZEL, M. and MOSER, P., *Helv. chim. Acta*, **53**, 15 (1970)
85. KARLE, I. L. and KARLE, J., *Acta crystallogr.*, **16**, 969 (1963)
86. KARLE, I. L., GIBSON, J. W. and KARLE, J., *J. Am. chem. Soc.*, **92**, 3755 (1970)
87. SCHWYZER, R., CARRIÓN, J. P., GORUP, B., NOLTING, H. and TUN-KYI, A., *Helv. chim. Acta*, **47**, 441 (1964)
88. SCHWYZER, R. and LUDESCHER, U., *Helv. chim. Acta*, **52**, 2033 (1969)
89. KOPPLE, K. D., OHNISHI, M. and GO, A., *J. Am. chem. Soc.*, **91**, 4264 (1969)
90. KOPPLE, K. D., OHNISHI, M. and GO, A., *Biochemistry*, **8**, 4087 (1969)
91. PRELOG, V. and GERLACH, H., *Helv. chim. Acta*, **47**, 2288 (1964)
92. GERLACH, H., OVCHINNIKOV, YU. A. and PRELOG, V., *Helv. chim. Acta*, **47**, 2294 (1964)
93. GERLACH, H., HAAS, G. and PRELOG, V., *Helv. chim. Acta*, **49**, 603 (1966)
94. IVANOV, V. T., SHILIN, V. V. and OVCHINNIKOV, YU. A., *J. gen. Chem. U.S.S.R.*, **40**, 902 (1970)
95. YANG, CHEN-SU, BLÁHA, K. and RUDINGER, J., *Colln Czech. Chem. Commun.*, **29**, 2633 (1964)
96. MLADENOVA-ORLINOVA, L., BLÁHA, K. and RUDINGER, J., *Colln Czech. Chem. Commun.*, **32**, 4070 (1967)
97. See WARNER, D. T., *Nature*, **190**, 120 (1961); WARNER, D. T., *J. Theoret. Biol.*, **1**, 514 (1961)
98. BLÁHA, K. and FRIČ, I., *Peptides 1968*, ed, Bricas, E., North-Holland Publishing Co., Amsterdam, 40 (1968)
99. BLÁHA, K., FRIČ, I. and RUDINGER, J., *Colln Czech. Chem. Commun.*, **34**, 3497 (1969)
100. OHNO, M. and IZUMIYA, N., *Bull. chem. Soc. Japan*, **38**, 1831 (1965)
101. ABE, O., TAKIGUCHI, H., OHNO, M., MAKISUMI, S. and IZUMIYA, N., *Bull. chem. Soc. Japan*, **40**, 1945 (1967)
102. KENNER, G. W. and LAIRD, A. H., *Chem. Commun.*, 305 (1965)
103. IZUMIYA, N., KATO, T., FUJITA, Y., OHNO, M. and KONDO, M., *Bull. chem. Soc. Japan*, **37**, 1809 (1964)
104. STUDER, R. O. and LERGIER, W., *Helv. chim. Acta*, **48**, 460 (1965)
105. KOPPLE, K. D. and NITECKI, D. E., *J. Am. chem. Soc.*, **83**, 4103 (1961); ibid., **84**, 4457 (1962)
106. KOPPLE, K. D., JARABAK, R. R. and BHATIA, P. L., *Biochemistry*, **2**, 958 (1963)
107. SHEEHAN, J. C. and MCGREGOR, D. N., *J. Am. chem. Soc.*, **84**, 3000 (1962)
108. SHEEHAN, J. C., GOODMAN, M. and RICHARDSON, W. L., *J. Am. chem. Soc.*, **77**, 6391 (1955)
109. SCHWYZER, R., *Ciba Foundation Symposium on Amino-acids and Peptides with Antimetabolic Activity*, Churchill, London, 171 (1958); *Record Chem. Prog.*, **20**, 147 (1959)
110. PAULING, L. and COREY, R. B., *Proc. natn. Acad. Sci. U.S.A.*, **39**, 247 (1953)
111. MOORE, A. T. and RYDON, H. N., *Acta Chim. Acad. Sci. Hung.*, **44**, 103 (1965)
112. KLOTZ, I. M. and FRANZEN, J. S., *J. Am. chem. Soc.*, **82**, 5241 (1960); ibid., **84**, 3461 (1962)
113. ROTHE, M., STEFFEN, K.-D. and ROTHE, I., *Angew. Chem.*, **75**, 1206 (1963); *Angew. Chem. Int. Edn*, **3**, 64 (1964)
114. ROTHE, M., ROTHE, I., TOTH, T. and STEFFEN, K.-D., *Peptides*, eds, Beyerman, H. C., van de Linde, A. and Maassen van den Brink, W., North-Holland Publishing Co., Amsterdam, 8 (1967)
115. SCHWYZER, R. and TUN-KYI, A., *Helv. chim. Acta*, **45**, 859 (1962)
116. BROCKMANN, H. and SPRINGORUM, M., *Tetrahedron Lett.*, 837 (1965); BROCKMANN, H. and ZELLERHOFF, K., *Tetrahedron Lett.*, 2291 (1965)
117. HAMILL, R. L., HIGGENS, C. E., BOAZ, N. E. and GORMAN, M., *Tetrahedron Lett.*, 4255 (1969)

118. OVCHINNIKOV, YU. A., IVANOV, V. T. and MIKHALEVA, I. I., *Tetrahedron Lett.*, 159 (1971)
119. SHEMYAKIN, M. M., OVCHINNIKOV, YU. A. and IVANOV, V. T., *Angew. Chem. Int. Edn*, **8**, 492 (1969)
120. SHEMYAKIN, M. M., OVCHINNIKOV, YU. A., IVANOV, V. T. and EVSTRATOV, A. V., *Nature*, **213**, 413 (1967)
121. SHEMYAKIN, M. M., OVCHINNIKOV, YU. A., IVANOV, V. T., ANTONOV, V. K., VINOGRADOVA, E. I., SHKROB, A. M., MALENKOV, G. G., EVSTRATOV, A. V., LAINE, I. A., MELNIK, E. I. and RYABOVA, I. D., *J. Membrane Biol.*, **1**, 402 (1969)
122. SHEMYAKIN, M. M., OVCHINNIKOV, YU. A., IVANOV, V. T., ANTONOV, V. K., SHKROB, A. M., MIKHALEAVA, I. I., EVSTRATOV, A. V. and MALENKOV, G. G., *Biochem. biophys. Res. Commun.*, **29**, 779 (1967)
123. MOORE, C. and PRESSMAN, B. C., *Biochim. biophys. Res. Commun.*, **15**, 562 (1964)
124. PRESSMAN, B. C., *Proc. natn. Acad. Sci. U.S.A.*, **53**, 1076 (1965)
125. MUELLER, P. and RUDIN, D. O., *Biochem. biophys. Res. Commun.*, **26**, 398 (1967)
126. ŠTEFANAC, Z. and SIMON, W., *Microchem. J.*, **12**, 125 (1967)
127. WIPF, H.-K., PIODA, L. A. R., ŠTEFANAC, Z. and SIMON, W., *Helv. chim. Acta*, **51**, 377 (1968)
128. OVCHINNIKOV, YU. A., IVANOV, V. T., EVSTRATOV, A. V., BYSTROV, V. F., ABDULLAEV, N. D., POPOV, E. M., LIPKIND, G. M., ARKHIPOVA, S. F., EFREMOV, E. S. and SHEMYAKIN, M. M., *Biochem. biophys. Res. Commun.*, **37**, 668 (1969)
129. DOBLER, M., DUNITZ, J. D. and KRAJEWSKI, J., *J. Molec. Biol.*, **42**, 603 (1969)
130. KELLER-SCHIERLEIN, W., PRELOG, V. and ZÄHNER, H., *Fortschr. Chem. org. NatStoffe*, **22**, 279 (1964)
131. EMERY, T. and NEILANDS, J. B., *J. Am. chem. Soc.*, **82**, 3658 (1960)
132. EMERY, T. and NEILANDS, J. B., *J. Am. chem. Soc.*, **83**, 1626 (1961)
133. ROGERS, S. J., WARREN, R. A. J. and NEILANDS, J. B., *Nature*, **200**, 167 (1963)
134. KELLER-SCHIERLEIN, W. and DEÉR, A., *Helv. chim. Acta*, **46**, 1907 (1963)
135. KELLER-SCHIERLEIN, W., *Helv. chim. Acta*, **46**, 1920 (1963)
136. WARREN, R. J. and NEILANDS, J. B., *J. biol. Chem.*, **240**, 2055 (1965)
137. ZALKIN, A., FORRESTER, J. D. and TEMPLETON, D. H., *J. Am. chem. Soc.*, **88**, 1810 (1966)
138. EMERY, T. and NEILANDS, J. B., *J. Am. chem. Soc.*, **82**, 4903 (1960)
139. KELLER-SCHIERLEIN, W. and MAURER, B., *Helv. chim. Acta*, **52**, 603 (1969)
140. LLINÁS, M., KLEIN, M. P. and NEILANDS, J. B., *J. Molec. Biol.*, **52**, 399 (1970)
141. EMERY, T. F., *Biochemistry*, **6**, 3858 (1967)
142. GAUZE, G. F. and BRAŽNIKOVA, M. G., *Novosti Med.*, **23**, 1 and 3 (1951)
143. BRAŽNIKOVA, M. G., MIKEŠ, O. and LOMAKINA, N. N., *Biokhimiya*, **22**, 111 (1959)
144. TURKOVÁ, J., MIKEŠ, O. and ŠORM, F., *Antibiotiki*, **10**, 878 (1962)
145. MIKEŠ, O. and TURKOVÁ, J., *Colln Czech. Chem. Commun.*, **27**, 581 (1962)
146. TURKOVÁ, J., MIKEŠ, O. and ŠORM, F., *Colln Czech. Chem. Commun.*, **27**, 591 (1962)
147. MIKEŠ, O., TURKOVÁ, J. and ŠORM, F., *Colln Czech. Chem. Commun.*, **28**, 1747 (1963)
148. PODDUBNAYA, N. A., LAVRENOVA, G. I., KRYSIN, E. P. and MAKEVNINA, L. G., *J. gen. Chem. U.S.S.R.*, **31**, 3565 (1961); LAVRENOVA, G. I. and PODDUBNAYA, N. A., *J. gen. Chem. U.S.S.R.*, **34**, 2896 (1964)
149. TURKOVÁ, J., MIKEŠ, O. and ŠORM, F., *Experientia*, **19**, 633 (1963); *Colln Czech. Chem. Commun.*, **29**, 280 (1964)
150. EL'NAGGAR, A. M. and PODDUBNAYA, N. A., *J. gen. Chem. U.S.S.R.*, **38**, 444 (1968)
151. TURKOVÁ, J., MIKEŠ, O. and ŠORM, F., *Colln Czech. Chem. Commun.*, **30**, 118 (1965)
152. LAVRENOVÁ, G. I. and PODDUBNAYA, N. A., *J. gen. Chem. U.S.S.R.*, **34**, 2896 (1964)
153. TURKOVÁ, J., MIKEŠ, O. and ŠORM, F., *Colln Czech. Chem. Commun.*, **31**, 2444 (1966)
154. NESVADBA, H. and YOUNG, G. T., *Tetrahedron Lett.*, 361 (1963)
155. LAW, H. D., MILLAR, I. T. and SPRINGNALL, H. D., *Proc. chem. Soc.*, 198 (1958); *J. chem. Soc.*, 279 (1961)
156. STEWART, F. H. C., *Aust. J. Chem.*, **22**, 2663 (1969)
157. SCHWYZER, R. and SIEBER, P., *Helv. chim. Acta*, **40**, 624 (1957)
158. WAKI, M. and IZUMIYA, N., *J. Am. chem. Soc.*, **89**, 1278 (1967); *Bull. chem. Soc. Japan*, **40**, 1687 (1967)
159. KONDO, M. and IZUMIYA, N., *Bull. chem. Soc. Japan*, **40**, 1975 (1967)
160. KONDO, M. and IZUMIYA, N., *Bull. chem. Soc. Japan*, **43**, 1850 (1970)
161. AOYAGI, H., KATO, T., OHNO, M., KONDO, M. and IZUMIYA, N., *J. Am. chem. Soc.*, **86**, 5700 (1964); AOYAGI, H., KATO, T., OHNO, M., KONDO, M., WAKI, M., MAKISUMI, S. and IZUMIYA, N., *Bull. chem. Soc. Japan*, **38**, 2139 (1965)

162. KONDO, M., AOYAGI, H., KATO, T. and IZUMIYA, N., *Bull. chem. Soc. Japan.*, **39**, 2234 (1966)
163. NAGATA, R., WAKI, M., KONDO, M., AOYAGI, H., KATO, T., MAKISUMI, S. and IZUMIYA, N., *Bull. chem. Soc. Japan*, **40**, 963 (1967)
164. KATO, T., KONDO, M., OHNO, M. and IZUMIYA, N., *Bull. chem. Soc. Japan*, **38**, 1202 (1965)
165. ABE, O., KUROMIZIU, K., KONDO, M. and IZUMIYA, N., *Bull. chem. Soc. Japan*, **43**, 914 (1970)
166. MATSUURA, S., WAKI, M., MAKISUMI, S. and IZUMIYA, N., *Bull. chem. Soc. Japan*, **43** 1197 (1970)
167. SCHWYZER, R. and SIEBER, P., *Helv. chim. Acta*, **41**, 1582 (1958)
168. WAKI, M., ABE, O., OKAWA, R., KATO, T., MAKISUMI, S. and IZUMIYA, N., *Bull. chem. Soc. Japan*, **40**, 2904 (1967)
169. ABE, O. and IZUMIYA, N., *Bull. chem. Soc. Japan*, **43**, 1202 (1970)
170. ROTHE, M. and EISENBEISS, F., *Angew. Chem. Int. Edn*, **7**, 883 (1968)
171. AOYAGI, H., KATO, T., WAKI, M., ABE, O., OKAWA, R., MAKISUMI, S. and IZUMIYA, N., *Bull. chem. Soc. Japan*, **42**, 782 (1969)
172. LEE, S., OHKAWA, R. and IZUMIYA, N., *Bull. chem. Soc. Japan*, **44**, 158 (1971)
173. AOYAGI, H. and IZUMIYA, N., *Bull. chem. Soc. Japan*, **39**, 1747 (1966)
174. HALSTROM, J. and KLOSTERMEYER, H., *Justus Liebigs Annln Chem.*, **715**, 208 (1968); OVCHINNIKOV, YU. A., KIRYUSHKIN, A. A. and KOZHEVNIKOVA, I. V., *J. gen. Chem. U.S.S.R.*, **38**, 2636 (1968)
175. SHEMYAKIN, M. M., OVCHINNIKOV, YU. A., IVANOV, V. T. and RYABOVA, I. D., *Experientia*, **23**, 326 (1967)
176. SHEMYAKIN, M. M., OVCHINNIKOV, YU. A. and IVANOV, V. T., *Angew. Chem. Int. Edn*, **8**, 492 (1969)
177. WAKI, M. and IZUMIYA, N., *Tetrahedron Lett.*, 3083 (1968); WAKI, M. and IZUMIYA, N., *Bull. chem. Soc. Japan*, **41**, 1909 (1968)
178. KATO, T., WAKI, M., MATSUURA, S. and IZUMIYA, N., *J. Biochem. Japan*, **68**, 751 (1970)
179. BALASUBRAMANIAN, D., *J. Am. chem. Soc.*, **89**, 5445 (1967)
180. POLLARD, L. W., BHAGAVAN, N. V. and HALL, J. B., *Biochemistry*, **7**, 1153 (1968)
181. MAKISUMI, S., MATSUURA, S., WAKI, M. and IZUMIYA, N., *Bull. chem. Soc. Japan*, **44**, 210 (1971)
182. MAKISUMI, S., WAKI, M. and IZUMIYA, N., *Bull. chem. Soc. Japan*, **44**, 143 (1971)
183. SCHMIDT, G. M. J., HODGKIN, D. C. and OUGHTON, B. M., *Biochem. J.*, **65**, 744 (1957)
184. HODGKIN, D. C. and OUGHTON, B. M., *Biochem. J.*, **65**, 752 (1957)
185. *See for example* VENKATCHALAM, C. H. and RAMACHANDRAN, G. N., *A. Rev. Biochem.*, **168**, 400 (1969)
186. SCHERAGA, H. A., LEACH, S. J., SCOTT, R. A. and NÉMETHY, G., *Discuss. Faraday Soc.*, **40**, 268 (1965)
187. VANDERKOOI, G., LEACH, S. J., NÉMETHY, G., SCOTT, R. A. and SCHERAGA, H. A., *Biochemistry*, **5**, 2991 (1966)
188. LIQUORI, A. M., DE SANTIS, P., KOVACS, A. L. and MAZZARELLA, L., *Nature*, **211**, 1039 (1966)
189. SCOTT, R. A., VANDERKOOI, G., TUTTLE, R. W., SHAMES, P. M. and SCHERAGA, H. A., *Proc. natn. Acad. Sci. U.S.A.*, **58**, 2204 (1967)
190. MOMANY, F. A., VANDERKOOI, G., TUTTLE, R. W. and SCHERAGA, H. A., *Biochemistry*, **8**, 744 (1969)
191. DE SANTIS, P. and LIQUORI, A. M., *Biopolymers*, **10**, 699 (1971)
192. LIQUORI, A. M. and CONTI, F., *Nature*, **217**, 635 (1968)
193. CONTI, F., *Nature*, **221**, 777 (1969)
194. STERN, A., GIBBONS, W. A. and CRAIG, L. C., *Proc. natn. Acad. Sci. U.S.A.*, **61**, 734 (1968)
195. LAIKEN, S. L., PRINTZ, M. P. and CRAIG, L. C., *Biochemistry*, **8**, 519 (1969)
196. ABBOTT, N. B. and AMBROSE, E. J., *Proc. R. Soc. A*, **219**, 17 (1953)
197. FEW, A. V., *Trans. Faraday Soc.*, **53**, 848 (1957)
198. OHNISHI, M. and URRY, D. W., *Biochem. biophys. Res. Commun.*, **36**, 194 (1969)
199. SCHWYZER, R. and LUDESCHER, U., *Biochemistry*, **7**, 2519 (1968)
200. SCHWYZER, R. and LUDESCHER, U., *Biochemistry*, **7**, 2514 (1968)
201. CAMILLETTI, G., DE SANTIS, P. and RIZZO, R., *Chem. Commun.*, 1073 (1970)
202. RUTTENBERG, M. A., KING, TE PIAO and CRAIG, L. C., *Biochemistry*, **4**, 11 (1965)
203. RUTTENBERG, M. A., KING, TE PIAO and CRAIG, L. C., *Biochemistry*, **3**, 758 (1964)

204. RUTTENBERG, M. A. and MACH, B., *Biochemistry* **5**, 2864 (1966)
205. FUJIKAWA, K., SAKAMOTO, Y., SUZUKI, T. and KWAHASHI, K., *Biochim. biophys. Acta*, **169**, 520 (1968)
206. OHNO, M. and IZUMIYA, N., *J. Am. chem. Soc.*, **88**, 376 (1966)
207. OHNO, M., KATO, T., MAKISUMI, S. and IZUMIYA, N., *Bull. chem. Soc. Japan*, **39**, 1738 (1966)
208. KUROMIZU, K. and IZUMIYA, N., *Experientia*, **26**, 587 (1970); KUROMIZU, K. and IZUMIYA, N., *Bull. chem. Soc. Japan*, **43**, 2199 (1970)
209. KUROMIZU, K. and IZUMIYA, N., *Bull. chem. Soc. Japan*, **43**, 2944 (1970)
210. MITZUYASU, N. and IZUMIYA, N., *Experientia*, **26**, 476 (1970); MITZUYASU, N., MATSUURA, S., WAKI, M., OHNO, M., MAKISUMI, S. and IZUMIYA, N., *Bull. chem. Soc. Japan*, **43**, 1829 (1970)
211. RUTTENBERG, M. A., KING, TE PIAO and CRAIG, L. C., *Biochemistry*, **5**, 2857 (1966)
212. LAIKEN, S., PRINTZ, M. and CRAIG, L. C., *J. biol. Chem.*, **244**, 4454 (1969)
213. STERN, A., GIBBONS, W. A. and CRAIG, L. C., *J. Am. chem. Soc.*, **91**, 2794 (1969)
214. GISIN, B. F., MERRIFIELD, R. B. and TOSTESON, D. C., *J. Am. chem. Soc.*, **91**, 2691 (1969)
215. SHEMYAKIN, M. M., SHCHUKINA, L. A., VINOGRADOVA, E. I., RAVDEL, G. A. and OVCHINNIKOV, YU. A., *Experientia*, **22**, 535 (1966)
216. HARRIS, E. J., CATLIN, G. and PRESSMAN, B. C., *Biochemistry*, **6**, 1360 (1967)
217. SHEMYAKIN, M. M., OVCHINNIKOV, YU. A., IVANOV, V. T., ANTONOV, V. K., SHKROB, A. M., MIKHALEVA, I. I., ESTRATOV, A. V. and MALENKOV, G. G., *Biochem. biophys. Res. Commun.*, **29**, 834 (1967)
218. PINKERTON, M., STEINRAUF, L. K. and DAWKINS, P., *Biochem. biophys. Res. Commun.*, **35**, 512 (1969)
219. HAYNES, D. H., KOWALSKY, A. and PRESSMAN, B. C., *J. biol. Chem.*, **244**, 502 (1969)
220. MAJUMDER, S. K. and BOSE, S. K., *Nature*, **181**, 134 (1958)
221. BANERGEE, A. B. and BOSE, S. K., *Nature*, **200**, 471 (1963)
222. MAJUMDER, S. K. and BOSE, S. K., *Biochem. J.*, **74**, 596 (1960)
223. SENGUPTA, S., BANERGEE, A. P. and BOSE, S. K., *Biochem. J.*, **121**, 843 (1971)
224. MEYER, C. E. and REUSSER, F., *Experientia*, **23**, 85 (1967)
225. PAYNE, J. W., JAKES, R. and HARTLEY, B. S., *Biochem. J.*, **117**, 757 (1970)
226. HASSALL, C. H. and MAGNUS, K. E., *Nature*, **194**, 1223 (1959)
227. BEVAN, K., DAVIES, J. S., HASSALL, C. H., MORTON, R. B. and PHILLIPS, D. A. S., *J. chem. Soc. C*, 514 (1971)
228. HASSALL, C. H., OGIHARA, Y. and THOMAS, W. A., *J. chem. Soc. C*, 522 (1971)
229. HASSALL, C. H., MORTON, R. B., OGIHARA, Y. and PHILLIPS, D. A. S., *J. chem. Soc. C*, 526 (1971)
230. BODANSZKY, M. and PERLMAN, D., *Science*, **163**, 352 (1969)
231. GODTFREDSON, W. O. and VANGEDAL, S., *Tetrahedron*, **26**, 4931 (1970)
232. BYCROFT, B. W., CAMERON, D., CROFT, L. R., HASSANALI-WALJI, A., JOHNSON, A. W. and WEBB, T., *Tetrahedron Lett.*, 5901 (1968)
233. KITAGAWA, T., SAWADA, Y. and MUIRA, T., ibid., 109
234. BYCROFT, B. W., CAMERON, D., CROFT, L. R., JOHNSON, A. W., WEBB, T. and COGGAN, P., *Tetrahedron Lett.*, 2925 (1968). See also BÜCHI, G. and RALEIGH, J. A., *J. org. Chem.*, **36**, 873 (1971)
235. FLOYD, J. C., BERTRAND, J. A. and DYER, J. R., *Chem. Commun.*, 998 (1968); KOYAMA, G., NAKAMURA, N., OMOTO, S., TAKITA, T., MAEDA, K. and IITAKA, Y., *J. Antibiot. Japan*, **22**, 34 (1969)
236. BYCROFT, B. W., CAMERON, D., CROFT, L. R., HASSANALI-WALJI, A., JOHNSON, A. W. and WEBB, T., *Nature*, **231**, 301 (1971)
237. WAKAMIYA, T., SHIBA, T., KANEDO, T., SAKAKIBARA, S., TAKC, T. and ABE, J., *Tetrahedron Lett.*, 3497 (1970)
238. BYCROFT, B. W., CAMERON, D., CROFT, L. R. and JOHNSON, A. W., *Chem. Commun.*, 1301 (1968)
239. LECHOWSKI, L., *Tetrahedron Lett.*, 479 (1969)
240. BYCROFT, B. W., CAMERON, D., HASSANALI-WALJI, A. and JOHNSON, A. W., *Tetrahedron Lett.*, 2539 (1969)
241. BODANSZKY, M., IZDEBSKI, J. and MURAMATSU, I., *J. Am. chem. Soc.*, **91**, 2351 (1969); BODANSZKY, M., MARCONI, G. and BODANSZKY, A., *J. Antibiot. Tokyo*, **22**, 40 (1969)
242. SHEEHAN, J. C., MANIA, D., NAKAMURA, S., STOCK, J. D. and MAEDA, K., *J. Am. chem. Soc.*, **90**, 462 (1968)
243. KOYAMA, G., IITAKA, Y., MAEDA, K. and UMEZAWA, M., *Tetrahedron Lett.*, 3587 (1967)

244. OGAWARA, H., MAEDA, K., KOYAMA, G., NAGANAWA, H. and UMEZAWA, H., *Chem. Pharm. Bull.*, **16**, 679 (1968)
245. OGAWARA, H., MAEDA, K. and UMEZAWA, H., *Biochemistry*, **7**, 3296 (1968)
246. See ref. 1 p. 405 (Vol. 2)
247. BOGDANOVA, I. A., KIRYUSHKIN, A. A., ROZYNOV, B. and BURIKOV, V. M., *Zh. Obsch. Khim.*, **39**, 891 (1969)
248. ONDETTI, M. and THOMAS, P., *J. Am. chem. Soc.*, **87**, 4373 (1965)
249. JOLLES, G. and BOUCHAUDON, J., *Peptides 1967*, eds, BEYERMAN, H. C., VAN DEN LINDE, A. and MAASSEN VAN DEN BRINK, W., North-Holland Publishing Co., Amsterdam, 258 (1967)
250. BODANSZKY, M., U.S. Pat. 3 420 816 (1969)
251. KELLER-SCHIERLEIN, W., MIHAILOVIC, M. LJ. and PRELOG, V., *Helv. Chim. Acta*, **42**, 305 (1959)
252. OTSUKA, H. and SHOJI, J., *Tetrahedron*, **23**, 1535 (1967)
253. YOSHIDA, Y. and KATAGIRI, K., *Biochemistry*, **8**, 2645 (1969)
254. BROCKMANN, H. and LACKNER, H., *Chem. Ber.*, **100**, 353 (1967)
255. BROCKMANN, H. and BOLDT, P., *Chem. Ber.*, **101**, 1940 (1968)
256. BROCKMANN, H. and MANEGOLD, J. H., *Chem. Ber.*, **100**, 3814 (1967)
257. MEIENHOFER, J. *Experientia*, **24**, 776 (1968)
258. BROCKMANN, H. and LACKNER, H., *Chem. Ber.*, **101**, 2231 (1968)
259. BROCKMANN, H., LACKNER, H., MECKE, G., TROEMEL, G. and PETRAS, H-S., *Chem. Ber.*, **99**, 717 (1966)
260. BROCKMANN, H. and SCHULZE, E., *Chem. Ber.*, **102**, 3205 (1969)
261. BROCKMANN, H. and LACKNER, H., *Chem. Ber.*, **101**, 1312 (1968)
262. LACKNER, H., *Chem. Ber.*, **103**, 2476 (1970)
263. BROCKMANN, H. and PETRAS, H-S., *Naturwissenschaften*, **48**, 218 (1961)
264. BROCKMANN, H. and SEELA, F., *Tetrahedron Lett.*, 161 (1968)
265. MAUGER, A. B. and WADE, R., *J. chem. Soc. C*, 1406 (1966)
266. BODANSZKY, M., FRIED, J., SHEEHAN, J. T., WILLIAMS, N. J., ALICINO, J., COHEN, P. I., KEELER, B. J. and BIRKHIMER, C. A., *J. Am. chem. Soc.*, **86**, 2478 (1964)
267. CROSS, D. F. W., KENNER, G. W., SHEPPARD, R. C. and STEHR, C. E., *J. chem. Soc. C*, 2143 (1963)
268. ANDERSON, B., HODGKIN, D. C. and VISWAMITRA, M. A., *Nature*, **225**, 233 (1970)
269. BARTON, M. A., KENNER, G. W. and SHEPPARD, R. C., *J. chem. Soc. C*, 2115 (1966)
270. JAMES, M. N. G. and WATSON, K. J., *J. chem. Soc. C*, 13 61 (1966)
271. RESSLER, C. and KASHELIKAR, D. V., *J. Am. chem. Soc.*, **88**, 2025 (1966)
272. ARIYOSHI, Y., SHIBA, T. and KANEKO, T., *Bull. chem. Soc. Japan*, **40**, 2648 (1967); *see also* 2654
273. CRAIG, L. C., PHILLIPS, W. F. and BURACHIK, M., *Biochemistry*, **8**, 2348 (1969)
274. WIELAND, TH., *Fortschr. Chem. org. NatStoffe*, **25**, 214 (1967); WIELAND, TH., *Science*, **159**, 946 (1968)
275. WIELAND, TH. and GEBERT, U., *Ann.*, **700**, 157 (1966)
276. WIELAND, TH. and DE VRIES, DE, J. X., *Ann.*, **700**, 174 (1966)
277. WIELAND, TH., REMPEL, D., GEBERT, U., BAKU, A. and BOEHRINGER, H., *Ann.*, **704**, 226 (1967); GEBERT, U., WIELAND, TH. and BOEHRINGER, H., ibid., **705**, 227
278. WIELAND, TH. and BAKU, A., *Justus Liebigs Annln Chem.*, **717**, 215 (1968)
279. WIELAND, TH. and WEHRT, H., *Justus Liebigs Annln Chem.*, **700**, 120 (1966); PFAENDER, P. and WIELAND, TH., ibid., 126; WIELAND, TH. and GEORGI, V., ibid., 133 and 149; WIELAND, TH., HASSAN, M. and PFAENDER, P., ibid., **717**, 205 (1968)
280. WIELAND, TH. and DÖLLING, J., *Naturwissenschaften*, **53**, 526 (1966)
281. FAULSTICH, M. and WIELAND, TH., *Justus Liebigs Annln Chem.*, **713**, 186 (1968)
282. WIELAND, TH. and JECK, R., *Justus Liebigs Annln Chem.*, **713**, 196 (1968)
283. FAHRENHOLZ, F., FAULSTICH, H. and WIELAND, TH., *Justus Liebigs Annln Chem.*, **743**, 83 (1971)
284. WIELAND, TH., JOCHUM, C. and FAULSTICH, H., *Justus Liebigs Annln Chem.*, **727**, 138 (1969)
285. WIELAND, TH., LÜBEN, G., OTTENHEYM, H., FAESEL, J., DE VRIES, J. X., PROX, A. and SCHMID, J., *Angew. Chem. Int. Edn*, **7**, 204 (1968)

286. KONIG, W. and GEIGER, R., *Justus Liebigs Annln Chem.*, **727**, 125 (1969); WIELAND, TH., BIRR, C. and FLOR, F., *Ann.*, **727**, 130 (1969); WIELAND, TH., FAESEL, J. and KONZ, W., *Justus Liebigs Annln Chem.*, **722**, 197 (1969)
287. TSCHESCHE, R. and REUTEL, I., *Tetrahedron Lett.*, 3817 (1968)
288. PAIS, M., MARCHAND, J., JARREAU, F.-X. and GOUTAREL, R., *Bull. Soc. chim. Fr.*, 1145 (1968)
289. SERVIS, R. E., KOSAK, A. I., TSCHESCHE, R., FROHBERG, E. and FEHLHABER, H. W., *J. Am. chem. Soc.*, **91**, 5619 (1969)
290. KLEIN, F. K. and RAPAPORT, H., *J. Am. chem. Soc.*, **90**, 2398 (1968)
291. TSCHESCHE, R., BEHRENDT, L. and FEHLHABER, H. W., *Chem. Ber.*, **102**, 50 (1969)
292. TSCHESCHE, R., FROHBERG, E. and FEHLHABER, H. W., *Chem. Ber.*, **103**, 250 (1970)
293. WARNHOFF, E. W., PRAHAM, S. K. and MA, J. C. N., *Can. J. Chem.*, **43**, 2594 (1965); WARNHOFF, E. W., MA, J. C. N. and REYNOLDS-WARNHOFF, P., *J. Am. chem. Soc.*, **87**, 4198 (1965); KLEIN, F. and RAPAPORT, H., ibid., **90**, 3576 (1968); SERVIS, R. E. and KOSAK, A. I., ibid., **90**, 4179 (1968)
294. TSCHESCHE, R. and LAST, H., *Tetrahedron Lett.*, 2993 (1968)
295. TSCHESCHE, R., LAST, H. and FEHLHABER, H. W., *Chem. Ber.*, **100**, 3937 (1967)
296. TSCHESCHE, R., RHEINGAUS, J., FEHLHABER, H. W. and LEGLER, G., *Chem. Ber.*, **100**, 3924 (1967)
297. TSCHESCHE, R., FROHBERG, E. and FEHLHABER, H. W., *Tetrahedron Lett.*, 1311 (1968)
298. TSCHESCHE, R., WALTERS, A. and FEHLHABER, H. W., *Chem. Ber.*, **100**, 323 (1967)
299. MARCHAND, J., PAIS, M., MONSEUR, X. and JARREAU, F-X., *Tetrahedron*, **25**, 937 (1969)
300. PAIS, M., MONSEUR, X., LUSINCHI, X. and GOUTAREL, R., *Bull. chem. Soc. Fr.*, 817 (1964); *Ann. Chim.*, 83 (1966)
301. PAIS, M., MARCHAND, J., RATLE, G. and JARREAU, F.-X., *Bull. Soc. chim. Fr.*, 2979 (1968)
302. WIELAND, TH., CORDS, H. and KECK, E., *Chem. Ber.*, **17**, 1312 (1954)
303. ZBIRAL, E., MÉNARD, E. L. and MULLER, J. M., *Helv. chim. Acta.*, **48**, 404 (1965)
304. MARUMO, S. and CURTIS, R. W., *Phytochemistry*, **1**, 245 (1962)
305. TAKEUCHI, S., SENN, M., CURTIS, R. W. and MCLAFFERTY, F. W., *Phytochemistry*, **6**, 287 (1967)
306. IRIUCHIJIMA, S. and CURTIS, R. W., *Phytochemistry*, **9**, 1199 (1970)
307. SCHÖBERL, A., RIMPLER, M. and CLAUSS, E., *Naturwissenschaften*, **56**, 516 (1969); *Justus Liebigs Annln Chem.*, **742**, 68 (1970)
308. GEIGER, R., *Angew. Chem. Int. Edn*, **10**, 152 (1971)
309. MANNING, M., COY, E. and SAWYER, W. H., *Biochemistry*, **9**, 3925 (1970)
310. JOŠT, K., and SŎRM, F., *Colln Czech. Chem. Commun.*, **36**, 234 (1971)
311. YAMASHIRO, D., HOPE, D. B. and DU VIGNEAUD, V., *J. Am. chem. Soc.*, **90**, 3857 (1968)
312. LANDE, S., WITTER, A. and DE WIED, D., *J. biol. Chem.*, **246**, 2058 (1971)
313. BEYCHOK, S. and BREWSTER, E., *J. biol. Chem.*, **243**, 151 (1968)
314. LORD, R. C. and YU, N., *J. Molec. Biol.*, **50**, 509 (1970)
315. NEHER, R., RINIKER, B., ZUBER, H., RITTEL, W. and KAHNT, F. W., *Helv. chim. Acta.*, **51**, 917 (1968); POTTS, J. T., NIALL, H. D., KEUTMANN, H. T., BREWER, H. B. and DELFTOS, L. J., *Proc. natn. Acad. Sci. U.S.A.*, **59**, 1321 (1968); BELL, P. H., BARG, W. F., COLUCCI, D. F., DAVIES, M. C., DZIOBOWSKI, C., ENGLERT, M. E., HEYDER, E., PAUL, R. and SNEDEKER, E. H., *J. Am. chem. Soc.*, **90**, 2704 (1968)
316. RITTEL, W., BRUGGER, M., KAMBER, B., RINIKER, B. and SIEBER, P., *Helv. chim. Acta.*, **51**, 924 (1968); GUTTMANN, ST., PLESS, J., SANDRIN, ED., JACQUENOUD, P.-A., BOSSERT, H. and WILLEMS, H., *Helv. chim. Acta.*, **51**, 1155 (1968)
317. KAMBER, B. and RITTEL, W., *Helv. chim. Acta*, **51**, 2061 (1968)
318. BELL, P. H., COLUCCI, D., DZIOBOWSKI, C., SNEDEKER, E. H., BARG JR., W. F. and PAUL, R., *Biochemistry*, **9**, 1655 (1970)
319. BREWER, H. B. and EDELHOCH, H., *J. biol. Chem.*, **245**, 2402 (1970)
320. GUTTMAN, ST., PLESS, J., HUGUENIN, R. L., SANDRIN, ED., BOSSERT, H. and ZEHNDER, K., *Helv. chim. Acta.*, **52**, 1789 (1969)
321. NEHER, R., RINIKER, B., RITTEL, W. and ZUBER, H., *Helv. chim. Acta*, **51**, 1900 (1968)
322. HARDY, P. M., RIDGE, B., RYDON, H. N. and SERRÃO, F. O. DOS S. P., *J. chem. Soc C*, 1722 (1971) and earlier papers in this series
323. HISKEY, R. G., DAVIS, G. W., SAFDY, M. E., INUI, T., UPHAM, R. D. and JONES JR., W. C., *J. org. Chem.*, **35**, 4148 (1970)
324. ANFINSON, C. B., *Harvey Lecture Series*, Academic Press, New York, 95 (1966)

325. GUTTE, B. and MERRIFIELD, R. B., *J. Am. chem. Soc.*, **91**, 501 (1969)
326. HIRSCHMANN, R., NUTT, R. F., WEBER, D. F., VITALI, R. A., VARGA, S. L., JACOB, T. A., HOLLEY, F. W. and DENKEWALTER, R. G., *J. Am. chem. Soc.*, **91**, 507 (1969)
327. ANFINSON, C. B., *J. biol. Chem.*, **221**, 405 (1956); TANIUCHI, H., *J. biol. Chem.*, **245**, 5459 (1970)
328. SPERLING, R., BURSTEIN, Y. and STEINBERG, I. Z., *Biochemistry*, **8**, 3810 (1969)
329. SPERLING, R. and STEINBERG, I. Z., *J. biol. Chem.*, **246**, 715 (1971)
330. YUNG-CONG, DU., RONG-QUING, J. and CHEN-LU, T., *Sci. Sinica*, **14**, 229 (1965)
331. ZAHN, H., DANHO, W. and GUTTE, B., *Z. Naturforsch.*, **21b**, 763 (1966); ZAHN, H., GUTTE, B., PFEIFFER, E. P. and AMMON, J., *Justus Liebigs Annln Chem.*, **691**, 225 (1966); KATSOYANNIS, P. G., TOMETSKO, A. M., GINOS, J. Z. and TILAK, M. A., *J. Am. chem. Soc.*, **88**, 164 (1966); KATSOYANNIS, P. G., TAMETSKO, A. and ZALUT, C., ibid., p. 166; CHING-I, N., YEUH-TING, K., WEI-TEH, H., LIN-TSUN, K., CHAN-CHIN, C., YUAN-CHUNG, C., YU-CANG, D., RONG-QING, J., CHEN-LU, T., SHIH-CHUAN, H., SHANG-QUAN, C. and KEH-ZHEN, *Sci. Sinica*, **15**, 231 (1966) and refs. therein
332. ZAHN, H. and SCHMIDT, G., *Justus Liebigs Annln Chem.*, **731**, 91 (1970)
333. WEBER, U., *Hoppe-Seylers Z. Physiol. Chem.*, **350**, 1421 (1969)
334. VEBER, D. F., MILKOWSKI, J. D., DENKEWALTER, R. G. and HIRSCHMANN, R., *Tetrahedron Lett.*, 3057 (1968)
335. WIELAND, T. and SIEBER, A., *Justus Liebigs Annln Chem.*, **722**, 222 (1969)
336. WIELAND, T. and SIEBER, A., *Justus Liebigs Annln Chem.*, **727**, 121 (1969)
337. INUKAI, N., NAKANO, K. and MURAKAMI, M., *Bull. chem. Soc. Japan*, **40**, 2913 (1967)
338. INUKAI, N., NAKANO, K. and MURAKAMI, M., *Bull. chem. Soc. Japan*, **41**, 182 (1968)
339. KONIG, W., GEIGER, R. and SIEDEL, W., *Chem. Ber.*, **101**, 681 (1968)
340. ZAHN, H., DANHO, W., KLOSTERMEYER, H., GATTNER, H. G. and REPIN, J., *Z. Naturforsch.*, **24b**, 1127 (1969)
341. SAKAKIBARA, S. and SHIMONISKI, Y., *Bull. chem. Soc. Japan*, **38**, 1412 (1965)
342. HISKEY, R. G. and SMITH, R. L., *J. Am. chem. Soc.*, **90**, 2677 (1968)
343. JOŠT, K., RUDINGER, J., KLOSTERMEYER, H. and ZAHN, H., *Z. Naturforsch*, **23b**, 1059 (1968)
344. STEINER, D. F. and OYER, P. E., *Proc. natn. Acad. Sci. U.S.A.*, **57**, 473 (1967)
345. CHANCE, R. E., ELLIS, R. M. E. and BROMER, W. M., *Science*, **161**, 165 (1968)
346. STEINER, D. F., CHO, S., OYER, P. E., TERRIS, S., PETERSON, J. D. and RUBINSTEIN, A. H., *J. biol. Chem.*, **246**, 1365 (1971)
347. STEINER, D. F. and CLARK, J. L., *Proc. natn. Acad. Sci. U.S.A.*, **60**, 622 (1968)
348. FRANK, B. H. and VEROS, A. H., *Biochem. biophys. Res. Commun.*, **32**, 155 (1968)
349. EDELMAN, G. M., CUNNINGHAM, B. A., GALL, W. E., GOTTLIEB, P. D., RUTISHAUSER, V. and WAXDAL, M. J., *Proc. natn. Acad. Sci. U.S.A.*, **63**, 78 (1969)
350. *See* reviews by EDELMAN, G. M. and GALL, W. E., *A. Rev. Biochem.*, **38**, 415 (1969); PUTNAM, F. W., *Science*, **163**, 633 (1969)
351. FREEDMAN, M. H. and SELA, M., *J. biol. Chem.*, **241**, 5225 (1966)
352. JATON, J.-C., KLINMAN, N. R., GIVOL, D. and SELA, M., *Biochemistry*, **7**, 4185 (1968)

6
CHEMILUMINESCENCE OF ORGANIC COMPOUNDS
F. McCapra

INTRODUCTION	231
GENERAL MECHANISMS	232
METHODS OF INVESTIGATION	235
ELECTRON TRANSFER LUMINESCENCE	236
Theoretical Considerations	237
ELECTROCHEMICALLY GENERATED CHEMILUMINESCENCE	238
OTHER ELECTRON TRANSFER REACTIONS	239
Oxidation of Radical Anions	239
Reduction Reactions	242
CHEMILUMINESCENCE FROM SINGLET OXYGEN	243
REACTIONS OF PEROXIDES	247
Hydrazide Chemiluminescence	248
Dioxetane Chemiluminescence	251
Imine Peroxides	255
Active Esters and Related Compounds	257
Derivatives of Oxalic Acid	257
Acridine Esters and Related Compounds	259
Acridan Esters	262
Lucigenin	263
BIOLUMINESCENCE	267
The Firefly	267
Cypridina Hilgendorfii	269
Latia Neritoides	271

INTRODUCTION

INTEREST in visible chemiluminescence ('cold light') has recently been increasing although its investigation is not yet widespread. Only in the last decade has it been possible to consider the mechanisms of the phenomenon in a meaningful way. There are still many problems to solve, but these are now at the same level of development as the study of photochemistry and, say, organic fragmentation reactions. Chemiluminescence occurs when the product of a reaction appears in an electronically excited state, light being emitted when the molecule reaches the ground state by emission of a photon. A characteristic of this form of luminescence is the exceedingly wide range of efficiency of reaction which is observed. Indeed, while the upper limit of 100% is almost attainable (in the firefly) the lower limit is set only by the sensitivity of detecting devices. Photon counting apparatus can detect luminescence from a great variety of chemical reactions, but the efficiency as measured by the quantum yield in many cases is 10^{-15} or less. Although the monitoring of the emission by kinetic and spectroscopic methods allows the study of even the weakest chemiluminescent reaction, there are obvious dangers in suggesting mechanisms for reactions whose products are present in nearly undetectable amounts. Characterisation by conventional chemical means is certainly impossible. Accordingly, this review is largely concerned with those reactions which produce fairly easily visible light, although very often the excited state involved may be formed in high yield but is quenched by other species. A particular case of visible luminescence, bioluminescence, is perhaps the most intriguing of all, and it is a little surprising that the chemistry of such systems has so seldom been investigated in detail. Almost by virtue of the progress made in understanding the chemiluminescence of organic compounds, bioluminescence is now in a situation where useful experiments can be devised and the details of the excitation process described in chemical terms.

GENERAL MECHANISMS

There is one sense in which the emission of bright light from a chemical reaction is not surprising. (The popularity of such demonstrations suggests, however, that most observers find the spectacle intriguing.) This is in the sense that, given a reaction of sufficient exothermicity, one is simply observing the partitioning of the energy in a non-equilibrium fashion. We agree to some extent with this view, particularly since the energy required for visible luminescence is around 50–80 kcal/mol. The paucity of reactions of this exothermicity may indeed be the most important factor limiting the frequency of occurrence of chemiluminescence. However, such an approach constitutes no answer to the questions concerning bright luminescence, the essence of which must be the nature of the effects responsible for the lack of equilibrium distribution. Although gas-phase chemiluminescence is outside the scope of this review, it does provide the simplest examples, at least in terms of the transition states involved. In the absence of an equilibrating solvent, vibrational energy can be preserved until a suitable electronic state becomes accessible among the reactants. The classical example[1] here is that of the sodium 'flame', in the reaction of sodium with halogens (*Scheme 1*):

$$Na + X_2 \longrightarrow NaX + X$$
$$X + Na_2 \longrightarrow NaX\dagger + Na$$
$$NaX\dagger + Na \longrightarrow NaX + Na^*$$
or
$$KX\dagger + Na \longrightarrow NaX + K^*$$

Scheme 1

The excitation step is actually an atom transfer reaction as can be seen by the use of a 'cross beam' of potassium. A variety of other chemiluminescent atom recombinations[2], such as

$$O + NO + M^3 \quad \text{and} \quad N + N + M \rightarrow N_2^{*\ 4}$$

where M is an inert third body, are known. The extreme reactivity of the atoms and the usually very narrow range of pressures required for luminescence are two of the reasons why such reactions provide little analogy for reactions in solution. However, related molecular reactions are closer in character to the systems discussed later. The oxidation of nitric oxide and sulphur monoxide by ozone[5] are two particularly well understood chemiluminescent reactions of this type.

$$O_3 + NO \rightarrow NO_2^* + O_2$$
$$O_3 + SO \rightarrow SO_2^* + O_2$$

The relative simplicity of the process allows a detailed study of the kinetics, and the adiabatic correlation rules applied to both sides of the equation lead to the reasonable potential energy surfaces shown for the sulphur monoxide reaction (*Scheme 2*). It is naturally impossible to apply the conclusions obtained from these diagrams directly to polyatomic molecules, but some points of general application emerge. The activation energy for formation of a given state is, as expected, slightly higher for the more energetic states. In accordance with the correlation rules, the probability of populating any state is lower if a surface crossing is involved (this is usually the case for excited state formation). The geometry in the transition state will strongly

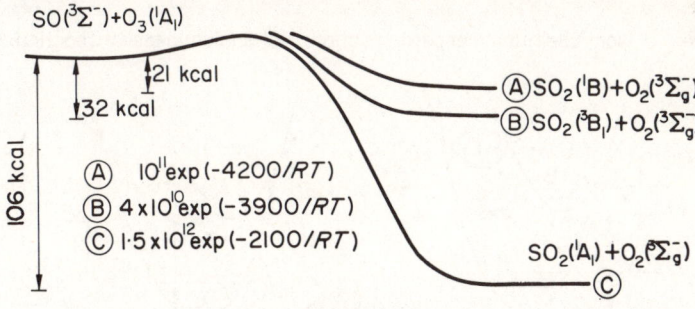

Scheme 2
After Thrush, B. A., *Chemistry in Britain*, 1966, p. 287 by courtesy of The Chemical Society

influence such a surface crossing, and the two states of NO_2 (ground and first excited states) can be traced to two different geometries of approach of the reactants. Similar considerations must hold in solution with more complex molecules but it is unlikely that such a detailed analysis would ever be possible. Nevertheless, these observations provide a useful basis for a description in the less easily understood reactions of large organic molecules.

By far the most common reaction involved in organic chemiluminescence is oxidation. Given the rarity of chemiluminescence, it would seem reasonable to expect some general mechanism for the process. However, the variety of reactions involved strongly suggests that several mechanisms obtain, and no one mechanistic type can cover all cases. A necessary feature in all proposals, of course, is that the energy of the light emitted must not be greater than that of the total exothermicity of the reaction including activation energy, if any (*Scheme 3*). This energy is necessarily large (50–80 kcal) for visible emission and its dissipation by bond fragmentation, vibration or equilibrium with the solvent might well be expected rather than formation of an excited state. Electron transfer reactions, with their extreme rapidity (similar in fact to that of an electronic transition), would seem to offer the best opportunity for avoidance of 'dark' routes. Nevertheless, bond rearrangements and fragmentations are also important reactions with strong evidence supporting their occurrence. Because of the unique character of oxygen and its excited states, it has been suggested[6] that singlet oxygen formation (to be discussed below) provides a general explanation for chemiluminescence involving oxygen or peroxides. Although this is an important route to luminescence, it is probably not as widespread as might have been expected.

Another general observation often made is that of 'sensitised chemiluminescence'. This refers to the excitation of a fluorescent molecule (which can be a reactant, product, by-product or additive) by a molecule in an excited state which itself does not emit or does so less effectively than the acceptor. The routes by which this may occur are naturally related to those of energy transfer processes much studied in connection with excitation experiments in solution[7]. At the present time, most of the established transfer mechanisms can be observed in chemiluminescence, and attention will be drawn to them as they occur.

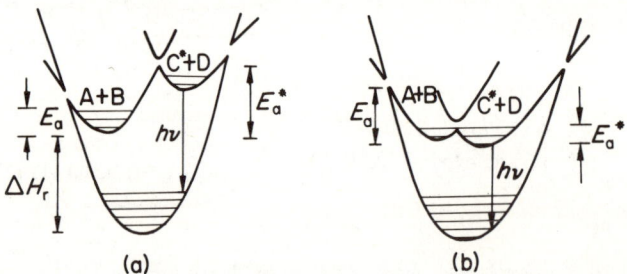

Scheme 3. *These curves apply most properly to electron transfer reactions where there is little change in atomic coordinates during the reaction. In all cases* $h\nu \leqslant \Delta H_r + E_a^*$

The problem of efficiency in chemiluminescence can be examined, as in photochemical investigations generally, in terms of the quantum yield. In this case it is useful to consider it as $\phi = \phi_C \times \phi_E \times \phi_F$ where ϕ_C is the normal chemical yield of the excited molecule, ϕ_E is the number of such molecules formed in the excited state and ϕ_F is its fluorescence efficiency. If sensitised chemiluminescence is involved then a further factor for the efficiency of the energy transfer must be included. Not much control can be exercised over the efficiency of fluorescence or energy transfer, and one would normally expect to arrange for the formation of a moderately fluorescent product. Investigation of the mechanism will perhaps suggest means of obtaining the optimum chemical yield and indicate reasons for efficient population of the excited rather than the ground state. Reference to *Scheme 3* suggests that means of enhancing the probability of entry into the excited state and of lowering the activation energy should be sought. Unfortunately it is rarely clearly understood how this may be achieved.

CHEMILUMINESCENCE OF ORGANIC COMPOUNDS
METHODS OF INVESTIGATION

Chemiluminescence is of course unique in that there are many of the problems associated with photochemical investigations superimposed upon a complex reaction in solution. The usual techniques of organic chemistry are used to define reaction conditions, the nature of products and a rough outline of the possible mechanism. The excited product must then be identified by its isolation and the comparison of its fluorescence spectrum with the spectrum of chemiluminescence. Commercial spectrophotofluorometers are usually used for fluorescence studies, and sometimes for the chemiluminescence measurements also. However, largely for reasons of sensitivity and inconvenient reaction conditions, separate monochromators, photomultiplying tubes with associated electronic devices and recorders, are usually necessary.

Measurement of the absolute quantum yield presents as many problems as does that of the more familiar fluorescence. The methods and criteria involved have been discussed in detail[8]. Relative quantum yields, however, are more easily obtained by using reactions of known yield as standard[8] or standardised radioactive scintillation sources[9]. Although it is impossible to give details here of the techniques used, it is crucial to any investigation of chemiluminescence that the excitation step be accurately identified. Most of the emissions involved consist of broad structureless bands and identification of the emitter may be in some doubt. Transient intermediates are likely as are radical chain processes. Further complications are introduced by energy transfer from a product in the main excitation route to an acceptor which can be the starting material, impurity or secondary product. These difficulties must be borne in mind while considering the mechanisms suggested for the reactions which follow, and specific reference will be made to their occurrence where appropriate.

Kinetic investigations of chemiluminescent reactions can be greatly facilitated, in the simpler cases, by the great ease of monitoring the decay of the emission. The following general and simple scheme serves to illustrate the relationship between intensity and order in reactant:

$$nR \xrightarrow{k_1} P^* \tag{1}$$

$$P^* \xrightarrow{k_f} P + h\nu \tag{2}$$

$$P^* \xrightarrow{k_Q} P \tag{3}$$

Invariably $k_f > k_1$ and

$$\frac{dP^*}{dt} = 0 = k_1 R^n - (k_f + k_Q) P^* \tag{4}$$

$$P^* = \frac{k_1}{k_f + k_Q} \cdot R^n \tag{5}$$

Since from (2) and (3)

$$\frac{dP}{dt} = (k_f + k_Q) P^* \tag{6}$$

$$\frac{dP}{dt} = k_1 R^n \text{ from (5) and (6)} \qquad (7)$$

Now $I = \dfrac{dh\nu}{dt}$ (I is the observed intensity at time t)

$$= k_f P^* \text{ from (2)} \qquad (8)$$

$$= \frac{k_1 k_f}{k_f + k_Q} \cdot R^n \text{ from (5)} \qquad (9)$$

For a first-order or pseudo first-order reaction:

(9) becomes $\qquad R = \dfrac{k_f + k_Q}{k_1 k_f} \cdot I$

and, on differentiation with respect to t,

$$-\frac{dR}{dt} = -\frac{k_f + k_Q}{k_1 k_f} \frac{dI}{dt}$$

$$= k_1 \frac{(k_f + k_Q)}{k_1 k_f} I \text{ from (9)}$$

Thus $-\dfrac{dI}{dt} = k_1 I$ and the rate constant for the excitation step can be readily obtained by plotting $\log I$ against t in the usual way. The assumption here is that excitation is the rate-determining step but in general any rate-determining step for a first-order or pseudo first-order reaction is obtained in this way. Rate-determining steps of higher order are also recognisable, for example, as in the case of the red emission from two oxygen singlets.

It is possible to use the maximum intensity of the reaction as a measure of the rate. However, whilst this is convenient it can lead to complex and confusing results since any reaction which does not show simple exponential decay of intensity with time must involve consecutive reactions whose kinetic analysis may be difficult. Measurement of initial intensities is still easier, but naturally much less information can be obtained and inconsistencies in the data are observed. These last two techniques are to be considered most cautiously when changes in rate constants with substitution (e.g. Hammett plots) are considered. Changes in substitution invariably alter fluorescence efficiencies, not always in a linear or otherwise predictable manner, and a single datum such as initial or maximum intensity is evidently misleading.

The reactions to be discussed now in some detail are selected on the basis of belonging to a specific mechanistic category. More comprehensive reviews are available[10], the most extensive being that by Gunderman[10a]. The approach used here is intended to serve as a guide to those wishing to collate the better understood examples of organic chemiluminescence.

ELECTRON TRANSFER LUMINESCENCE[11-13]

The final step in any chemiluminescence is identical to emission by fluorescence or, very rarely, by phosphorescence. Recombination luminescence[14] occurs when a glassy solution (at 77 K) of a potentially phosphorescent molecule is allowed to rise in temperature after irradiation. This result can, in a certain sense, be considered as a chemiluminescence, and as such has implications for more complex examples. The electron ejected by the photoionisation is trapped in the rigid medium which on being warmed allows the electron to return to the radical cation produced with the formation, in this case, of the first excited triplet state. Phosphorescence from this state then occurs in the usual manner. The first observation of a chemical counterpart was made by Hoijtink, as reported by Chandross and co-workers[15], in their further investigation of the reaction. Chemiluminescence was observed during electrolysis of solutions of fluorescent aromatic hydrocarbon, a result simultaneously reported by Rauhut et al.[16] and Hercules[17]. This method of generating a system capable of electron transfer is much more easily controlled than other oxidative (electron transfer) reactions, and the results from electrochemical experiments are accordingly more easily interpreted.

Theoretical Considerations

Electron transfer reactions should provide an opportunity for a deeper theoretical understanding of chemiluminescence since the excitation step involves only electron transfer. An attempt has been made to examine the reaction theoretically on the basis of the scheme shown where [ox] and [red] are oxidised and reduced forms of an electron acceptor.

Antibonding   ↑   A⁻ + [ox]   ⟶   ↑   A* + [red]
Bonding   ↑↓   ↑

            A*   ⟶   A + hν

At first sight, it may be surprising that the electron transferred is not the more energetic one in the more diffuse antibonding orbital (in so far as electrons can be identified in this rough fashion). The explanation suggested for this effect[18] is based on a calculation which shows that the activation energy for entry into a particular state is directly proportional to the energy difference between reactants and products (excited *or* ground state). Thus, if this analysis is correct, the rate of removal of a *bonding* electron from the radical anion could well be competitive with the removal of the antibonding electron, since the reactants and *excited* state of the hydrocarbon are closer in energy. It might be expected that the exothermicity of the reaction would influence the extent of population of the excited state. Hence the situation is as depicted in *Scheme 3b* rather than as in *3a*. Diagrams of this sort can only be illustrative for large organic molecules but nevertheless provide some basis for discussion. Obviously the activation energy depends very largely on the assumption that differences in the atomic coordinates of the reactant and product are small since the height of the crossing above the ground state of reactant would be markedly increased by a greater separation of the curves. For electron transfer reactions in polynuclear aromatic hydrocarbons and related compounds this assumption is probably justified.

In chemiluminescent reactions with large structural differences between reactant and product, the meaning of the curves is obviously not the same. Another feature of electron transfer is the rapidity with which the reaction is completed, in contrast to the time required for the considerable bond reorganisation occurring in a typical organic reaction. Since experience in photochemistry[19] strongly suggests that changes in bonding in the excited state promote internal conversion, electron transfer is *in principle* more likely to result in radiation.

ELECTROCHEMICALLY GENERATED CHEMILUMINESCENCE

As indicated above, the experimental techniques in this form of chemiluminescence allow both control of the chemiluminescence and measurement of the energy of the reaction and of the species generated. Following Hercules[11], the scheme which describes the situation obtaining in alternating current electrolysis is shown below. For an outline description of the apparatus used the reviews[11,12] and references cited should be consulted.

(1) $A + e \rightarrow A^-$ Reduction of A at electrode surface
(2) $A \rightarrow A^+ + e$ Oxidation of A at electrode surface
(3) $A^+ + A^- \rightarrow {}^1A^* + A$ Reaction of radical cation and anion to form excited singlet state
(3a) $A^+ + A^- \rightarrow {}^3A^* + A$ Formation of lowest triplet state of A
(4) ${}^1A^* \rightarrow A + h\nu$ Fluorescence of A
(4a) ${}^3A^* + {}^3A^* \rightarrow {}^1A^* + A$ Triplet–triplet annihilation
(5) ${}^3A + Q \rightarrow A + Q$ Triplet quenching
(6) $A^+ + A^- \rightarrow {}^1A_2^*$ Reaction of cation and anion to give excimer
(6a) ${}^1A_2^* \rightarrow 2A + h\nu_e$ Excimer fluorescence
(7) $A^- + P^+ \rightarrow A^3 + P$ Reaction of radical anion with electrochemically generated oxidant to give triplet state of A
(8) $A^+ + D^- \rightarrow A^3 + D$ Reaction of radical cation with electron donor to give triplet state of A

A particular advantage of the use of alternating current is that short-lived species may still be present on reversal of the current. The rubrene radical anion is less stable than the radical cation. The quantum yield (based on a coulomb of electricity consumed) if the anion is first generated is 0·006 while the efficiency is more than twice as great at 0·015 if the more stable radical cation is first formed. The cation–anion reaction of Equation (3) can be viewed as the oxidation of the radical anion by an electron-deficient species. Both singlet and triplet states are possible and although phosphorescence from the triplet is rare, indirect evidence for its formation exists. For example, wherever there is insufficient energy for the direct formation of the singlet state, triplet–triplet annihilation is the most likely source of the observed fluorescence (Equations (4) and (4a)). The difference in electrode potential for oxidation and reduction gives (in volts) the energy available for population of the excited state. If the O–O energy for the ground–excited state transition is known, it is immediately apparent whether or not there is sufficient energy for the direct formation of the singlet. By assuming that quenching of the electrochemically generated triplet occurs readily in the solvents used, Chang et al.[20] have obtained evidence which indicates that, in addition to direct singlet formation, fluorescence as a result of triplet–triplet annihilation also occurs[21]. Since the excited state is formed at the instant of collision of radical anion and cation, it is reasonable to suppose[22] that an excimer would result. These complexes of excited and ground state molecules are of some theoretical interest, and are usually characterised by a broadened emission at a wavelength longer than that of the uncomplexed excited state. Although Parker and Short[23] obtained evidence for excimer formation in the electrolysis of dimethylanthracene, Werner et al.[24] suggest that the *direct* formation (Equation (6)) is not necessarily proved, and that the excimer may well result from collision of a singlet excited anthracene with a ground state molecule—the normal radiative route. In fact, these workers and others[25] have shown that emission from oxidation products of unsubstituted anthracene can be mistaken for excimer emission. Energy from excited anthracene is transferred to such an oxidation product (anthranol), which is strongly fluorescent at longer wavelengths (460 nm). A similar conclusion is reached concerning perylene[26], where the direction of the current flow is important. Acting as an anode, the electrode oxidises perylene to (presumably) hydroxylated fluorescent impurities, but chemiluminescence as emission from the singlet state of perylene occurs normally if the electrode is first negatively charged. The reactions of Equations (7) and (8) can be inferred from the fact that chemiluminescence is still observed even when the energy, as represented by the difference between the potential required to generate, say, the radical anion and the more positive potential generated by the electrode as the current sweep continues, is insufficient for singlet state formation. The reactants P and D in these equations appear to be generated electrochemically from the solvent or supporting electrolyte[27]. Good evidence for triplet formation was also obtained[28] for the oxidation of the phenanthrene radical anion, phosphorescence being observed, substantiated by quenching of the light by triplet quenchers of the appropriate energy.

OTHER ELECTRON TRANSFER REACTIONS

Earlier reviews[11–13] give details of other reactions which can be classified as involving an electron transfer in the excitation step. They can be remarkably general, and although the more efficient examples usually require electron abstraction from a pre-formed radical anion, similar reactions may occur in the course of oxidations proceeding by radical chains.

Oxidation of Radical Anions

Sodium 9,10-diphenylanthracenide, for example, reacts with a variety of electron acceptors to form the first excited singlet of diphenylanthracene[29]. The oxidants used include bromine, chlorine, benzoyl peroxide, oxaloyl chloride and *p*-toluenesulphonyl chloride. The reaction of the radical anion with 9,10-dichloro-9,10-diphenyl-9,10-dihydroanthracene is also chemiluminescent, and is thought to proceed as shown in *Scheme 4*.

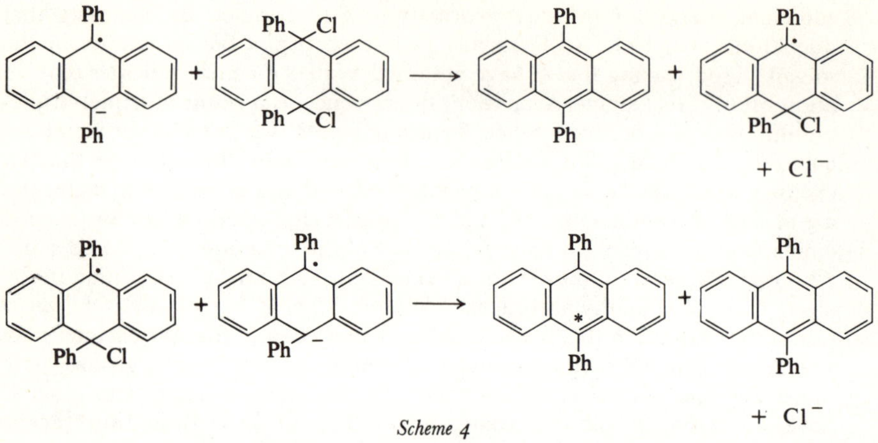

Scheme 4

Similar reactions occur between the radical anions of *N*-phenylcarbazole and *N*-methylacridone (the ketyl) and the electron acceptor benzoyl peroxide or *p*-toluenesulphonyl chloride.

Grignard reagents have long been known to emit light on oxidation by dry oxygen[30,31] or other electron acceptors such as chloropicrin[30,32]. A recent re-investigation[33] of the reaction suggests the following sequence (*Scheme 5*). The isolation of *p*-terphenyl as a principal fluorescent product accounts for the emission of visible light from a reaction involving compounds which absorb and emit in the ultraviolet region. Since the quantum yield[31b,32] is about 10^{-8} to 10^{-9}, interpretation is obviously difficult. Indeed it is likely that this is not a complete explanation since independent work[34] on the rather more

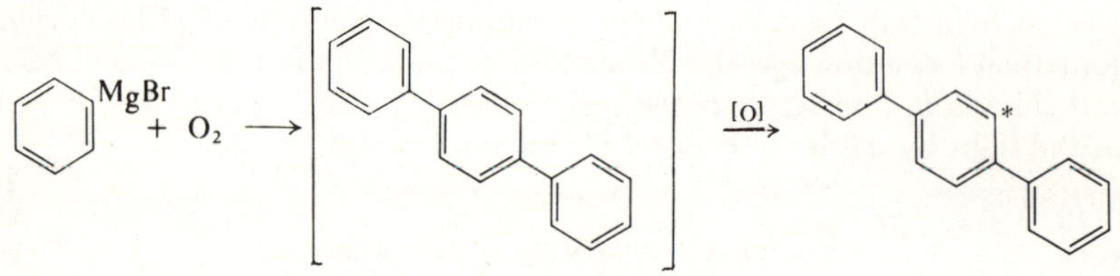

Scheme 5

efficient *p*-chlorophenylmagnesium bromide, while confirming certain aspects of the scheme shown, throws some doubt on the nature of the excitation step. For example, *p,p'*-dichloro-*p*-terphenyl is isolated as the major fluorescent product, while the emission observed occurs at slightly longer wavelength. In addition, the oxidation of the *p*-terphenyl anion (another isomer is obviously necessary in the chlorosubstituted case) is not the best candidate for the excitation step. Similar anions are easily generated by other means and are not chemiluminescent in proportion to the now much increased concentration (the yield of the *p*-terphenyls is less than 1%). Nevertheless, these terphenyls are virtually the only fluorescent products and it seems that energy transfer from an undiscovered excited product is not occurring since added terphenyl does not enhance the quantum yield in the expected manner. The oxidation of a radical anion as shown is thus the most attractive explanation[33].

The oxidation of sodium naphthalene by alkyl halides[35] is another example of an electron transfer reaction, and the scheme shown may have some implications for the Grignard chemiluminescence (Scheme 6).

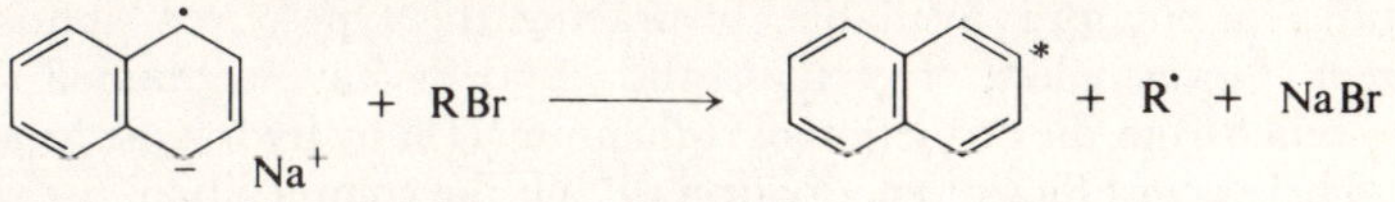

Scheme 6

A particularly interesting reaction of this type is the oxidation of perylene, chrysene and other aromatic radical anions by Wursters Blue cation[36] (Scheme 7).

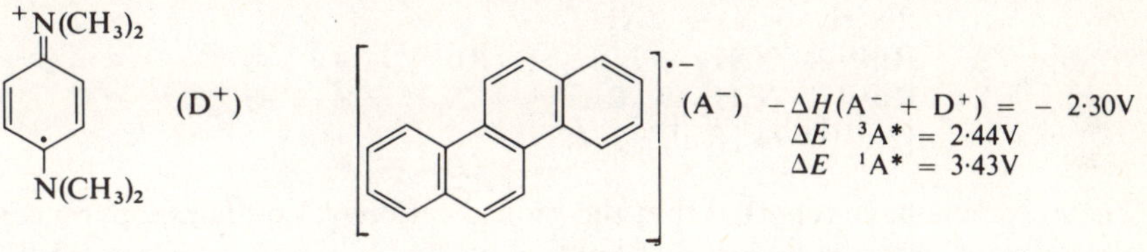

$(A^-) \quad -\Delta H(A^- + D^+) = -2.30V$
$\Delta E \ ^3A^* = 2.44V$
$\Delta E \ ^1A^* = 3.43V$

Scheme 7

It is easily established by electrochemical measurements that there is insufficient energy available from the oxidation to populate the first excited singlet state of the hydrocarbon, yet fluorescence from this state is observed. It is thus almost certain that the singlet state is the result of the triplet–triplet annihilation. In the case of perylene $-\Delta H$ for the oxidation is $+2.01$ V while ΔE for the singlet state–ground state is 2.83 V, and for the triplet state–ground state 1.55 V. There is thus only sufficient energy for triplet state formation. This view is confirmed by the observation of chrysene phosphorescence from the reaction with the chrysene radical anion.

A two-electron oxidation giving rise to chemiluminescence has recently been reported[37]. Oxidation of acridan by benzoyl peroxide results in fluorescence from the acridinium cation (E 58.6 kcal mol^{-1}) rather than from acridine itself (E 67.9 kcal mol^{-1}), probably the result of the difference in energy of the transitions involved (Scheme 8). Since the quantum yield is low (3×10^{-7}) this explanation should be accepted with caution, particularly since other possible two-electron chemiluminescent reactions are not observed[36].

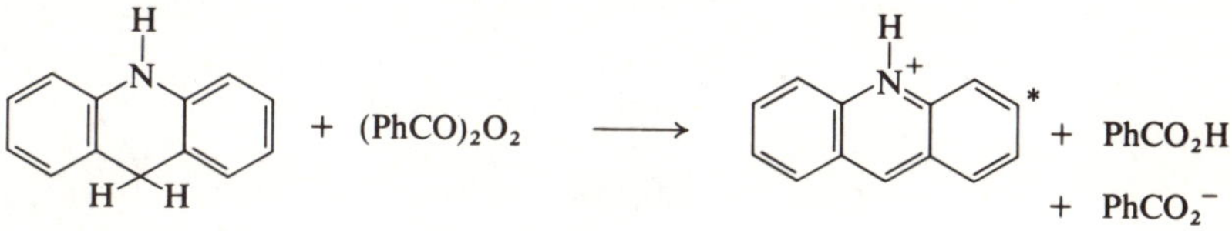

Scheme 8

Reduction Reactions

Although the electron transfers discussed above involve reduction, their nature is such that the excited state of the molecule from which the electron is removed is formed. In principle, reduction of an electron-deficient molecule by completion of its valency shell could be achieved by addition of the required electron to the lowest unoccupied molecular orbital, providing that sufficient energy is available. In practice this type of reaction is rarely observed. Nevertheless, Hercules and Lytle[38] have obtained chemiluminescence from the reduction of ruthenium(III) by hydrazine or aqueous alkali. There is insufficient energy overall, but the complexity of the reaction may result in certain of the intermediate steps (involving very unstable species) having the energy required to populate the excited state observed, that of the complexed ruthenium(II) ion. Their scheme is shown below, without ligands, which can be bipyridine or substituted phenanthrolines.

$$Ru^{III} + N_2H_4 \rightarrow [Ru^{II}]^* + N_2H_3 \quad \phi = 1\%$$
$$2N_2H_3 \rightarrow N_2 + 2NH_3$$
$$Ru^{III} + N_2H_3 \rightarrow N_2H_2 + [Ru^{II}]$$
$$Ru^{III} + N_2H_2 \rightarrow [Ru^{II}]^* + N_2 \quad \phi = 99\%$$
$$[Ru^{II}]^* \rightarrow [Ru^{II}] + h\nu$$

These workers have reported that the radical cation of 1,6-diaminopyrene is also reduced with the emission of light.

Another reaction which has been formulated[40] in a way which suggests a similar interpretation is that of tetralin peroxide and zinc tetraphenylporphine, the last step in the scheme shown being considered the reduction of the zinc tetraphenylporphine (ZnTPP) radical cation (*Scheme 9*).

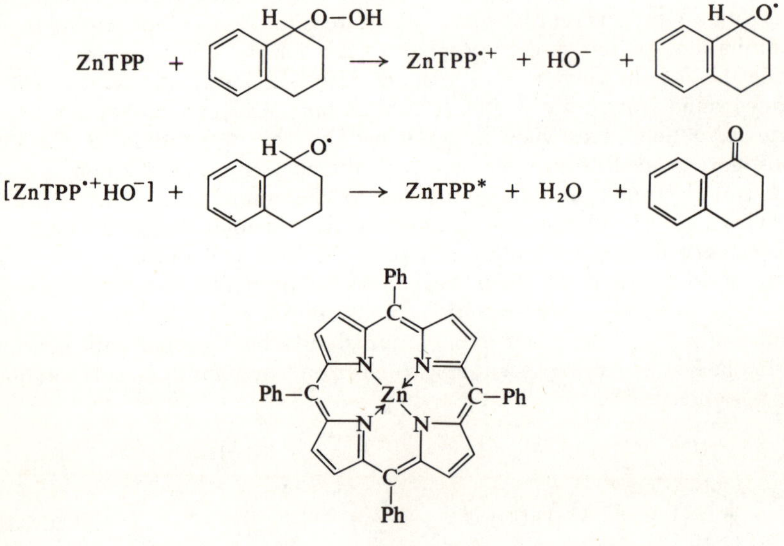

ZnTPP + [tetralin-O-OH] $\longrightarrow$ ZnTPP$^{\cdot+}$ + HO$^-$ + [tetralin-O$^\cdot$]

[ZnTPP$^{\cdot+}$HO$^-$] + [tetralin-O$^\cdot$] $\longrightarrow$ ZnTPP* + H$_2$O + [tetralone]

ZnTPP

Scheme 9

CHEMILUMINESCENCE FROM SINGLET OXYGEN[41]

The evolution of oxygen which occurs when sodium hypochlorite solution and hydrogen peroxide are mixed is accompanied by the emission of red light. Although some of the maxima in the emission had been observed by earlier workers[42], the assignment of the bands was not immediately apparent. Khan and Kasha[43] suggested that the prominent bands at 635 nm and 703 nm were the solvent shifted (0·0) and (0·1) bands of the $^1\Delta_g \leftrightarrow {}^3\Sigma_g^-$ transition. Ogryzlo and his co-workers[44] made the now fully accepted assignment of the band system to a simultaneous transition from *two* $^1\Delta_g$ excited singlet oxygen molecules, such an event being spin allowed in this case, and forbidden for single molecules. Though Khan and Kasha[45], and more recently Khan[46], have suggested that the formation of singlet oxygen is a general mechanism for chemiluminescence, most efficient reactions studied recently do not belong to this class. In the reaction of chlorine with alkaline hydrogen peroxide emission from the $^1\Delta_g \cdot {}^1\Sigma_g^+$ pair is seen. It is then possible to construct a 'manifold' of oxygen excited states capable of exciting a variety of fluorescent acceptors (*see Scheme 10*). An objection to this scheme for acceptors which require the participation of $^1\Sigma_g^+$ oxygen is that in view of the short lifetime of 'double molecules' containing the $^1\Sigma_g^+$ species, the probability of energy transfer approaches that of termolecular collisions.

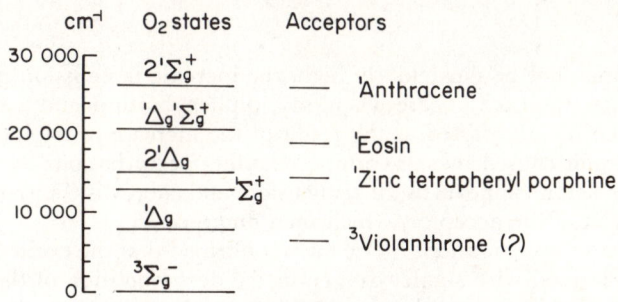

Scheme 10

Chemiluminescent reactions with direct emission from the 'dimol' are relatively uncommon but some interesting examples are shown in *Scheme 11*. Probably the most important of these is the chain termination of autoxidation reactions by the 'Russell mechanism'[47] (Equation 1). This reaction is particularly significant in the many cases of relatively weak chemiluminescence accompanying hydrocarbon and polymer oxidation[48]. Vasil'ev[49] had earlier suggested that the autoxidation of a variety of aliphatic and aromatic hydrocarbons, especially those such as ethylbenzene, tetralin and diphenylethane, resulted in the formation of the major product (e.g. acetophenone, *see Scheme 11*) in an excited state.

The reaction is accompanied by a blue–green emission (420–450 nm), but the quantum yield is very low (10^{-8}–10^{-10}) and the exact nature of the emitter must remain in some doubt. Radical initiation is obviously helpful, but the most valuable observation is the particular effect of heavy atom 'activators' such as 9,10-dibromoanthracene. Although the fluorescence

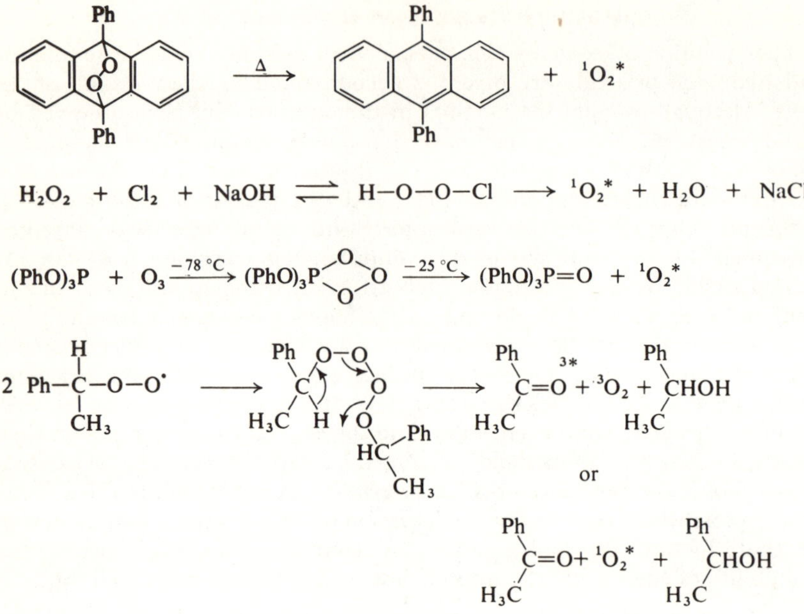

Scheme 11

efficiency may not be particularly high, the increase in emission intensity is much greater than is the case when, say, diphenylanthracene is used. This enhancement is interpreted as the result of the increase in triplet to singlet energy transfer caused by spin–orbital coupling perturbations by the heavy atoms. The ketone appears in the triplet state and energy is transferred to the lowest singlet of the acceptor, which then fluoresces.

More recent investigations[48,50–52] have confirmed that the excited carbonyl product is formed with singlet oxygen in the decomposition of the unstable tetra-oxide. Evidence for this is provided by the following observations: (a) the termination rate constant for secondary radicals (as shown) is 10^3 less than that for tertiary radicals; (b) K_H/K_D for termination is 1·9; (c) the characteristic products[53] of singlet oxygen can be obtained in the Ce^{IV} catalysed decomposition of secondary peroxides. It is likely that emission from both excited molecules is occurring, but efficiency is low. It seems that the excited carbonyl product is quenched in the reaction cage by the oxygen formed, leading to inefficient *free* triplet formation. Triplet *formation* may well occur with near unit efficiency—a strong indication that spin conservation is a controlling factor. However, a recent investigation of the autoxidation of dimedone[54] contradicts some of the assumptions made by Kellogg, and the author concludes that the reaction is considerably less efficient in populating the excited state than was at first thought. A value of $\phi_E = 4·5 \times 10^{-8}$ is derived from this study, which is remarkably low. It is obviously difficult to establish whether both excited states are formed in the decomposition, or whether only triplet carbonyl products result, the singlet oxygen observed being produced by triplet carbonyl to triplet (ground state) oxygen energy transfer.

Another reaction with very similar characteristics is the decomposition of alkylidene peroxides[55,56]. Kurz originally suggested[55] for the sensitised reaction using dibenzanthrone (violanthrone) that oxygen produced in the decomposition formed the endoperoxide (I, Scheme 12). The reaction was carried out by adding dibenzal diperoxide to a solution of dibenzanthrone in paraffin oil at 200°C, and the resulting endoperoxide was assumed to dissociate rapidly into ground state oxygen and triplet dibenzanthrone. However, it has now been shown[56] that this rather unlikely sequence of events does not occur. Other acceptors more likely to form endoperoxides (e.g. 9,10-diphenylanthracene) are less effective sensitisers, and the emission observed corresponds to the fluorescence of dibenzanthrone, not the phosphorescence. There is a very weak emission without acceptor, but a spectrum was unobtainable even when an image intensifier spectrograph was used. After a detailed consideration of possible mechanisms, the later workers

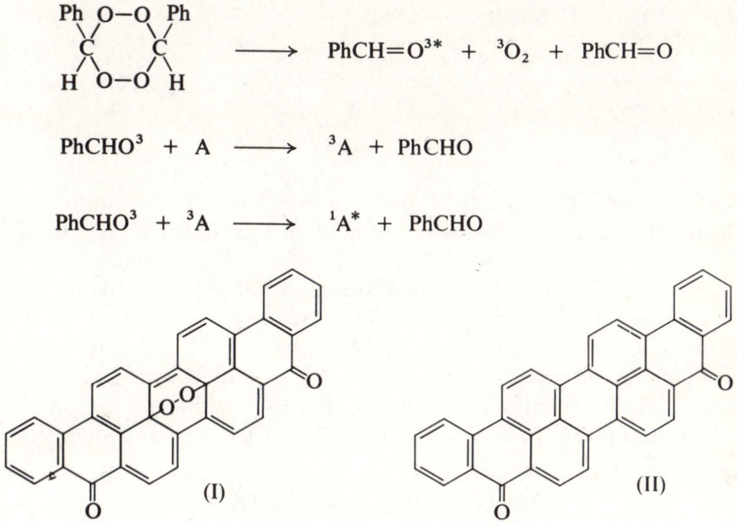

$$\text{Ph}_2\text{C(OO)}_2\text{CHPh} \longrightarrow \text{PhCH=O}^{3*} + {}^3\text{O}_2 + \text{PhCH=O}$$

$$\text{PhCHO}^3 + \text{A} \longrightarrow {}^3\text{A} + \text{PhCHO}$$

$$\text{PhCHO}^3 + {}^3\text{A} \longrightarrow {}^1\text{A}^* + \text{PhCHO}$$

Scheme 12

conclude that long range dipole–dipole energy transfer from triplet benzaldehyde to give the excited singlet acceptor is a major source of the light seen. Evidence was also obtained for the $^1\Delta_g\text{O}_2$ pair, but these authors feel that an energy pooling process involving *successive* energy transfer from single $^1\Delta_g\text{O}_2$ molecules is more likely. Triplet–triplet annihilation then provides the singlet excited state of the acceptor, a fairly common final step in moderately strong chemiluminescence.

These reactions of peroxides form singlet oxygen and singlet carbonyl or triplet (ground state) oxygen and triplet (excited) carbonyl inevitably if the reaction is concerted. In fact most heterolytic reactions forming oxygen which are at least 22·4 kcal mol^{-1} exothermic will be likely sources of singlet oxygen, as pointed out by McKeown and Waters[57]. They observed weak chemiluminescence from the reaction of nitriles with alkaline hydrogen peroxide and, following Wiberg, depicted the formation of the singlet oxygen as shown in Equation (1) (*Scheme 13*). An alternative formulation has been given by House[58], but a third possibility, hitherto unconsidered, is also shown (2). A large number of heterolytic reactions may also give rise to a much weaker chemiluminescence which may be associated with singlet oxygen formation[59]. However, as in so many cases of weak chemiluminescence, certain identification of the mechanism is difficult and this type of reaction will not be discussed further.

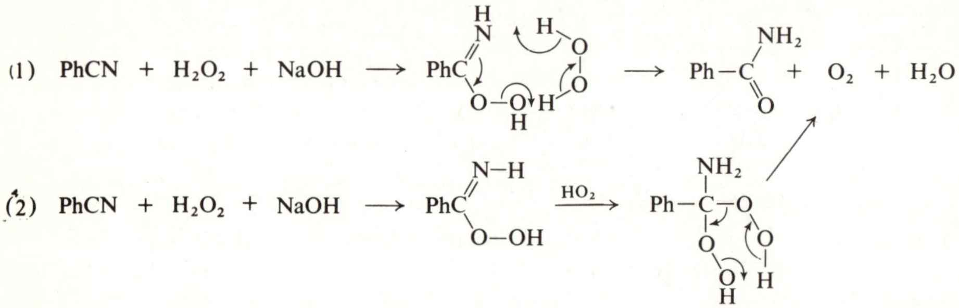

(1) $PhCN + H_2O_2 + NaOH \longrightarrow PhC\begin{smallmatrix}NH\\\\O-OH\end{smallmatrix} \cdots \longrightarrow Ph-C\begin{smallmatrix}NH_2\\\\O\end{smallmatrix} + O_2 + H_2O$

(2) $PhCN + H_2O_2 + NaOH \longrightarrow PhC\begin{smallmatrix}N-H\\\\O-OH\end{smallmatrix} \xrightarrow{HO_2} Ph-C\begin{smallmatrix}NH_2\\\\\cdots\end{smallmatrix}$

Scheme 13

Naturally enough, a study of singlet oxygen induced chemiluminescence is made much easier by the generation of the excited oxygen in quantity. This is most conveniently achieved by use of a radiofrequency or microwave discharge. Ogryzlo and co-workers[60] using a 2450 MHz discharge can obtain 6% $^1\Delta_g O_2$ in the effluent oxygen. They have made a study of the brightest of all such chemiluminescent oxygen reactions which uses violanthrone (II) as acceptor. As originally suggested by Khan and Kasha[45], singlet oxygen pairs provide a 'manifold' of energy donors capable of exciting a large variety of acceptors (*see Scheme 10*). However, Ogryzlo has shown that of the

(1) $2\,^1\Delta_g O_2 + {}^1V_0 \longrightarrow 2\,^3\Sigma_g^- O_2 + {}^1V_1$

(2) $^1\Delta_g O_2 + {}^1V_0 \longrightarrow {}^3\Sigma_g^- O_2 + {}^3V_1$
$\ {}^3V_1 + {}^1\Delta_g O_2 \longrightarrow {}^3\Sigma_g^- O_2 + {}^1V_1$

(3) $^1\Delta_g O_2 + {}^1V_0 \longrightarrow {}^3\Sigma_g^- O_2 + {}^3V_1$
$\ {}^3V_1 + {}^3V_1 \longrightarrow {}^1V_1 + {}^1V_0$

Common final step:
$$^1V_1 \longrightarrow {}^1V_0 + h\nu$$

(V is violanthrone)

Scheme 14

three possibilities given in outline in *Scheme 14* reaction (2) is the most likely, and that the direct transfer from $2\,^1\Delta_g O_2$ molecules to form singlet violanthrone is definitely excluded. Reactions (2) or (3) are further examples of the energy pooling process required to explain the observation of fairly energetic visible light from energy-deficient sources, in this case excited oxygen. Nevertheless, there can be doubts concerning this interpretation since it depends on the population of a hitherto unobserved triplet state of the acceptor of rather low (22·4 kcal) energy. Support for this view has come, however, from a similar study using a more extensive series of acceptors[61]. The reaction using rubrene was studied in some detail and the conclusions are outlined in *Scheme 15*. The author feels that energy transfer from $2^1\Delta_g O_2$

$$^1R_0 + {}^1\Delta_g O_2 \longrightarrow {}^3R_1 + {}^3\Sigma_g^- O_2$$
$$^3R_1 + {}^1\Delta_g O_2 \longrightarrow {}^1R_1 + {}^3\Sigma_g^- O_2$$
$$^1R_1 \longrightarrow {}^1R_0 + h\nu$$
$$^1R_0 + {}^1\Delta_g O_2 \longrightarrow RO_2 \text{ (endoperoxide)}$$

(R is rubrene)

Scheme 15

molecules is not ruled out completely. Although the triplet state energy of rubrene is not known, that of the parent naphthacene is 29 kcal and the existence of a rubrene triplet sufficiently low in energy to accept energy from the $^1\Delta_g O_2$ molecule is entirely possible. The direct formation of rubrene singlet (whose fluorescence constitutes the observed chemiluminescence) is not likely, there being a deficit of 8 kcal mol^{-1}. All reactions which produce singlet oxygen in good yield also show this type of chemiluminescence, although with the exception of violanthrone and the chlorine–hydrogen peroxide system they must be considered rather inefficient. This is largely a result of the need to generate the easily quenched $^1\Sigma_g^+$ oxygen or triplet ketone in order to excite higher energy acceptors.

A third reaction sometimes thought to be chemiluminescent as a result of singlet oxygen release is that of the aromatic endoperoxides. Singlet oxygen can indeed be trapped in the usual way[62] but, as will be discussed later, the emission seen most probably arises from another type of reaction.

REACTIONS OF PEROXIDES

The decomposition of organic peroxides has long been associated with extremely, sometimes violently, exothermic reactions. It must be remembered, however, that the large release of energy is not necessarily so dramatic when viewed as a single molecular event. Chain processes occurring in a short time account for the violence. Light corresponding to the release of about 85 kcal mol^{-1} [33], an amount of energy exceeding that in most chemical bonds, is in many ways more spectacular. The involvement of peroxides has been indicated earlier, but these examples all suffer from a preferential formation of triplet states or singlet oxygen. Neither of these species radiates efficiently, and although efficient energy transfer can undoubtedly be observed[56,63], in most cases it cannot be expected to lead to efficient population of excited states owing to spin restrictions. For bright chemiluminescence *direct* singlet state formation is obviously desirable and, of course, molecules so formed should be highly fluorescent. Although structural changes in this class of peroxide lead to a range of efficiencies, values of ϕ_E as high as 60% are possible. By including bioluminescence in the type (and there is excellent evidence in favour of this), overall quantum yields as high as 90% are to be found.

Hydrazide Chemiluminescence

Most of the recent reviews have described studies on these compounds in some detail[10,64,65], and only an outline together with more recent work will be included here. Many workers have added to the number of oxidants which catalyse the reaction or have prepared derivatives which do not significantly advance knowledge of the basic reactions involved. For reasons of space it is not possible to include these.

Luminol (5-amino-2,3-dihydro-1,4-phthalazinedione, (III)) was first reported by Albrecht[66] to emit a bright blue light on oxidation by alkaline ferricyanide in the presence of hydrogen peroxide. This oxidation can be catalysed by a wide variety of transition metals and oxidising agents such as hypochlorite and persulphate, and is usually carried out in aqueous solution, although co-solvents such as ethanol can be used. A more general and milder system was introduced by White[65,67], leading to a substantial increase in the information available from the reaction. Strong base (potassium t-butoxide

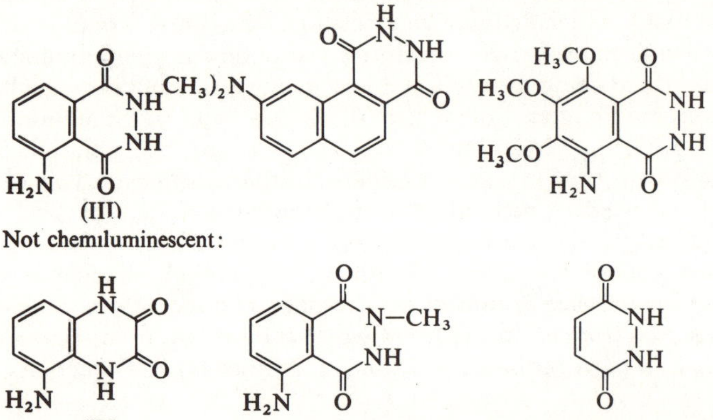

Scheme 16

for example) in aprotic solvents such as dimethylformamide or, more particularly, dimethyl sulphoxide allows direct reaction of the luminol anion with oxygen, with a high yield of product and few by-products. Cyclic hydrazides of this general class (*see Scheme 16*) are only chemiluminescent if the heterocyclic ring is completely unsubstituted. Other isomers, e.g. those based on quinoxalines (IV), are also totally ineffective. Although, as will be seen later, linear hydrazides can be strongly chemiluminescent, cyclic hydrazides have been more thoroughly studied.

It seems likely that the reaction paths in the aqueous and aprotic solvent systems converge, and that the early stages in water (which usually require transition metals) involve free radical oxidation[67]. In dimethyl sulphoxide and strong base luminol reacts as shown in *Scheme 17*. The principal features to note[65] are that: (a) the *di*anion is the species which reacts with oxygen, (b) both oxygen atoms appear virtually quantitatively in the product, (c) light emission at 485 nm corresponds exactly to the fluorescence of the aminophthalate dianion, (d) the reaction is first order in base, oxygen and

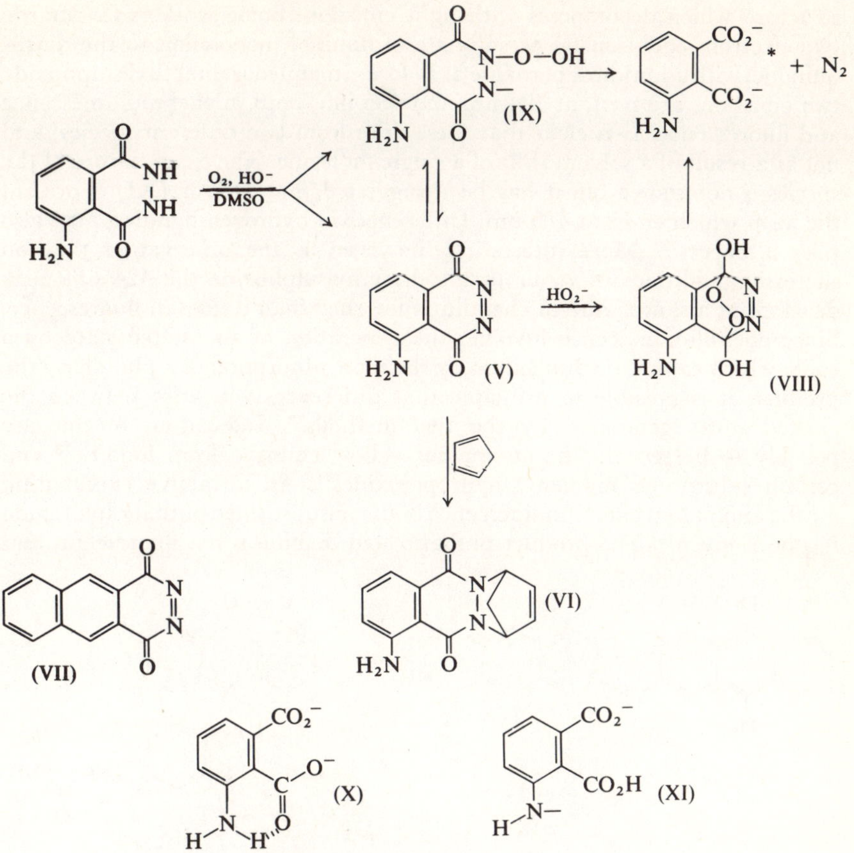

Scheme 17

luminol. Recent efforts to identify the actual excitation step have demonstrated that the diazaquinone (V) is an intermediate under certain reaction conditions. In what is essentially an aqueous system, Omote, Miyake and Sugiyama[68] found that addition of cyclopentadiene effectively quenched chemiluminescence, and they isolated the compound (VI) in confirmation of a diazaquinone intermediate. It is not clear whether the reaction of the dianion with oxygen in dimethyl sulphoxide occurs via the diazaquinone and the hydrogen peroxide so formed or whether it forms a peroxide directly. The isolation of stable diazaquinones[69] such as (VII) and addition of hydrogen peroxide provides a common point in the aqueous and aprotic systems. However, none of these findings provides proof of the nature of the intermediate which decomposes to the excited dicarboxylate dianion. Although (VIII) is a most attractive precursor[70,71] other formulations (e.g. (IX)) can not be ruled out. A study of the reaction of luminol with persulphate in aqueous solution shows that it is first order in luminol and persulphate and zero order in base and hydrogen peroxide, although the quantum yield is strongly dependent on the last reagent[70]. It seems reasonable to assume that there are several routes to the peroxide of unknown structure which decomposes with light emission. Some workers favour the two-electron oxidation by persulphate of luminol monoanion to the diazaquinone, with addition of peroxide as before. In aqueous dimethyl sulphoxide two emissions are seen, at 424 nm and 485 nm, both in chemiluminescence and fluorescence. It is clear that these arise from two different species, and not as a result of a solvent shift of a single molecule. The exact nature of the species is not known but it has been suggested that (X) or (XI) represent the form which emits at 485 nm. Differences in hydrogen bonding may also play a part[65,72]. More interesting, however, is the observation that on successive additions of water to the dimethylsulphoxide the 424 nm peak develops at a slower rate in chemiluminescence than it does in fluorescence. Since chemiluminescence involves the generation of an excited state by a route which cannot be duplicated by the mere absorption of a photon by the product, it is possible in principle that differences will arise between the excited states generated by the two methods[72]. Indications of this are possibly to be seen in the anomalous yellow emission from lophine[73] and certain substituted indolenyl hydroperoxides[74]. An attractive explanation for the anomalous chemiluminescence of the unsubstituted phthalic hydrazide has been given[75]. The product phthalic acid dianion is not fluorescent, and

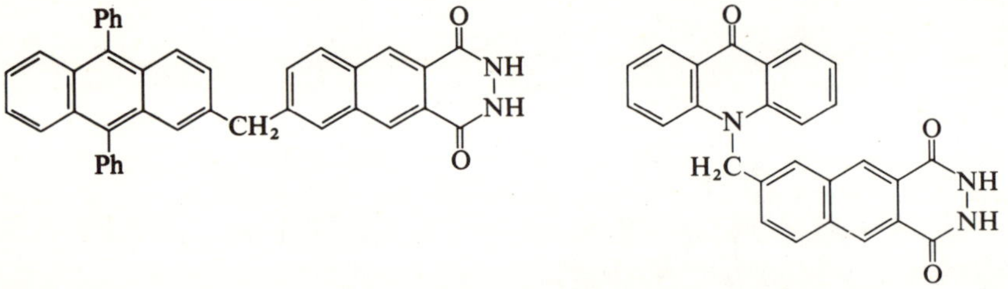

Scheme 18

moderately strong yellow light (525 nm) is observed only in fairly concentrated solutions. The intensity is dependent on concentration and the spectrum corresponds to the fluorescence of the monoanion of the starting hydrazide. It is therefore suggested that energy is transferred from excited phthalate dianion to the monoanion. The authors point out, however, that this explanation is not without problems since other potential energy acceptors do not enhance the emission.

The hydrazides have been used to investigate energy transfer in general with the excited phthalate dianion formed in the chemiluminescent reaction causing an attached fluorescent molecule to emit. Examples are shown in *Scheme 18*. It appears that triplet–singlet transfer of a mixed dipole–dipole and exchange type is the most likely mechanism[63,76].

Linear hydrazides with rather low chemiluminescent efficiency have been known for some time[77]. Simple benzenoid hydrazides are much less efficient than luminol, but recent examples[78] ((XII) and (XIII)) approach luminol in efficiency, leading to the possibility that cyclic hydrazides are only a special case of hydrazide chemiluminescence. At the present time the mechanism of light emission is not clear but, as in the case of the cyclic hydrazides, the possibility of nitrogen release and the involvement of hydrogen peroxide seem essential. For example, (XIV) is not chemiluminescent[78]. An electron transfer mechanism has been suggested[79] for linear hydrazides, although this has as yet no unambiguous support. The enormous energy released by the simultaneous formation of nitrogen and a carbonyl group at the expense of weak bonds (it can be as high as 135 kcal) makes the

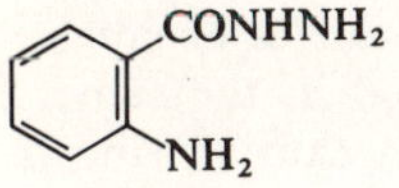

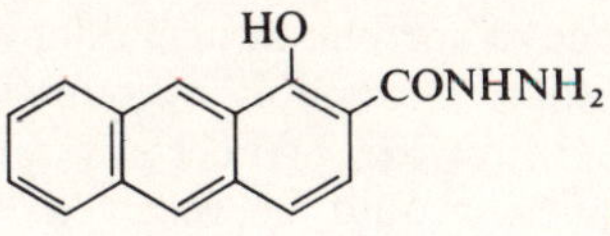

$$R \cdot CONHNH_2 \xrightarrow[\text{or } H_2O_2]{O_2} R \cdot CO_2^{-*} + N_2$$

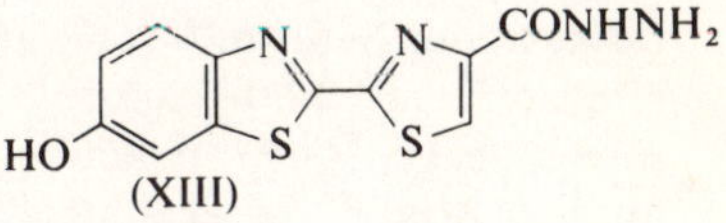

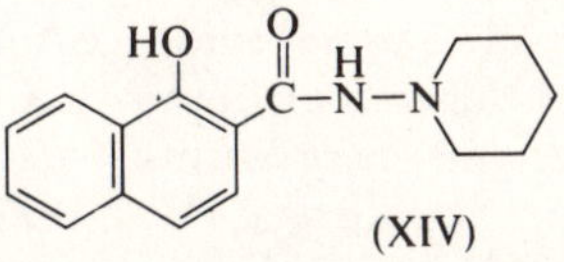

Scheme 19

alternative concerted decompositions (*Schemes 17* and *19*) attractive. Nevertheless, there is no convincing explanation for the fairly high population of the excited state and much work remains to be done.

Dioxetane Chemiluminescence

Interest in 1,2-dioxetanes (e.g. (XV)) as possible sources of chemiluminescence was aroused by the suggestion[80] that many existing chemiluminescent reactions may involve dioxetanes as intermediates. Although these compounds may provide a particular reason for discrimination against the ground state, a more basic property may well be the extremely exothermic generation of two carbonyl groups within a single transition state[78,81]. The exothermicity of the reaction, as roughly estimated from bond strengths, varies from about 80 to 110 kcal mol^{-1}, depending on substitution. Extended Hückel M.O. calculations give higher values[82] but the accuracy of these is not established. A more thorough examination of the thermochemistry of dioxetane decomposition[83] based on the group factors of Benson *et al.*[84] gives values of $\Delta H = -75.9$ kcal mol^{-1} for the parent compound and -93.5 kcal mol^{-1} for tetramethyl dioxetane. (R = H and R = CH$_3$ respectively.)

These values include an allowance for strain energy (26 kcal mol^{-1}) and activation energy (21·5–24·7 kcal mol^{-1}). The last figures agree rather well with the experimental value for the thermal decomposition of trimethyl dioxetane[85] (23·7 kcal mol^{-1}) as does that for log A (12·2 as against 12·4 for the calculated value). Since the calculations are based on a pathway involving a diradical (*see Scheme 20*), this mode of decomposition is indicated. This point will be taken up later.

Kopecky and Mumford[85], reporting the first isolation of a simple dioxetane (XVI), also noted that light was emitted when the compound was heated in benzene solution ($t_{\frac{1}{2}}$ at 60°C is about 20 min). Energy transfer to fluorescent acceptors was also observed, but no quantum yield was reported.

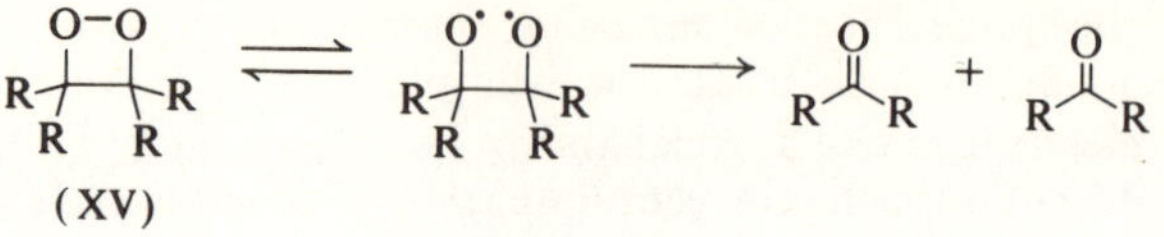

[Scheme 20 diagram]

Scheme 20

The emission from the pure compound consisted of a broad band, λ_{max} 430 nm, whereas the fluorescence of acetone or acetaldehyde was thought to be at about 345 nm[86]. The authors reasonably account for this as excimer emission since both carbonyl groups are of necessity formed in the same solvent cage. This attractive view should be received with caution since another study[87] of acetone fluorescence does not support the conclusion. There is in fact no evidence for acetone excimer formation in solution. Traces of fluorescent impurity are always likely in such highly exothermic reactions, and energy transfer to these may be occurring.

Bartlett and Schaap[88] and Foote[89] have prepared and isolated pure dioxetanes by the addition of singlet oxygen at low temperatures to oxygen-substituted ethylenes. Wilson and Schaap[90] in a thorough study of one of

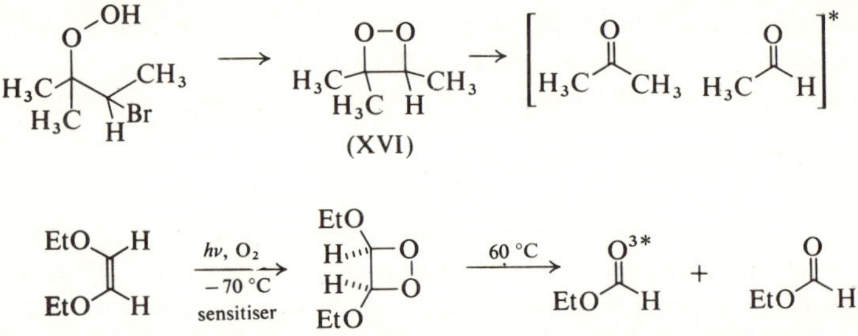

[Scheme for compound XVI]

Scheme 21

these, *cis*-diethoxy-1,2-dioxetane, have shown that it decomposes quantitatively to ethyl formate (*Scheme 21*). In the presence of an acceptor such as diphenylanthracene light is emitted ($\phi = 7 \times 10^{-6}$). Significantly, the quantum yield with 9,10-dibromoanthracene is much larger (5×10^{-4}) and, together with other evidence, strongly indicates a high yield of triplet formate. In fact the authors suggest that triplet formate is formed in nearly quantitative yield, the observed quantum yield being the result of inefficiencies in energy transfer. This intriguing conclusion may need acceptance with care since it depends in part on estimated rate constants. There is probably sufficient energy to achieve the n,π* singlet state of the

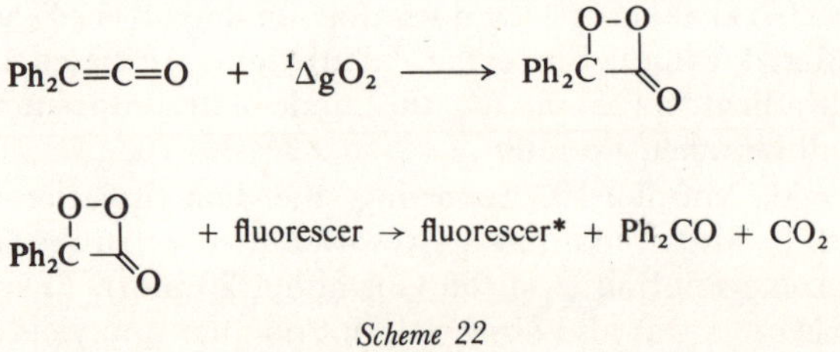

$$Ph_2C=C=O + {}^1\Delta_g O_2 \longrightarrow Ph_2C\underset{O}{\overset{O-O}{\diagup\!\!\!\diagdown}}$$

$$Ph_2C\underset{O}{\overset{O-O}{\diagup\!\!\!\diagdown}} + \text{fluorescer} \rightarrow \text{fluorescer*} + Ph_2CO + CO_2$$

Scheme 22

formate, but there is no evidence as to whether the observed triplet state results from intersystem crossing from this or is formed directly.

Other workers[82] have added singlet oxygen to ketenes (Scheme 22) and inferred the intermediacy of oxodioxetanes from the products formed. Light also results from energy transfer to fluorescent acceptors with moderate

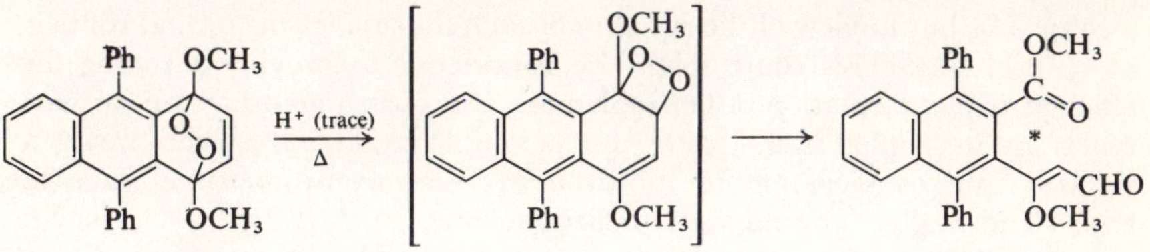

Scheme 23

efficiency ($\phi = 10^{-3}$–10^{-4}). Wilson[91] and Lundeen and Adelman[92] in an independent study of the Dufraisse[93] endoperoxide (XVII) conclude that a rearrangement is a key step in the chemiluminescence. The immediate product of the decomposition[94] is (XVIII), and it has been suggested by all the investigators that it proceeds through a dioxetane as shown (Scheme 23).

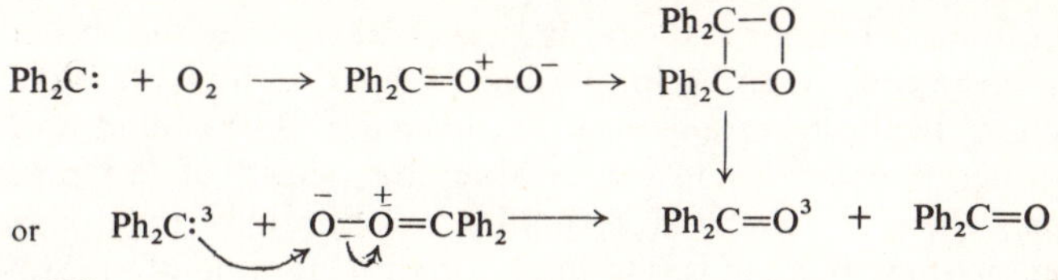

$$Ph_2C: + O_2 \longrightarrow Ph_2C=\overset{+}{O}-O^- \longrightarrow \begin{array}{c} Ph_2C-O \\ | \quad | \\ Ph_2C-O \end{array}$$

$$\downarrow$$

or $\quad Ph_2C:^3 + \overset{-}{O}-\overset{+}{O}=CPh_2 \longrightarrow Ph_2C=O^3 + Ph_2C=O$

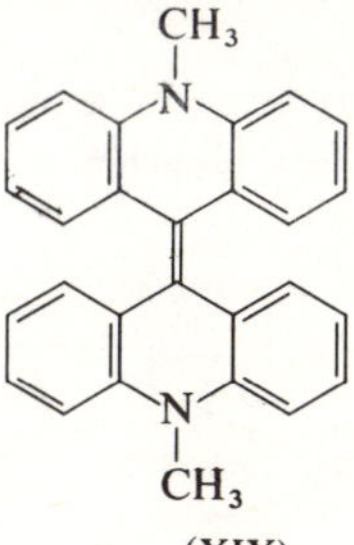

(XIX)

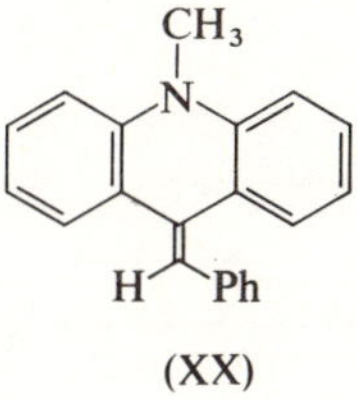

(XX)

Scheme 24

In particular, Wilson has demonstrated that emission is from the fully aromatic compound formed by loss of oxygen, the fluorescence of this compound corresponding to the emission. Such loss of oxygen is not the excitation step since the rearrangement shown is necessary for light emission.

Another probable dioxetane reaction which is highly chemiluminescent is that of dimethylbiacridene (XIX) or the related compound (XX)[95]. Since one of the products in each case (*N*-methylacridone) is highly fluorescent, efficient emission is possible. The presumed intermediate dioxetane cannot be isolated in these cases, although the reason for this lack of stability is not quite clear. The reaction of diphenylcarbene with oxygen at low temperatures[96] results in the phosphorescence of benzophenone. A possible reaction path is through the dioxetane formed from the intermediates shown (*Scheme 24*), but in view of the stability of such dioxetanes the second route is more probable. This route might be considered to provide a reason for efficient triplet formation of benzophenone if it is assumed that the carbene reacts in its triplet state, while the peroxide zwitterion is necessarily a singlet. Spin conservation in the atom transfer transition state leads to triplet and singlet (ground state) benzophenone.

Since other reactions discussed later may involve dioxetane intermediates, it is obvious that the reasons for their efficacy in chemiluminescence are worth studying. In the first place it is clear that an enthalpy change of greater than 100 kcal mol^{-1} is difficult to 'store' since it is larger than almost all bond strengths in organic compounds. Some kind of 'safety catch' is desirable and this is provided by symmetry prohibitions. The 'tiger in a paper cage' referred to by Woodward and Hoffmann[97] is in a similar position. If the reaction takes place in two steps then the activation energy provided by the prohibition ensures that the fragments are formed with an energy greater than that of an accessible excited state of one of them. If spin relaxation occurs at the diradical stage, then the resultant excited state must necessarily be a triplet. This may occur with simple dioxetanes. Whilst it may not be correct to assume a concerted decomposition in view of the prohibitions mentioned above, it is very likely that the second step would follow so rapidly as to be impossible to separate from the first. If the bond stretching in the transition state involves the O—O bond and this permits access to an excited state lower than the 'crest' of the potential energy surface, then a concerted formation of products may occur. Complex dioxetanes may react in this fashion. Finally, Hammond's postulate[98] may be

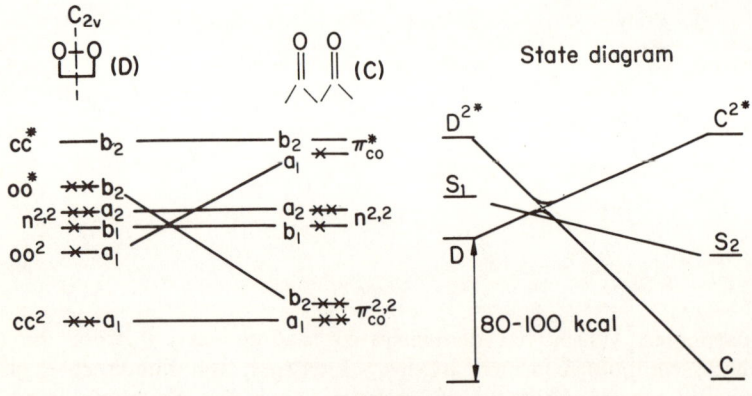

Scheme 25

S_2 is an n,π* state of one of the two carbonyl (C) compounds, either singlet or triplet, and S_1 is a correlated state of the dioxetane (D). It may be of interest that the n,π* transition is allowed in this case, and that early breakage of the O—O bond may lead to a lowering of the energy of S_1

considered as being of some interest in this context. Obviously this is an extreme example of an exothermic reaction, and as such the transition state should occur very early along the reaction co-ordinate. The geometry of the transition state will strongly resemble that of the reactant and in this case such a configuration is closer to the excited state of a carbonyl product than its ground state[99]. The correlation diagram serves merely to illustrate these ideas (Scheme 25). The state of the product chosen is a π,π^* or n,π^* state of low energy, and it is assumed that the localised decomposition of the dioxetane merges gradually with the 'hinterland' of orbitals in the complex fluorescent substrate. In simple dioxetanes the n,π^* state will be lower, and both singlet and triplet versions of this may be expected to cross the potential energy surface at a suitable point. It is probably not profitable to enquire as to the nature of such a crossing, or to view the diagram as more than an indication of events. Such surface crossings are notoriously difficult to understand in complex molecules. In essence, the dioxetane structure imposes an upward direction on the reaction leading to possible entry into an excited state which may be of lower energy than is usual. The lowering of energy may be the result of excimer formation or the simple consequence of the non-crossing rule.

Imine Peroxides

A sufficient number of examples of this type of chemiluminescent compound now exists to justify their description as a class. Lophine[100], 2,4,5-triphenylimidazole, is the earliest example of the type, and is indeed the earliest reported case of a chemiluminescent organic compound, having been discovered in 1877. Treatment of lophine with base in ethanol in

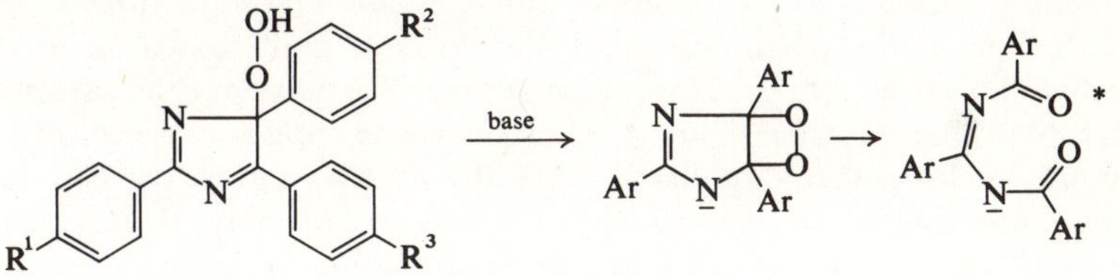

Scheme 26

air gives a long-lived yellow light. A peroxide (XXI) is undoubtedly involved and isolation[73,101] of the pure peroxide and its strong, brief chemiluminescence in basic ethanol confirms this (Scheme 26). The emission observed (λ_{max} 525 nm) is anomalous, and no material whose fluorescence corresponds to this has ever been found. However, a substituted lophine emits light whose spectrum is identical to that of the corresponding substituted dibenzoylbenzamidine. The mechanism shown is supported by the reaction of the peroxide in dry dimethyl sulphoxide.

Indolenyl peroxides[102,103] are similarly chemiluminescent, as are the parent indoles in basic solution in the presence of oxygen[104]. Detailed studies[102,103] of the peroxides of dialkyl substituted indoles (e.g. R = CH_3) suggest the sequence shown (Scheme 27).

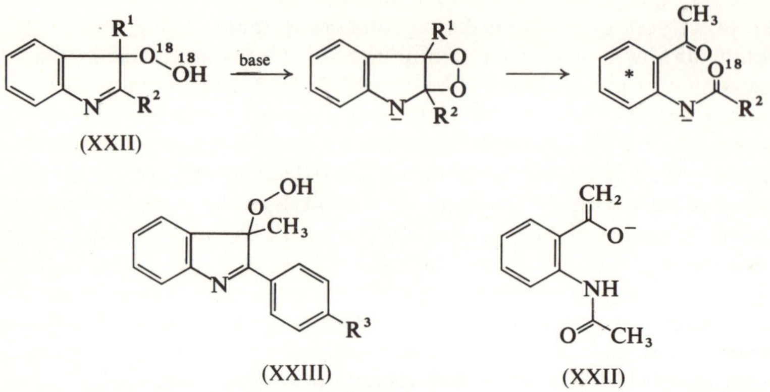

(XXII)

(XXIII)　　　　　　(XXII)

Scheme 27

When $^{18}O_2$ is used as a label and the reaction is carried out in highly aqueous dimethyl sulphoxide there is a strong indication[102] that the sole reaction path involves a transient dioxetane. The amide anion is certainly the emitter in this case since exact correspondence between its fluorescence spectrum and that of the chemiluminescence is obtained. The suggestion[103] that the anion is (XXII) is not reasonable since the amide has a pK_a at least two units lower than that of the methyl ketone[105]. Synthesis of 2-phenyl substituted indoles has made possible an examination of the effect of substitution on luminescence efficiency. Electron-donating groups slow the reaction moderately (XXIII, $R_3 = NMe_2$) and enhance the fluorescence efficiency of the anion. Total light yield is enhanced by a greater amount than can be accounted for merely by this increase, and some direct effect of substitution on the partitioning between ground and excited states seems possible. This is an extremely interesting area for study, but the substitution effects described earlier[106] are not as simple as previously thought. For example, *p*-chlorophenylindolenyl peroxides do not emit at the wavelength of the anion shown in *Scheme 27*, showing yellow light only. Reasons for this can be discerned but are too complex for inclusion here[105].

In response to the problem raised by the surprising simplicity of Latia luciferin (discussed later) certain Schiff bases were tested for chemiluminescence[107,108] (*Scheme 28*). Whilst only moderately efficient at best, the

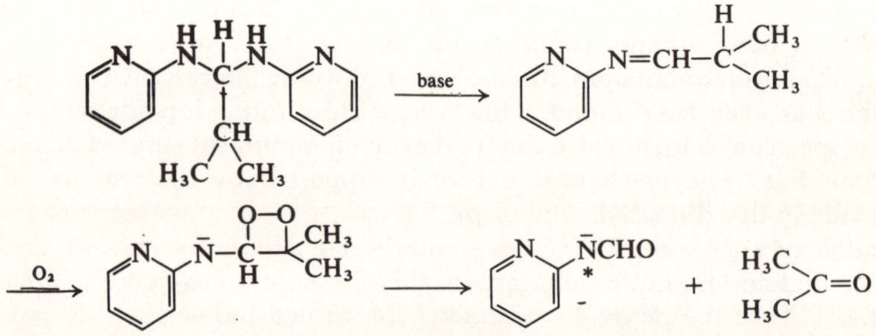

Scheme 28

reaction is simple and the excited molecule is clearly identifiable. The formamidopyridine emission at 395 nm is among the most energetic yet obtained from this type of reaction, and an estimate of the enthalpy available from a variety of possible mechanisms leaves that involving a dioxetane as the most likely.

The chemiluminescence seen when the liquid tetrakisdimethylaminoethylene (XXIV) is exposed to moist air is unique in being the only efficient

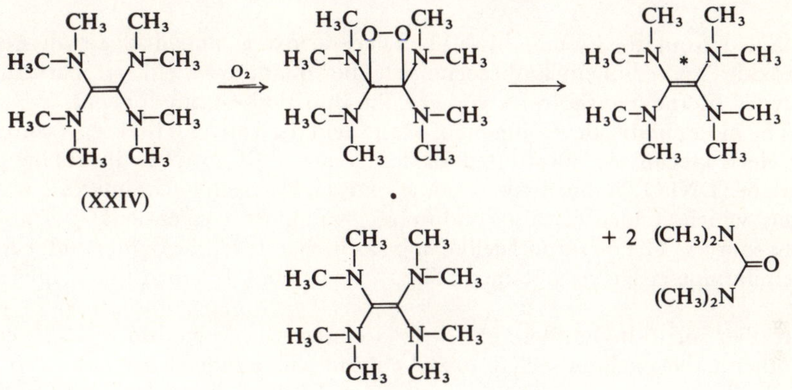

Scheme 29

luminescence which is spontaneous. This has resulted in a variety of potential commercial uses[109]. The mechanism[110] accepted at present is shown in *Scheme 29*, the excited starting material being the emitter. If the oxidation sequence is correct, this compound may reasonably be included in the imine peroxide class.

Active Esters and Related Compounds

This class of chemiluminescent compound is in many ways the most interesting, for three reasons. The range of structures is extensive, the mechanisms appear to be complementary to that of the well-studied ester hydrolysis and, not least, compounds of this type are excellent models for certain fairly well understood bioluminescent systems.

Derivatives of Oxalic Acid

On treatment with aqueous hydrogen peroxide in the presence of anthracene oxalyl chloride emits light corresponding to anthracene fluorescence. This first report suggested that excited oxygen was a product, but this was not substantiated by a more detailed study[112] which supported the reaction sequence shown in *Scheme 30*. There are very marked solvent effects, and although water is certainly required for strong chemiluminescence in ethers, in dimethylphthalate the effect is virtually absent. The quantum yield, depending on the fluorescent acceptor used, can be as high as 0·05 making this one of the most efficient compounds. A similar result ($\phi = 0.13$) is

$$Cl\cdot CO\cdot CO\cdot Cl + H_2O_2 + H_2O \rightarrow HO-O-\underset{\underset{O}{\|}}{C}-\underset{\underset{O}{\|}}{C}-OH$$

$$HO-O-\underset{\underset{O}{\|}}{C}-\underset{\underset{O}{\|}}{C}-OH + R^\cdot + \text{fluorescer} \rightarrow \text{fluorescer}^* + 2CO_2 + RH + HO^\cdot$$

Scheme 30

obtained from the system bis-biphenylacetic oxalic anhydride, hydrogen peroxide and diphenylanthracene[113]. The quantum yield is markedly reduced by the free radical scavenger, 2,6-dimethyl-4-t-butylphenol.

The most notable development of oxalic acid derivatives is that represented by electronegatively substituted oxalate esters[114,115]. Bis-2,4-dinitrophenyl oxalate (DNPO) is the basis of some extremely effective commercial products which, under certain conditions, rival more conventional portable light sources. The quantum yield can be as high as 0·23, the highest value for chemiluminescence yet obtained. They are particularly suited to emergency use.

Rather surprisingly, there is a distinct change in mechanism with respect to other active oxalates such as oxalyl chloride. Free radicals are not directly involved nor is the key intermediate in the earlier type, the monoperoxyoxalic acid. In fact water, far from being essential, lowers the quantum yield.

Spectral data for the seemingly unlikely and certainly unstable dioxetanedione (XXV) have proved impossible to obtain, with the major exception of a mass spectrum[116] where the parent peak at m/e 88 was observed to decay at the same rate as the chemiluminescence in a parallel experiment. However, most evidence, including volatility, supports the proposed structure and the aromatic hydrocarbons which eventually fluoresce in the reaction actually

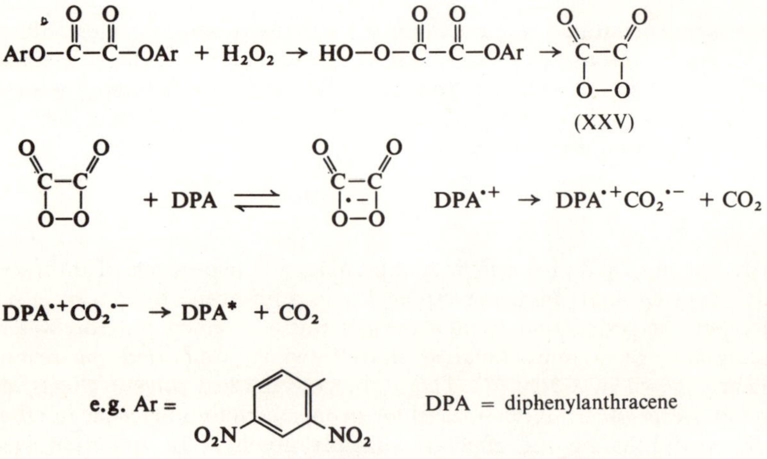

$$ArO-\underset{\underset{O}{\|}}{C}-\underset{\underset{O}{\|}}{C}-OAr + H_2O_2 \rightarrow HO-O-\underset{\underset{O}{\|}}{C}-\underset{\underset{O}{\|}}{C}-OAr \rightarrow \underset{O-O}{\overset{O=C-C=O}{|\ \ \ |}}$$

(XXV)

$$\underset{O-O}{\overset{O=C-C=O}{|\ \ \ |}} + DPA \rightleftharpoons \underset{O-O}{\overset{O=C-C=O}{|\cdot\ -|}} DPA^{\cdot+} \rightarrow DPA^{\cdot+}CO_2^{\cdot-} + CO_2$$

$$DPA^{\cdot+}CO_2^{\cdot-} \rightarrow DPA^* + CO_2$$

e.g. Ar = 2,4-dinitrophenyl (O_2N-, NO_2)   DPA = diphenylanthracene

Scheme 31

act as catalysts for the decomposition. Solutions of what seems to be the dioxetanedione can be collected at low temperatures and made to emit light at a characteristic and fast rate by the addition of a variety of fluorescent polynuclear aromatic hydrocarbons[117]. It seems reasonable to assume that the highly strained and potentially extremely energetic dioxetanedione is preserved for sufficiently long by orbital symmetry prohibitions as to react by a lower energy pathway involving formation of the easily accessible first excited singlet state of the fluorescer. The details of the actual excitation step are unavailable, but an interesting possibility which has certain features in common with electron transfer luminescence is shown in *Scheme 31*. Other

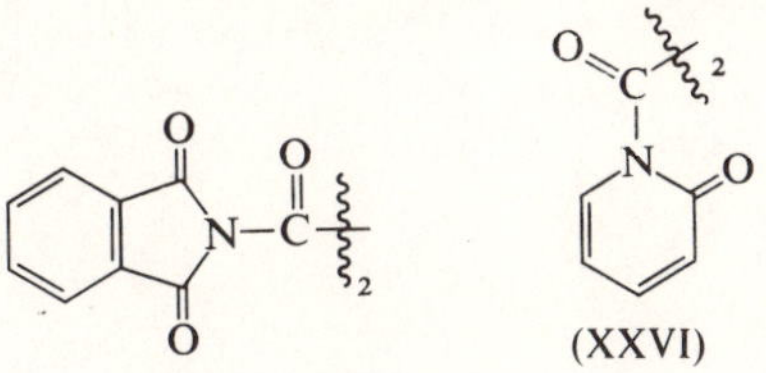

Scheme 32

compounds with structures[118] similarly based on their reactivity towards hydrogen peroxide are shown. They are also among the brightest of chemiluminescent compounds ($\phi = 0.15$). Interestingly, and as might be expected, compound (XXVI) reacts by acid catalysis rather than by the base catalysis seen in the ester series (*Scheme 32*).

Acridine Esters and Related Compounds

Compounds of this type are rewarding subjects for study since certain intermediates can be isolated and informative structural changes are easily made. In addition, at the oxidation level of the acridan they serve as excellent models of some bioluminescent reactions, while the acridinium salts provide a means of investigating intermediates which are not otherwise accessible. The first of the series was the nitrile (XXVII)[102,119], unusual in that most chemiluminescent compounds are discovered by accident whereas bright luminescence is expected in this case if the details to be discussed below obtain (*Scheme 33*).

The reaction is formulated as involving an intermediate four-membered peroxide ring since the sequence of events includes initial fast addition of peroxide to the acridinium nucleus. The (slow) addition of hydrogen peroxide to the nitrile group is unlikely since the rate of reaction is not sensitive to peroxide concentration. Moreover, attack of hydroxide on the nitrile, besides being notoriously slow, is not a likely chemiluminescent route since this must form the amide as an intermediate. The amide (XXVIII) is not chemiluminescent[120]. There is no question that *N*-methylacridone is the excited product since correspondence of the chemiluminescence and fluorescence spectra is exact and it is the only product. It is not difficult to construct other systems on this basis, with the added advantage that the dark route shown does not occur.

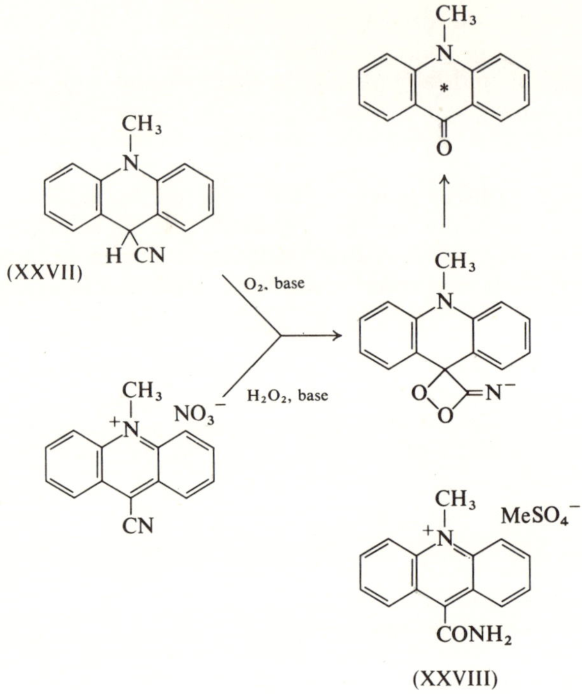

Scheme 33

The acid chloride (XXIX) (*Scheme 34*) was simultaneously investigated by two groups[102,121], one of which favoured a mechanism involving the peroxide (XXX). The main observations in support of this are that dilution of the initial solution of (XXIX) in 90% hydrogen peroxide with water markedly increased the intensity of emission. Other peroxides such as t-butyl peroxide and perlauric acid also produced light, but not as efficiently as hydrogen peroxide itself. Both groups commented on the failure of the peroxyacid (XXXI) to give light. The reaction is not quite as straightforward as suggested however. The increase of emission with addition of water is not simply the result of the addition of water to the acridinium nucleus, since the intensity increases proportionately with a volume increase from 0·01 to 3 litres. Obviously, changes in pH are occurring, an interpretation supported by the rate increase caused by base. Furthermore, the t-butylperoxide reaction does not emit light on simple dilution with water.

A related, more stable series[102,106] with greater opportunity for experimental investigation is provided by the phenyl esters (XXXII). Again the major product (90% yield) is *N*-methylacridone, the light emission from this series occurring with a quantum yield of about 2%. The most significant observations[106,122] are detailed below:

(a) The reaction is first order in base (over a limited pH range), first order in acridinium salt and zero order in hydrogen peroxide (at pH values above 8).

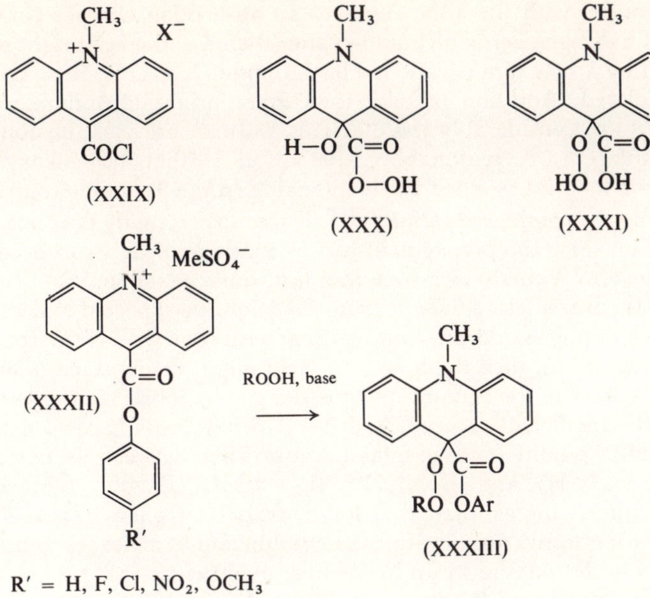

[Scheme 34 structures: (XXIX), (XXX), (XXXI), (XXXII), (XXXIII) — N-methylacridinium derivatives]

R' = H, F, Cl, NO$_2$, OCH$_3$

Scheme 34

(b) Hydrogen peroxide is essential, t-butyl and methyl hydroperoxide being considerably less effective (the small amount of light emitted almost certainly being caused by traces of hydrogen peroxide).

(c) Quantum yield is not significantly affected by substitution on the phenyl group, providing that the conjugate acid of the leaving group is stronger than hydrogen peroxide. The reaction constant (ρ) in aqueous ethanol is +4·6.

(d) The order of addition of the reagents, base and hydrogen peroxide, is important. In either case the acridinium absorption disappears virtually

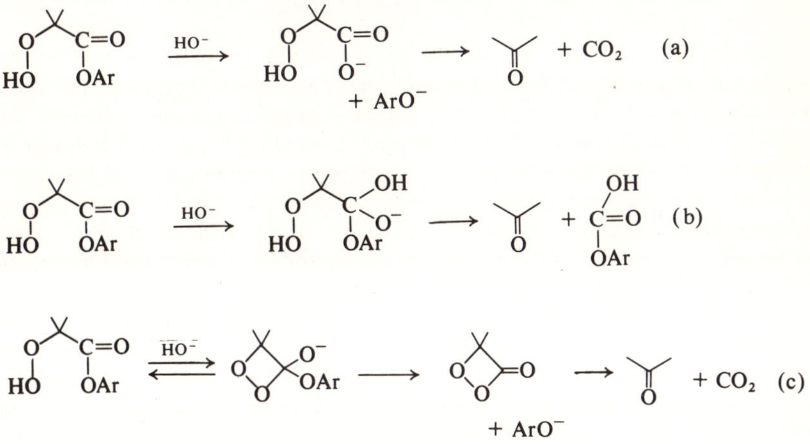

[Scheme 35: reaction pathways (a), (b), (c) showing peroxy intermediates decomposing to give CO$_2$ + ArO$^-$]

Scheme 35

instantaneously, with the appearance of an absorption characteristic of an acridan. If hydrogen peroxide is added after the base, the fast bright reaction is replaced by a slow dim emission. The peroxide (XXXIII, R = H) can in fact be isolated. Addition of base then gives bright chemiluminescence. Alkyl esters are considerably less efficient, and can only just be considered chemiluminescent. A reasonable explanation is that the alkoxy group (pK_a 16) is not easily expelled by the peroxide (pK_a 12). Three routes to the products may be considered (*Scheme 35*). Route (a) certainly does not lead to light emission since the peroxy acid can be made and shown to decompose quantitatively to N-methylacridone in a fast, dark, reaction.

The tetrahedral intermediate in route (b) might be expected to decompose by expulsion of phenoxide ion, thus merging with route (a). If this route does give rise to emission, then the quantum yield might be expected to fall with increasing acidity of the leaving (phenoxide) group. This is not observed. In addition, the inefficient reaction with alkyl peroxides is hard to explain—there should be no difference in final quantum yield between the compounds (XXXIII, R = H) and (XXXIII, R = alkyl). Finally, the value of $\rho(+4\cdot 6)$ is much higher than that for hydrolysis of phenyl esters ($+2\cdot 1$), probably owing to the need to shift the equilibrium in route (c) to the right, off-setting the developing strain in the four-membered ring.

It is in fact not surprising that route (c) is the most likely, given the high reactivity of peroxide anions towards carbonyl groups[123] and the intramolecular nature of the reaction. The extra strain energy (20–30 kcal mol^{-1}) of the dioxetane may in fact be required to populate the excited state (about 70 kcal mol^{-1}). If this is so, then intermediate dioxetanes are implicated in all chemiluminescent reactions which produce excited states of energy greater than about 70 kcal. It would be of interest to add substituents to the nucleus in order to lower the energy of the first excited singlet of the acridone, perhaps allowing open chain routes such as (a) or (b) to populate the excited state efficiently. Route (b), in particular, expelling the resonance stabilised carbonate grouping is about 15 kcal more exothermic than route (a).

Acridan Esters

Reduction of the acridinium esters in acetic acid with zinc gives the corresponding acridans. These react in strong base in polar aprotic solvents to give bright chemiluminescence[122]. The reaction is strongly base catalysed and the position of the equilibrium shown (*Scheme 36*) strongly influences the rate of emission. The reaction of the anion with oxygen is fast and appears to occur by the autoxidative route expected of such anions. It is not markedly catalysed by transition metals. This resemblance to the enzymic reactions of the luciferins is interesting.

The quantum yield is higher than in the case of the acridinium salts, perhaps owing to direct formation of the peroxide, excluding competition with other nucleophiles. These are among the brightest of all chemiluminescent organic compounds ($\phi = 10\%$), and serve as useful models for bioluminescent processes. After the autoxidation the sequence is probably identical to that of the acridinium salts. Fairly obvious extensions of the type

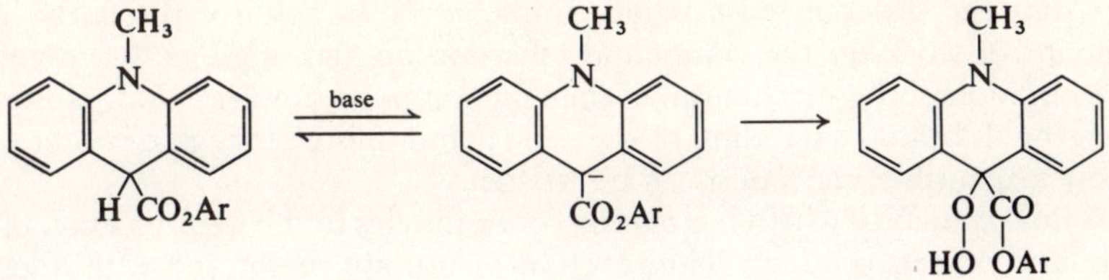

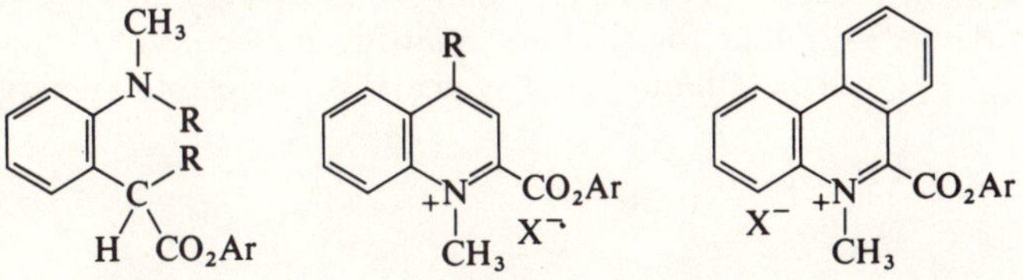

Scheme 36

can be made. The relatively simple series represented by (XXXIV) where R is alkyl or a ring residue is almost as efficient as the acridans. Quaternary salts such as (XXXV) or (XXXVI) are less efficient, but the emission from (XXXVI) is remarkably energetic, with λ_{max} 385 nm. The 0,0 level of the emitter, phenanthridone, is not observable and the exact energy of the excited state is not known. However, it is likely that at least 80 kcal mol^{-1} is available in the reaction. It is very fast at room temperature ($t_{\frac{1}{2}} = 2$ sec at pH 12) so that no great increment is included for activation energy. Again, a dioxetanone intermediate seems necessary to provide this large amount of energy.

Lucigenin

Standing in relative isolation (with lophine and luminol) for so long, lucigenin (XXXVII) attracted a fair amount of experimental effort, but with many contradictory results. The chemiluminescent reaction is very sensitive to medium effects but in essence requires a base and hydrogen peroxide[125]. It proceeds best in hydroxylic solvents, and is rather sensitive to the structure of the alcohol used[126,127].

The nature of the base is also relevant, some amines being considered more effective than others[126,128]. However, it is probable that these effects are the result of changes in medium and rate caused by the added compounds (a change to general base catalysis, for example) and that the quantum yield is not affected[126]. In fact, the influence of a variety of additives[126,128,130] has been noted, but seldom investigated in a way which would lead to any conclusion about the involvement of the additive in the mechanism. The reaction is complex, and the major product, *N*-methylacridone, is not formed quantitatively. There is some indication that the initial intensity of the reaction has a linear relationship to the hydrogen peroxide concentration, but that the order in base is not simple[128,130]. In spite of the fairly large amount of work on the compound, there is no thorough and convincing examination of the mechanism of the reaction in its simplest form. However, the growth in understanding of organic chemiluminescence and recent work allow a plausible mechanism to be written.

A final difficulty which has caused confusion lies in the identification of the primary excited product. The reaction as usually performed with alkaline hydrogen peroxide in aqueous alcohol displays a green emission. Since lucigenin itself is green fluorescent (λ_{max} c. 500 nm) it was suggested[131] that lucigenin is re-formed (as a biradical for example) with subsequent emission. A related view has been expressed recently[132]. However, there is now conclusive evidence[133,134] that the primary emitting molecule is N-methylacridone (λ_{max} 442 nm in ethanol) and, since this is the most energetic

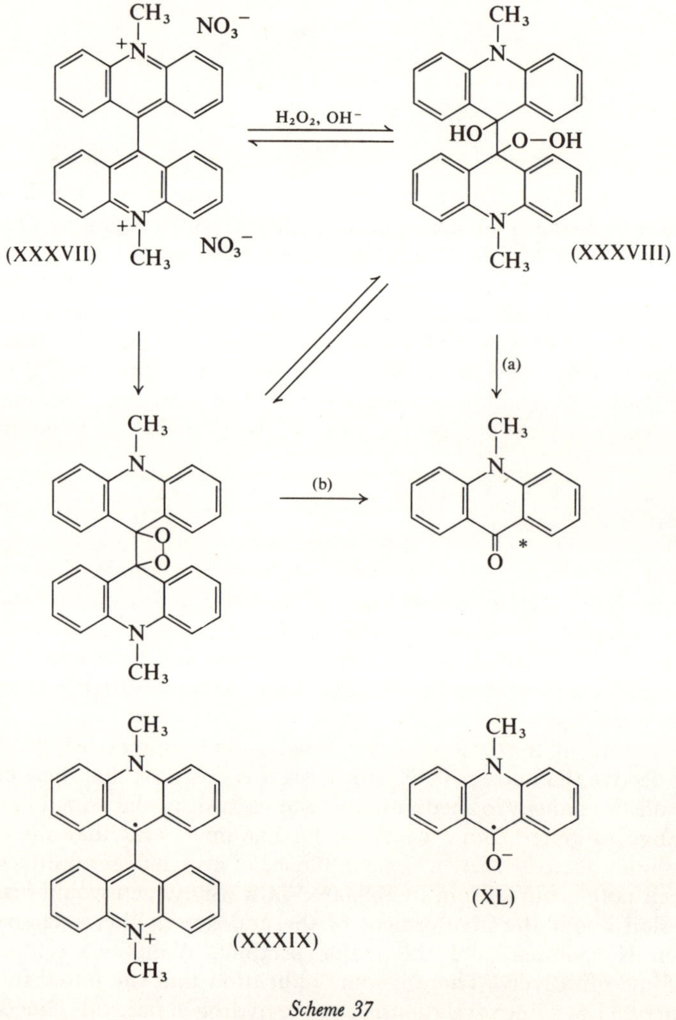

Scheme 37

emission observed, energy transfer to lucigenin or a degradation product is occurring. Such transfer to added fluorescent molecules is easily demonstrated[131a,135].

It is then necessary to provide a mechanism which produces N-methylacridone directly in an excited state. The details of this process are not entirely clear, but it must occur by a route similar to those shown. Although a radical mechanism involving a radical derived from lucigenin (XXXVII) or its pseudo base (XXXVIII) (*Scheme 37*) is possible, recent extensive investigations[134,136] have shown that such intermediates do not lead to efficient chemiluminescence. For example, Janzen and co-workers[136] have identified DBA·+ (XXXIX), formed by the addition of hydroxyl ion to lucigenin by e.s.r. However, subsequent reaction with oxygen produced a quantum yield 10^{-4} times that of the standard reaction using hydrogen peroxide. Moreover no radicals were detectable during the latter reaction. Weak chemiluminescence, also requiring oxygen, was observed by the addition of cyanide ion to lucigenin. Hercules and co-workers repeated earlier work[129] which seemed to indicate that electrogenerated chemiluminescence was possible from lucigenin. However, it is apparent[134] that direct reduction in the absence of oxygen, at a platinum electrode, merely reduces lucigenin to dimethylbiacridene, and that a mercury electrode at potentials more negative than -0.15 V reduces oxygen which then reacts with lucigenin to give light. In fact, at pH 7 in water the reduction of oxygen gives both the reagents required for the classical chemiluminescence.

$$O_2 + 2H_2O + 2e \rightarrow H_2O_2 + 2HO^-$$

Using non-aqueous solvents, the superoxide ion $O_2^{·-}$ is produced from oxygen, and lucigenin is also reduced to dimethylbiacridene. With the solvents ethanol, dimethylformamide, dimethyl sulphoxide and acetonitrile, it is possible to have varying solubilities and fluorescence efficiencies for the two principal energy acceptors. These are lucigenin itself (λ_{max} 500 nm) and dimethylbiacridene (λ_{max} 510 nm). Because of this it can be shown that N-methylacridone is the primary emitter, and that at concentrations of acceptor greater than 10^{-4} M, singlet to singlet energy transfer of the Förster type[137] occurs. Thus, the arguments in favour of nucleophilic attack of hydrogen peroxide or its conjugate base as a principal step in all cases of lucigenin luminescence are reinforced.

Various other mechanisms[131,132,138] have been advanced, some accompanied by considerable experimental work. However, almost all can be excluded on the grounds that the excitation step either does not form N-methylacridone, or equally important, does not release the large amount of energy required to populate the observed excited state.

To date it has not been possible to distinguish between routes (a) and (b), but in so far as alkyl peroxides are considerably less efficient (if effective at all) then route (b) is implied. Lucigenin is a relatively efficient compound although the quantum yield of 1.6% obtained[133] on the basis of the N-methylacridone produced is very variable, and the quantum yield based on lucigenin consumed may be only 0.08%.

A variety of lucigenin analogues have also been prepared[139-142]. By alkylating the potassium salt of acridone and treating the N-alkylacridone with zinc in methanolic hydrochloric acid, the N,N-dialkylbiacridenes are formed[141]. Oxidation with nitric acid gives the dialkyl, dipropyl and dibutyl analogues of lucigenin. Some N-aryl derivatives have also been prepared via the N-arylacridone, made by the condensation of an N-arylanthranilic acid with cyclohexanone. They are all chemiluminescent, and it is not expected that important differences in behaviour will be observed. However, the electron-withdrawing effect of the aryl group may have some influence on the efficiency of the light reaction[142]. Substituents on the nucleus have a more pronounced effect[140] since the reaction may now take a different course. Changes in colour of the emission are observed but no detailed investigation has been made.

An interesting example of this class of chemiluminescent compound was uncovered during a study of overcrowded aromatic systems[143]. These compounds (XLI–XLIII, Scheme 38) and the unbridged parent compound

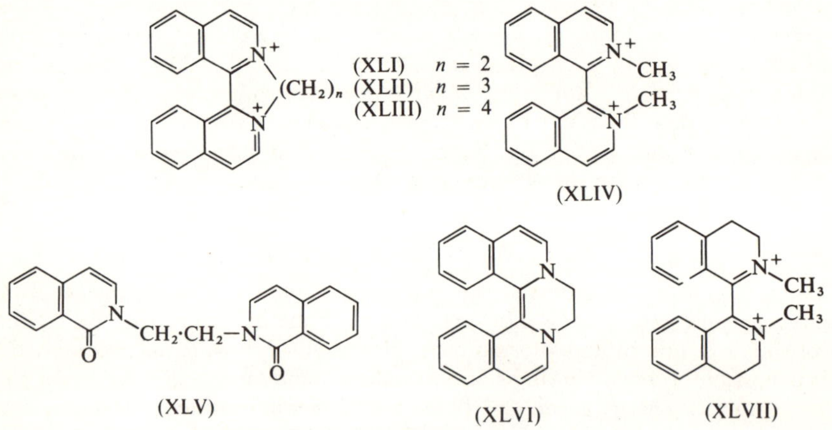

(XLI)   n = 2
(XLII) n = 3
(XLIII) n = 4

(XLIV)

(XLV)   (XLVI)   (XLVII)

Scheme 38

(XLIV) are brightly chemiluminescent under conditions very similar to those used for lucigenin. In the case of (XLI) the compound (XLV) was identified as the emitter and, although the investigators did not suggest a mechanism, it seems possible that it is related to that for the lucigenin reaction. Similar solvent effects are observed, and the efficiency of (XLI) is comparable ($\phi = 1.5 \times 10^{-3}$). The most intriguing aspect of these compounds is that efficiency falls through nine powers of ten in the series (XLI–XLIV). The reasons for this effect are not clear but would obviously repay investigation. Treatment of the quaternary salts in degassed methanol with base gave a red compound which reacted subsequently with oxygen with chemiluminescence. Structure (XLVI) was suggested for this intermediate, and if correct would in fact be expected to react as an electron-rich olefin such as tetrakisdimethylaminoethylene (TMAE). An investigation[144] of a tetrahydroderivative of (XLIV), i.e. (XLVII), shows that, although complex, the chemiluminescent reaction with oxygen in the presence of base has much in common with that of TMAE.

BIOLUMINESCENCE

Only those aspects of this dramatic and intriguing subject which relate directly to chemiluminescence can be discussed here. Several excellent reviews of the more biological material are available[145]. There are many indications that bioluminescent systems do not conform to a single chemical reaction type. Indeed, some of the more intriguing examples cannot be described in the terms applicable to the few fairly well understood reactions discussed below. The difficulty of collecting sufficient organisms, small amounts of material and very complex molecules are all factors which impede progress in this area.

The reactions discussed all involve a discrete small organic molecule, although the individual structures are fairly complex and all rather unusual. These molecules are called luciferins[14] and interact with an enzyme called a luciferase producing, by oxidation, a product in the first excited singlet state. Understandably, such products are highly fluorescent, although only one has been adequately characterised as a product of the natural system. The reactions can be duplicated to a greater or lesser extent by model compounds and much of the information available is derived from this source.

The Firefly

The isolation of firefly luciferin and luciferase came as the result of the excellent work by McElroy and co-workers[147] over many years. It was

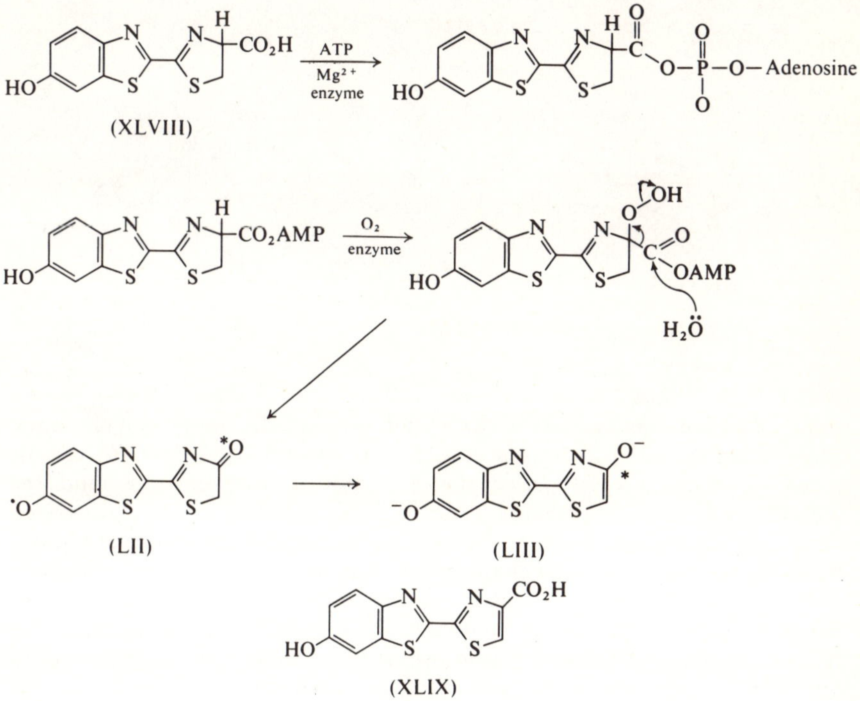

Scheme 39

followed eventually by the elucidation of the structure of the luciferin (XLVIII, *Scheme 39*) and its synthesis[148], the first luciferin to be successfully so characterised. It is also the only naturally occurring example of a benzothiazole. All fireflies so far examined[149,150], some twelve or so species, utilise the same luciferin and probably all systems operate by the synthesis of the adenylate before oxidation. Thus it would seem that the actual substrate for the oxidation is analogous to the active ester series, and particularly to the acridan phenyl esters. Following the successful use of a model compound in the case of Cypridina luciferin, two groups independently reported the chemiluminescent reaction of derivatives of firefly luciferin. Dehydroluciferin (XLIX) is an oxidation product both *in vivo* and *in vitro* and in

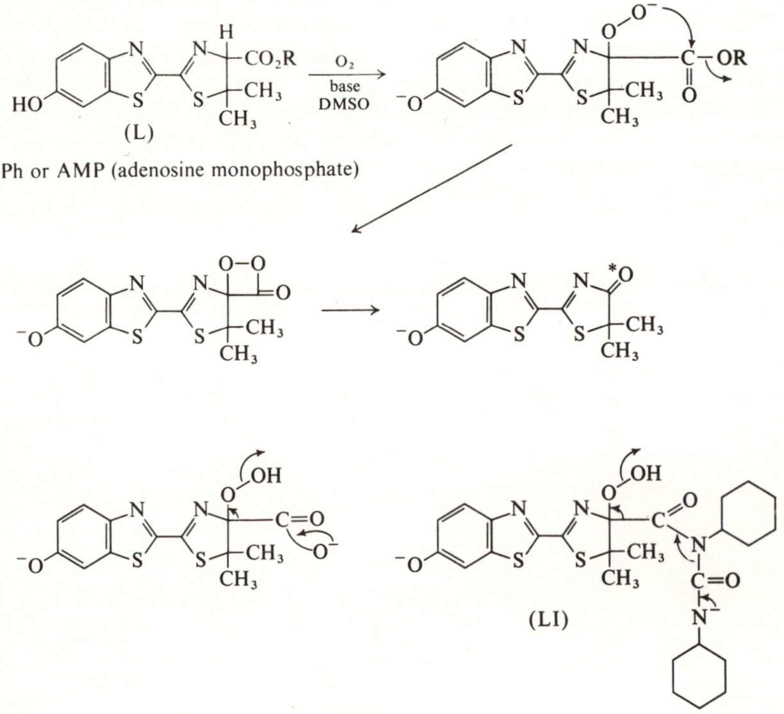

Scheme 40

most oxidative experiments is the major product. By using penicillamine rather than cysteine in the synthesis the compound (L, *Scheme 40*) can be made. The first group[151] used the adenylate as the substrate, and the second synthesised the phenyl ester[152]. Interpretation in the latter case should be less ambiguous since the possible nucleophilic reactions are fewer. Both groups related the mechanisms to that of the active esters discussed previously (*see Scheme 35*), involving a dioxetanone as intermediate. Recent labelling experiments using luciferase show clearly that a linear route (either (a) or (b) of *Scheme 35*) is in operation. Although exchange reactions are ruled out by this work such reactions could possibly account for the observation. Nevertheless this is rather unlikely and the experiment should be taken as indicating the operation of routes (a) or (b). The model compounds, on the other hand, do not seem to react in this way, the free carboxylic acid and compound (LI) decomposing as shown without the emission of light. Decomposition of the acid certainly provides sufficient energy to populate the excited state of the product.

The use of the blocked substrate (L) whilst strongly indicating the type of reaction being dealt with obscures the nature of the actual emitter. The firefly light is yellow (λ_{max} 565 nm) whereas the model compound (L) emits red light (λ_{max} 625 nm) on oxidation. White, Seliger and their co-workers[154] have shown that deprotonation of the excited anion (LII) (*Scheme 39*) in the excited state causes a shift to shorter wavelength. Thus the emitter is in fact the dianion (LIII) in the normal yellow luminescence. Enzyme binding might be expected to alter this since the position of the fluorescence of the free compound does not quite correspond. Synthesis of the presumed *in vivo* emitter[155] and addition to the enzyme in solution, however, did not result in any significant alteration in the spectrum. Binding to the enzyme need not be reversible, and in any case the proposed emitter is formed by oxidation from a previously bound precursor.

Cypridina Hilgendorfii

The quantum yield of bioluminescence from this organism in the luciferin–luciferase reaction with oxygen is estimated at 0·28[156]. Emission occurs with λ_{max} 460 nm. Cypridina is a small ostracod crustacean found mainly in the Sea of Japan, and the light occurs under natural conditions as the reaction mentioned above takes place in sea water. The flash of light is thought to be a means of distracting predators. As in the case of the firefly, the enzyme does not contain cofactors normally associated with oxidations by oxygen (e.g. haeme), but is even simpler in that ATP or other activating molecules are not required.

The structure and synthesis of the luciferin (LIV, *Scheme 41*) is the work of Goto, Johnson and their co-workers[157]. The molecule was known to be

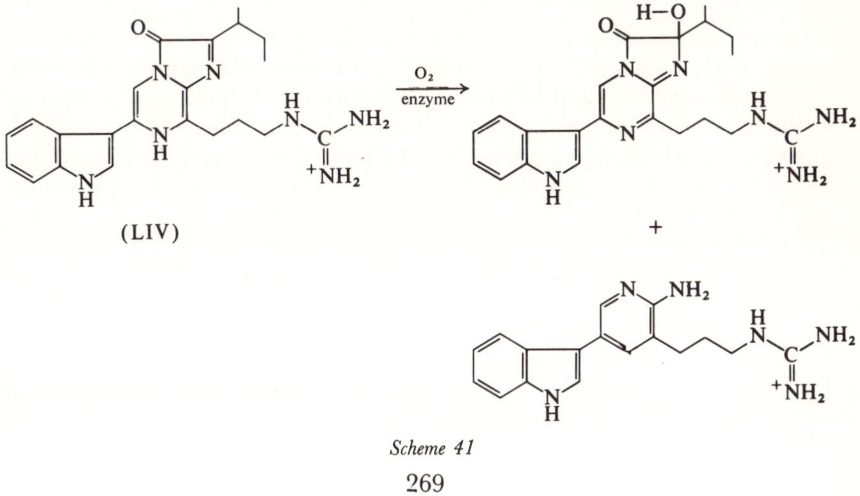

Scheme 41

chemiluminescent in dimethyl sulphoxide[158] but no product was isolated, nor was there any indication of the mechanism of the luminescent reaction. Although Cypridina luciferin may fairly be considered a multifunctional molecule, there is in fact an analogy with the structure of the acridan phenyl esters. McCapra and Chang[159] using a model compound (LV) were able to suggest a satisfactory mechanism for the chemiluminescent reaction (*Scheme 42*). Hitherto the suggested reaction sequence[157] (*Scheme 41*) *in vivo* did not include the generation of CO_2 as a fragment. A further study of the enzymic reaction confirmed that the route shown for the model compound did in fact operate *in vivo* also[160]. The structures of the luciferin and the several model compounds are such that the intermediate peroxide can react as an active acyl pyrazine, most probably via a dioxetanone. The other routes discussed for active esters in general are also possible, but the high energy of the emission (λ_{max} 460 nm) must again be considered. Also of interest is the observation that under mild (acetate) base catalysis the emitter

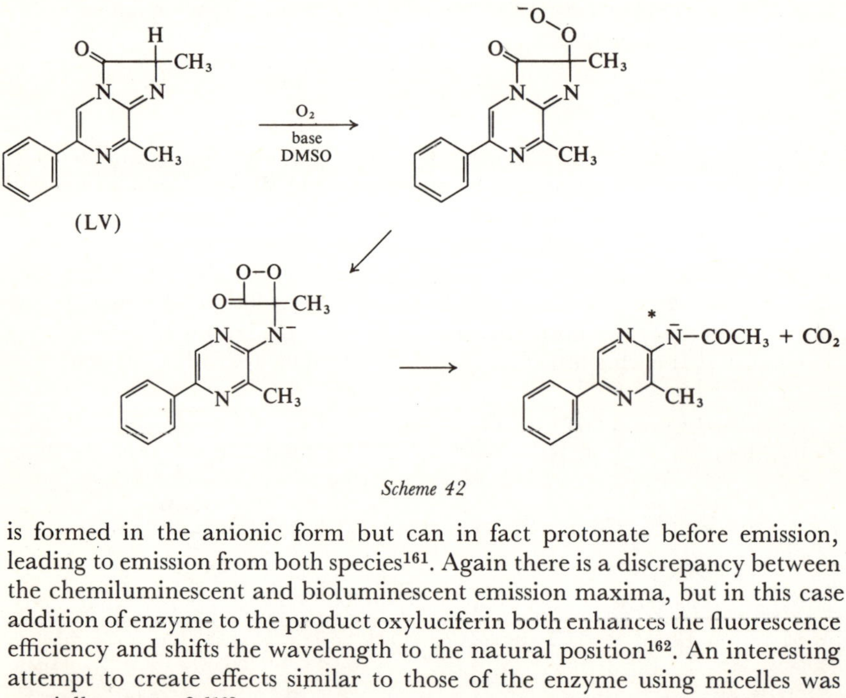

Scheme 42

is formed in the anionic form but can in fact protonate before emission, leading to emission from both species[161]. Again there is a discrepancy between the chemiluminescent and bioluminescent emission maxima, but in this case addition of enzyme to the product oxyluciferin both enhances the fluorescence efficiency and shifts the wavelength to the natural position[162]. An interesting attempt to create effects similar to those of the enzyme using micelles was partially successful[163].

A chemiluminescent system based on the luciferin structure is that of the extremely simple imidazopyridinone (LVI, *Scheme 43*)[164]. With R = alkyl the light is blue (λ_{max} 420–440 nm) but with R = H the emission is that of the yellow fluorescence of the reactant anion. Examination of the spectra of reactant and product in the two cases suggests that alkyl substitution shifts the emitter (LVII) fluorescence to a position in which there is little overlap with the absorption of the reactant anion. Hence energy transfer in this case is not observed. In the case of (LVII, R = H) the much more energetic

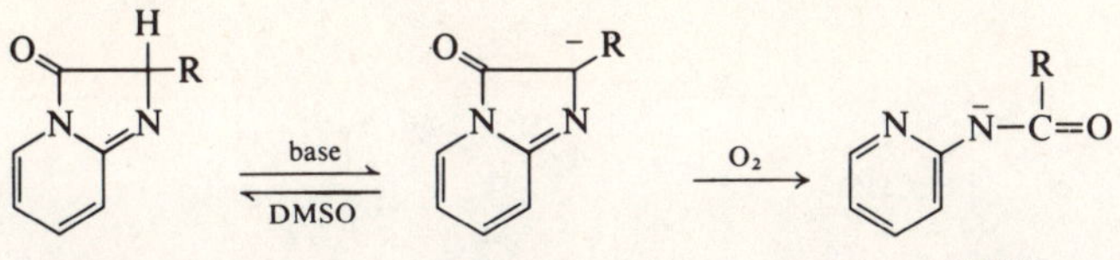

Scheme 43

emission (395 nm) overlaps effectively with the reactant absorption. The fluorescence maxima of the amide anions are seemingly much affected by steric hindrance to solvation.

Latia Neritoides
This organism is unique in being the only example of a luminescent fresh water limpet—another case in fact of isolated evolution from New Zealand.

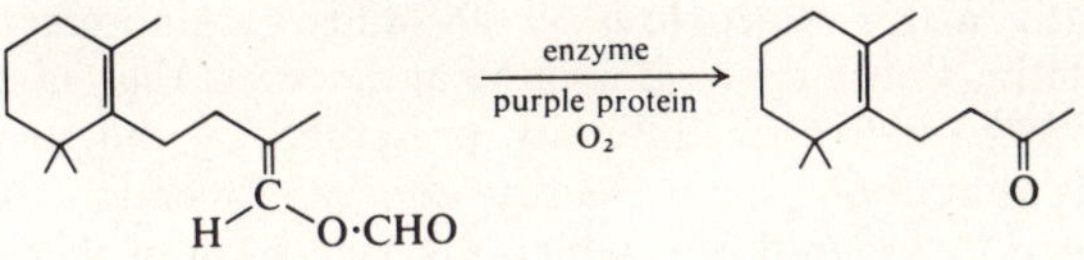

Scheme 44

The function of the luminescence is unknown. Although a luciferin–luciferase can be demonstrated, certain features prevent direct classification with the previous examples. Isolation of the luciferin[165] was followed by the elucidation[166] of the structure of both the luciferin (LVIII) and its oxidation product (LIX, *Scheme 44*). This unusual enol formate has been synthesised[167,168], the later synthesis[168] being particularly elegant and perhaps biogenetically significant as it uses a molecule of classic sesquiterpenoid type as the starting material (*Scheme 45*).

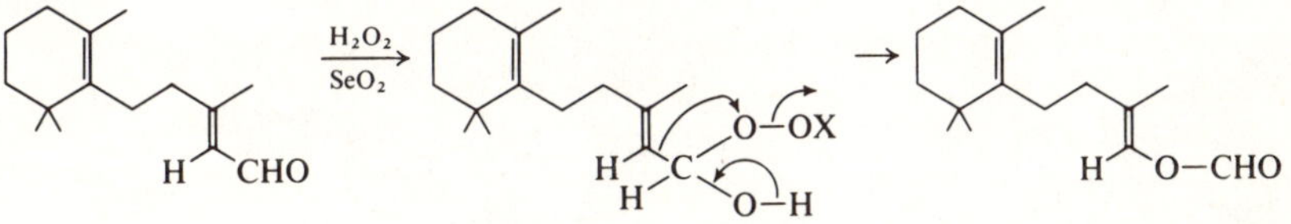

Scheme 45

It is fairly certain that the ketone (LIX) which is the only product of the oxidation of the luciferin found is not directly luminescent. Although the enzyme might be expected to facilitate energy transfer, it is unlikely that the rapid intersystem crossing to the triplet ketone would be affected. Triplet–singlet energy transfer is then a possible mechanism for emission of light. Associated with the enzyme and luciferin is a third essential molecule (not including oxygen of course) which has been called a 'purple protein'. The

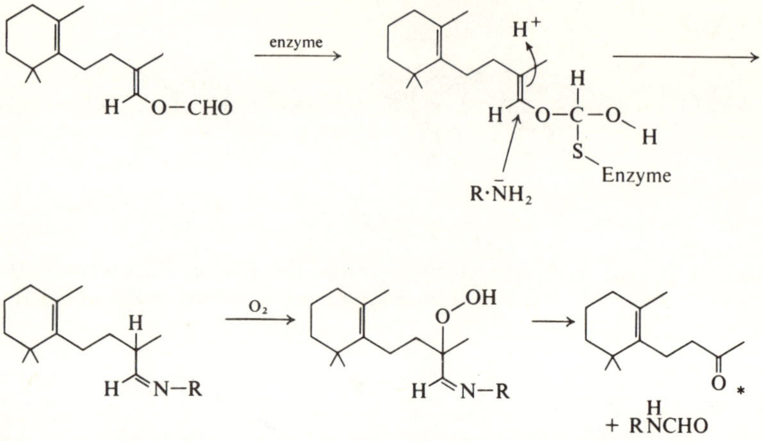

Scheme 46

molecular weight of this molecule is 39 000 and it is fluorescent. However, the position of the fluorescence does not agree with that of the observed chemiluminescence. A further difficulty presented by the structure of the enol formate is that there is no readily apparent means of oxidation by molecular oxygen. A suggestion[169] which takes account of these factors—i.e. inefficient emission from a simple carbonyl compound, the role of the purple protein, the discrepancy in emission maxima and the reaction with oxygen— is shown in *Scheme 46*. The association of the luciferin with either the protein or the enzyme may result in a Schiff's base. Oxidation leads to a formyl

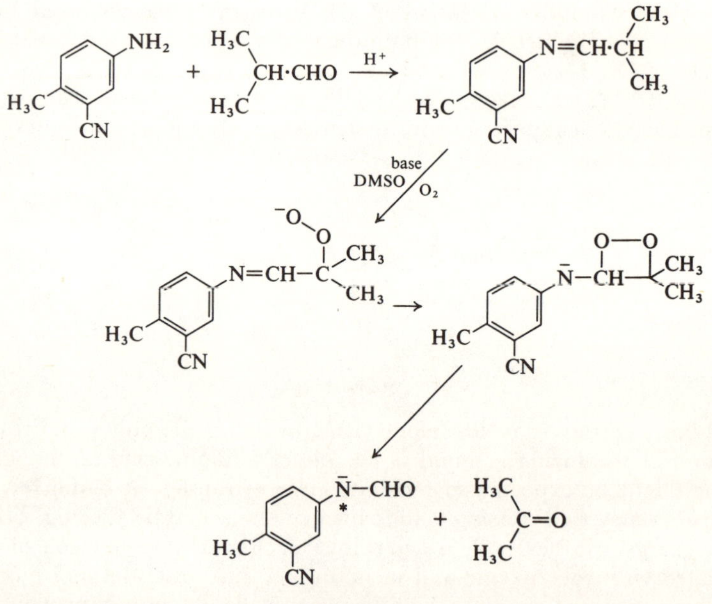

Scheme 47

derivative in an excited state by the route shown. Various model compounds (*see Scheme 47* for an example) are moderately chemiluminescent, and evidence for the reaction path has been obtained.

The luciferin–luciferase reaction has been obtained for a wide variety of organisms so that the study of chemiluminescence of discrete organic molecules as described in this article is obviously relevant. However, there are examples of bioluminescence where a more complex situation obtains, and these must await further biochemical investigation before a correlation with chemiluminescent reactions can be made.

Manuscript received August 1971.

REFERENCES

1. HABER, F. and ZISCH, W., *Z. Phys.*, **9**, 302 (1922); BEUTLER, H. and POLANYI, M., *Naturwissenschaften*, **13**, 711 (1925)
2. WAYNE, R. P., 'Luminescence in the Gas Phase', in *Luminescence in Chemistry*, ed, BOWEN, E. J., Van Nostrand, London, chap. 4 (1968)
3. CLYNE, M. A. A. and THRUSH, B. A., *Proc. R. Soc.*, **A269**, 404 (1962)
4. LEWIS, P., *Astrophys. J.*, **12**, 8 (1900); STRUTT, R. J., *Proc. R. Soc.*, **A176**, 1 (1940)
5. THRUSH, B. A., *Chem. Br.*, 287 (1966)
6. KHAN, A. U. and KASHA, M., *J. Am. chem. Soc.*, **88**, 1574 (1966)
7. WILKINSON, F., *Advances in Photochemistry*, eds, Noyes, W. A., Hammond, G. S. and Pitts, J. N., Interscience, New York, **3** (1964)
8. LEE, J. and SELIGER, H. H., *Photochem. Photobiol.*, **4**, 1015 (1965)
9. HASTINGS, J. W. and WEBER, G., *J. opt. Soc. Am.*, **53**, 1410 (1963)
10. (a) GUNDERMAN, K. D., *Chemilumineszenz Organischer Verbindungen*, Springer-Verlag, Berlin (1968); (b) MCCAPRA, F., *Q. Rev.*, **20**, 485 (1968); (c) GUNDERMAN, K. D., *Angew. Chem. Int. Edn*, **4**, 566 (1965)
11. HERCULES, D. M., *Accts Chem. Res.*, **2**, 301 (1969)
12. ZWEIG, A., *Advances in Photochemistry*, **6**, 425 (1968)
13. CHANDROSS, E. A., *Trans. N.Y. Acad. Sci.*, **31**, 571 (1969)
14. LEWIS, G. N. and BIGELEISEN, J., *J. Am. chem. Soc.*, **65**, 2424 (1943)
15. CHANDROSS, E. A. and SONNTAG, F. I., *J. Am. chem. Soc.*, **86**, 3178 (1964); VISCO, R. E. and CHANDROSS, E. A., ibid., **86**, 5350 (1964)
16. RAUHUT, M. M., MARICLE, D. L., KENNEDY, G. W. and MOHNS, J. P., 'Chemiluminescent Materials', *American Cyanamid Co. Tech. Rep.*, No. 5 (1964)
17. HERCULES, D. M., *Science, N.Y.*, **145**, 808 (1964)
18. MARCUS, R. A., *J. chem. Phys.*, **43**, 2654 (1965); ibid., **52**, 2803 (1970)
19. TURRO, N. J., *Molecular Photochemistry*, W. A. Benjamin Inc., New York (1967)
20. CHANG, J., HERCULES, D. M. and ROE, D. K., *Electrochim. Acta*, **13**, 1197 (1968)
21. FELDBERG, S. W., *J. phys. Chem.*, **70**, 3928 (1966)
22. CHANDROSS, E. A., LONGWORTH, J. W. and VISCO, R. E., *J. Am. chem. Soc.*, **87**, 3259 (1965)
23. PARKER, C. A. and SHORT, G. D., *Trans. Faraday Soc.*, **63**, 2618 (1967)
24. WERNER, T. C., CHANG, J. C. and HERCULES, D. M., *J. Am. chem. Soc.*, **92**, 763 (1970)
25. FAULKNER, L. J. and BARD, A. J., *J. Am. chem. Soc.*, **90**, 6284 (1968)
26. WERNER, T. C., CHANG, J. and HERCULES, D. M., *J. Am. chem. Soc.*, **92**, 5560 (1970)
27. VISCO, R. F. and CHANDROSS, E. A., *J. Am. chem. Soc.*, **86**, 5350 (1964); HERCULES, D. M., LANSBURG, R. C. and ROE, D. K., *J. Am. chem. Soc.*, **88**, 4578 (1966)
28. ZWEIG, A., MARICLE, D. L., BRUIEN, J. S. and MAURER, A. H., *J. Am. chem. Soc.*, **89**, 473 (1967)
29. RAUHUT, M. M., MARICLE, D. L., KENNEDY, G. W. and MOHNS, J. P., 'Chemiluminescent Materials', *American Cyanamid Co. Tech. Rep. No.* 5 (1964); CHANDROSS, E. A. and SONNTAG, F. I., *J. Am. chem. Soc.*, **86**, 3179 (1964); ibid., **88**, 1089 (1966)
30. WEDEKIND, E., *Z. wiss. Photogr.*, **5**, 29 (1907)
31. (a) DUFFORD, R. T., CALVERT, S. and NIGHTINGALE, D., *J. Am. chem. Soc.*, **45**, 2058 (1923); ibid., **47**, 95 (1925); (b) THOMAS, C. D. and DUFFORD, R. T., *J. opt. Soc. Am.*, **23**, 251 (1933); (c) EVANS, W. V. and DIEPENHORST, E. M., *J. Am. chem. Soc.*, **48**, 715 (1926); BACHMANN, W. E., *J. Am. chem. Soc.*, **56**, 1363 (1934)
32. BREMER, T. and FRIEDMANN, H., *Bull. Soc. chim. Belg.*, **63**, 415 (1954)

33. BARDSLEY, R. L. and HERCULES, D. M., *J. Am. Chem. Soc.*, **90**, 4545 (1968)
34. MCCAPRA, F. and WARD, P. J., unpublished observations
35. HAAS J. W. JR., and BAIRD, J. E., *Nature*, **214**, 1006 (1967)
36. WELLER, A. and ZACHARIASSE, K., *J. chem. Phys.*, **46**, 4984 (1967)
37. STEENKEN, S., *Photochem. Photobiol.*, **11**, 279 (1970)
38. HERCULES, D. M. and LYTLE, F. E., *J. Am. chem. Soc.*, **88**, 4745 (1966); LYTLE, F. E. and HERCULES, D. M., ibid., **91**, 253 (1969)
39. HERCULES, D. M. and LYTLE, F. E., U.S. Pat. 3 515 674 through *Chem. Abstr.*, **73**, 50662 (1970)
40. LINSCHITZ, H., *Light and Life*, Johns Hopkins Press, Baltimore, 173 (1961)
41. WILSON, T. and HASTINGS, J. W., *Photophysiology*, **5**, 49 (1970)
42. MALLET, L., *C.r. hebd. Séanc. Acad. Sci., Paris*, **185**, 352 (1927); GROH, P. and KIRMANN, A., ibid., **215**, 275 (1942); GATTOW, G. and SCHNEIDER, A., *Naturwissenschaften*, **41**, 116 (1954); SELIGER, H. H., *Analyt. Biochem.*, **1**, 60 (1960); *J. chem. Phys.*, **40**, 3133 (1964)
43. KHAN, A. and KASHA, M., *J. chem. Phys.*, **39**, 2105 (1963)
44. ARNOLD, S. J., OGRYZLO, E. A. and WITZKE, H., *J. chem. Phys.*, **40**, 1769 (1964); BROWNE, R. J. and OGRYZLO, E. A., *Proc. chem. Soc.*, 117 (1964); *Photochem. Photobiol.*, **4**, 963 (1965)
45. KHAN, A. U. and KASHA, M., *J. Am. chem. Soc.*, **88**, 1574 (1966)
46. KHAN, A. U., *Science, N.Y.*, **168**, 476 (1970)
47. RUSSELL, G. A., *J. Am. chem. Soc.*, **79**, 387 (1957)
48. VASIL'EV, R. F., *Prog. Reaction Kinetics*, **4**, 305 (1967)
49. VASIL'EV, R. F., *Nature*, **196**, 668 (1962)
50. KELLOGG, R. E., *J. Am. chem. Soc.*, **91**, 5433 (1969)
51. LUNDEEN, G. and LIVINGSTONE, R., *Photochem. Photobiol.*, **4**, 1085 (1965)
52. HOWARD, J. A. and INGOLD, K. V., *Can. J. Chem.*, **43**, 2737 (1965); *J. Am. chem. Soc.*, **90**, 1056 (1968)
53. FOOTE, C. S., *Accts Chem. Res.*, **1**, 104 (1968)
54. BEUTEL, J., *J. Am. chem. Soc.*, **93**, 2615 (1971)
55. KURZ, R. B., *Ann. N.Y. Acad. Sci.*, **16**, 399 (1954)
56. ABBOTT, S. R., NESS, S. and HERCULES, D. M., *J. Am. chem. Soc.*, **92**, 1128 (1970)
57. MCKEOWN, E. and WATERS, W. A., *Nature*, **203**, 1063 (1964)
58. HOUSE, H. O., *Modern Synthetic Reactions*, W. A. Benjamin Inc., New York, 119 (1965)
59. STAUFF, J., *Photochem. Photobiol.*, **4**, 1199 (1965); see also GUNDERMAN, K. D., *Chemilumineszenz organischer Verbindungen*, Springer-Verlag, Berlin, 9 (1968)
60. OGRYZLO, E. A. and PEARSON, A. E., *J. phys. Chem.*, **72**, 2913 (1968); OGRYZLO, E. A. and coworkers, *loc. cit.* (see e.g. ref. 17)
61. WILSON, T., *J. Am. chem. Soc.*, **91**, 2387 (1969)
62. WASSERMAN, H. H. and SCHEFFER, J. R., *J. Am. chem. Soc.*, **89**, 3073 (1967)
63. WHITE, E. H. and ROSWELL, D. F., *J. Am. chem. Soc.*, **89**, 3944 (1967); ROBERTS, D. R. and WHITE, E. H., *J. Am. chem. Soc.*, **92**, 4861 (1970); ROSWELL, D. F., PAUL, V. and WHITE, E. H., ibid., **92**, 4855 (1970)
64. MCCAPRA, F., *Q. Rev.*, **20**, 485 (1966)
65. WHITE, E. H. and ROSWELL, D. F., *Accts Chem. Res.*, **3**, 54 (1970)
66. ALBRECHT, H. O., *Z. phys. Chem.*, **136**, 321 (1928)
67. WHITE, E. H., in *Light and Life,* ed. McElroy, W. D. and Glass, B., Johns Hopkins Press Baltimore, 183 (1961)
68. OMOTE, Y., MIYAKA, T. and SUGIYAMA, N., *Bull. chem. Soc. Japan*, **40**, 2446 (1967)
69. WHITE, E. H., NASH, G. E., ROBERTS, D. R. and ZAFIRIOU, O. C., *J. Am. chem. Soc.*, **90**, 5932 (1968)
70. RAUHUT, M. M., SEMSEL, A. M. and ROBERTS, B. G., *J. org. Chem.*, **31**, 2431 (1966)
71. DREW, H. D. K. and GARWOOD, R. F., *J. chem. Soc.*, 791 (1938); DREW, H. D. K., *Trans Faraday Soc.*, **35**, 207 (1939); WILHELMSE, P. C., LUMRY, R. and EYRING, H., in *The Luminescence of Biological Systems*, ed, Johnson, F. H., Am. Ass. Adv. Science, Washington D.C., 75 (1955)
72. LEE, J. and SELIGER, H. H., *Photochem. Photobiol.*, **11**, 247 (1970)
73. WHITE, E. H. and HARDING, M. J. C., *J. Am. chem. Soc.*, **86**, 5686 (1964); *Photochem. Photobiol.*, **4**, 1129 (1965)
74. MCCAPRA, F., CHANG, Y. C. and LONG, P. V., unpublished results
75. WHITE, E. H., ROSWELL, D. F. and ZAFIRIOU, O. C., *J. org. Chem.*, **34**, 2462 (1969)

76. WHITE, E. H., ROBERTS, D. R. and ROSWELL, D. F., in *Molecular Luminescence*, ed, Lim, E. C., W. A. Benjamin Inc., New York, 479 (1969)
77. WITTE, A. A. M., *Recl. Trav. chim. Pays-Bas Belg.*, **64**, 471 (1935); WASSERMAN, J. S. and MIKLUCHIN, G. P., *Zh. obshch. Khim.*, **9**, 606 (1939); OJIMA, H., *Naturwissenschaften*, **48**, 600 (1961); KROH, J. and LUSZCZEWSKI, J., *Roczniki chem.*, **30**, 647 (1956)
78. WHITE, E. H., BURSEY, M. M., ROSWELL, D. F. and HILL, J. H. M., *J. org. Chem.*, **32**, 1198 (1967)
79. WHITE, E. H. and ROSWELL, D. F., Joint Conference Can. Inst. Chem. and Am. Chem. Soc., Toronto, May 1970
80. MCCAPRA, F., *Chem. Commun.*, 155 (1968)
81. MCCAPRA, F. and RICHARDSON, D. G., *Tetrahedron Lett.*, 3167 (1964)
82. BOLLYKY, L. J., *J. Am. chem. Soc.*, **92**, 3230 (1970)
83. O'NEAL, H. E. and RICHARDSON, W. H., *J. Am. chem. Soc.*, **92**, 6553 (1970); the figures for the calculated available energies have been amended in *J. Am. chem. Soc.*, **93**, 1828 (1971)
84. BENSON, S. W., CRUIKSHANK, F. R., GOLDEN, D. M., HAUGEN, G. R., O'NEAL, H. E., ROGERS, A. S., SHAW, R. and WALSH, R., *Chem. Rev.*, **69**, 279 (1969)
85. KOPECKY, K. R. and MUMFORD, C., *Can. J. Chem.*, **47**, 709 (1969)
86. SULLIVAN, M. O. and TESTA, A. C., *J. Am. chem. Soc.*, **90**, 6245 (1968)
87. RENKES, G. D. and WETTOCK, F. S., *J. Am. chem. Soc.*, **91**, 7514 (1969)
88. BARTLETT, P. B. and SCHAAP, A. P., *J. Am. chem. Soc.*, **92**, 3223 (1970)
89. MAZUR, S. and FOOTE, C. S., *J. Am. chem. Soc.*, **92**, 3225 (1970)
90. WILSON, T. and SCHAAP, A. P., *J. Am. chem. Soc.*, **93**, 4126 (1971)
91. WILSON, T., *Photochem. Photobiol.*, **10**, 441 (1970)
92. LUNDEEN, G. W. and ADELMAN, A. H., *J. Am. chem. Soc.*, **92**, 3914 (1970)
93. DUFRAISSE, C. and VELLUZ, L., *Bull. Soc. chim. Fr.*, **9**, 171 (1942); DUFRAISSE, C., RIGAUDY, J., BASSELIER, J. J. and CUONG, N. K., *C.r. hebd. Séanc Acad. Sci., Paris*, **260**, 5031 (1965)
94. BALDWIN, J. E., BASSON, B. H. and KRAUSS, H., *Chem. Commun.*, 984 (1968)
95. MCCAPRA, F. and HANN, R. A., *Chem. Commun.*, 443 (1969) and unpublished work
96. TROZZOLO, A. M., MURRAY, R. W. and WASSERMAN, E., *J. Am. chem. Soc.*, **84**, 4990 (1962)
97. WOODWARD, R. B. and HOFFMANN, R., *The Conservation of Orbital Symmetry*, Academic Press, London, 108 (1970)
98. HAMMOND, G. S., *J. Am. chem. Soc.*, **77**, 334 (1955)
99. ROBINSON, G. W. and DIGIORGIO, V. E., *Can. J. Chem.*, **36**, 31 (1958); RAYNES, W. T., *J. chem. Phys.*, **44**, 2755 (1966)
100. RADZIEZEWSKI, B., *Chem. Ber.*, **10**, 70 (1877)
101. DUFRAISSE, C., ETIENNE, A. and MARTEL, J., *C.r. hebd. Séanc. Acad. Sci., Paris,* 970 (1957); SONNENBERG, J. and WHITE, D. M., *J. Am. chem. Soc.*, **86**, 5685 (1964)
102. MCCAPRA, F., RICHARDSON, D. G. and CHANG, Y. C., *Photochem. Photobiol.*, **4**, 1111 (1965); MCCAPRA, F. and CHANG, Y. C., *Chem. Commun.*, 522 (1966)
103. SUGIYAMA, N., YAMAMOTO, H., OMOTE, Y. and AKUTAGAWA, M., *Bull. chem. Soc., Japan*, **41**, 1917 (1968)
104. JOHNSON, F. H., STACHEL, H. D., TAYLOR, E. C. and SHIMOMURA, O., *Bioluminescence in Progress,* ed, Johnson, F. H. and Haneda, Y., Princeton University Press, 67 (1966); SUGIYAMA, N., AKUTAGAWA, M., GASHA, T. and SAIGA, Y., ibid., 83; PHILBROOK, G. E., AYERS, J. B., GARST, J. F. and TOTTER, J. R., *Photochem. Photobiol.*, **4**, 869 (1965)
105. MCCAPRA, F. and LONG, P. V., to be published
106. MCCAPRA, F., *Pure appl. Chem.*, **24**, 611 (1970)
107. MCCAPRA, F. and WRIGGLESWORTH, R., *Chem. Commun.*, 1256 (1968)
108. MCCAPRA, F. and WRIGGLESWORTH, R., unpublished work
109. *See* for example, WINBERG, H. E., *Chem. Abstr.*, **73**, 5069 (1970); CUTTER, M., CARLOW, K. and SHERMAN, L. M., ibid., 104, 214
110. FLETCHER, A. N. and HELLER, C. A., *J. phys. Chem.*, **71**, 1507 (1967); URRY, W. H. and SHEETO, J., *Photochem. Photobiol.*, **4**, 1067 (1965)
111. CHANDROSS, E. A., *Tetrahedron Lett.*, 761 (1963)
112. RAUHUT, M. M., ROBERTS, B. G. and SEMSEL, A. M., *J. Am. chem. Soc.*, **88**, 3604 (1966)
113. BOLLYKY, L. J., WHITMAN, R. H., ROBERTS, B. G. and RAUHUT, M. M., *J. Am. chem. Soc.*, **89**, 6523 (1967)
114. RAUHUT, M. M., BOLLYKY, L. J., ROBERTS, B. G., ROY, M., WHITMAN, R. H., IANNOTTA, A. V., SEMSEL, A. M. and CLARKE, R. A., *J. Am. chem. Soc.*, **89**, 6515 (1967)

115. RAUHUT, M. M., *Accts Chem. Res.*, **2**, 80 (1969)
116. CORDES, H. F., RICHTER, H. P. and HELLER, C. A., *J. Am. chem. Soc.*, **91**, 7209 (1969)
117. MCCAPRA, F. and SEVERN, D. J., unpublished observations
118. MAULDING, D. R., CLARKE, R. A., ROBERTS, B. G. and RAUHUT, M. M., *J. org. Chem.*, **33**, 250 (1968); BOLLYKY, L. J., ROBERTS, B. G., WHITMAN, R. H. and LANCASTER, J. E., ibid., **34**, 836 (1969)
119. MCCAPRA, F. and RICHARDSON, D. G., *Tetrahedron Lett.*, 3167 (1964)
120. MCCAPRA, F. and RICHARDSON, D. G., unpublished observations
121. RAUHUT, M. M., SHEEHAN, D., CLARKE, R. A., ROBERTS, B. G. and SMALL, A. M., *J. org. Chem.*, **30**, 3587 (1965)
122. MCCAPRA, F., RICHARDSON, D. G. and HANN, R. A., to be published
123. JENCKS, W. P., *Catalysis in Chemistry and Enzymology*, McGraw-Hill Inc., New York (1968); BUNTON, C. A., in *Peroxide Reaction Mechanisms*, ed, Edwards, J. D., Interscience Publishers, New York, 11 (1962)
124. MCCAPRA, F. and FRANCOIS, V. P., to be published
125. GLEU, K. and PETSCH, P., *Angew. Chem.*, **48**, 57 (1935)
126. WEBER, K., *Z. phys. Chem.*, **B50**, 100 (1941)
127. ERDEY, L., TACKACS, J. and BURZAS, I., *Acta chim. Hung.*, **39**, 295 (1963)
128. SCHALES, O., *Chem. Ber.*, **72**, 1155 (1939)
129. TAMAMUSHI, B. and AKIYAMA, H., *Trans. Faraday Soc.*, **35**, 491 (1939)
130. WEBER, K. and OCHSENFELD, W., *Z. phys. Chem.*, **51**, 63 (1942)
131. (a) KARIAKIN, A. V., *Optics Spectrosc.*, **7**, 75 (1959); (b) RYZHIKOV, B. D., *Bull. Acad. Sci. USSR phys. Ser.*, **20**, 487 (1956)
132. MAEDA, K. and HAYASHI, T., *Bull. chem. Soc., Japan*, **40**, 169 (1967)
133. (a) TOTTER, J. R., *Photochem. Photobiol.*, **3**, 231 (1964); (b) SPRUIT, C. J. and SPRUIT VAN DER BURG, A., in *The Luminescence of Biological Systems*, ed, Johnson, F. H., Washington (1955); SPRUIT VAN DER BURG, A., *Compt. rend.*, **69**, 1525 (1950); SVESHNIKOV, B. YA., *Izv. Akad. Nauk. USSR, Ser. fiz.*, **9**, 341 (1945)
134. LEGG, K. D. and HERCULES, D. M., *J. Am. chem. Soc.*, **91**, 1902 (1969)
135. ERDEY, L., *Acta. chim. Acad. Sci. hung.*, **3**, 81 (1953)
136. JANZEN, E. G., PICKETT, J. B., HAPP, J. W. and DEANGELIS, W., *J. org. Chem.*, **35**, 88 (1970); HAPP, J. W. and JANZEN, E. G., ibid., 96
137. FÖRSTER, TH., *Z. Electrochem.*, **53**, 93 (1949)
138. KAUTSKY, H. and KAISER, H., *Naturwissenschaften*, **31**, 505 (1943); ERDEY, L., *Acta chim. hung.*, **3**, 95 (1953); TOTTER, J. R. and PHILBROOK, G., *Photochem. Photobiol.*, **5**, 177 (1966); TOTTER, J. R., in *Bioluminescence in Progress*, eds, Johnson, F. H. and Haneda, Y., Princeton University Press, 25 (1966)
139. GLEU, K. and SCHUBERT, A., *Ber. dt. chem. Ges.*, **73B**, 805 (1940)
140. GLEU, K. and SCHAARSCHMIDT, R., *Ber. dt. chem. Ges.*, **73B**, 909 (1940); GLEU, K. and NITZSCHE, S., *J. prakt. Chem.*, **153**, 233 (1939)
141. KORMENDY, K., *Acta chim. hung.*, **21**, 83 (1959), through *Chem. Abstr.*, **54**, 18524 (1960)
142. BRAUN, A., DORABIALSKA, A. and REIMSCHUESSEL, W., *Roczniki chem.*, **40**, 247 (1966), *Chem. Abstr.*, **65**, 1620 (1966); CHRZASZCZEWSKA, A., BRAUN, A. and NOWACZYK, M., *Soc. Sci. Lodziensis Acta Chim.*, **3**, 93 (1958), *Chem. Abstr.*, **53**, 13148 (1959)
143. MASON, S. F. and ROBERTS, D. R., *Chem. Commun.*, 476 (1967)
144. HENRY, R. A. and HELLER, C. A., Joint Meeting Can. Inst. Chem. and Am. Chem. Soc., Toronto, May 1969
145. HARVEY, E. N., *Bioluminescence*, Academic Press, New York (1952); *Light and Life*, eds, McElroy, W. D. and Gloss, B., The Johns Hopkins Press, Baltimore (1961); *Light—Physical and Biological Action*, eds, Seliger, H. H. and McElroy, W. D., Academic Press, New York (1965); *Bioluminescence in Progress*, eds, Johnson, F. H. and Haneda, Y., Princeton University Press (1966); HASTINGS, J. W., *A. Rev. Biochem.*, **37**, 597 (1968); JOHNSON, F. H., in *Comprehensive Biochemistry*, eds, Florkin, M. and Stotz, E. H., **27**, 79 (1967)
146. GOTO, T. and KISHI, Y., *Angew. Chem. Int. Edn*, **7**, 407 (1968)
147. MCELROY, W. D., *Fedn Proc. Fedn. Am. Socs exp. Biol.*, **19**, 941 (1960)
148. WHITE, E. H., MCCAPRA, F. and FIELD, W. F., *J. Am. chem. Soc.*, **85**, 337 (1963)
149. SELIGER, H. H. and MCELROY, W. D., *Light—Biological and Physical Action*, Academic Press, New York, 182 (1965)

150. GOTO, T., KISHI, Y., MATSURA, S., INOUE, S. and SHIMOMURA, O., *Tetrahedron Lett.*, 2847 (1968)
151. HOPKINS, T. A., SELIGER, H. H., WHITE, E. H. and CASS, M. W., *J. Am. chem. Soc.*, **89**, 7148 (1967)
152. MCCAPRA, F., CHANG, Y. C. and FRANCOIS, V. P., *Chem. Commun.*, **22** (1968)
153. DELUCA, M. and DEMPSEY, M. E., *Biochim. biophys. Res. Commun.*, **40**, 117 (1970)
154. WHITE, E. H., RAPOPORT, E., HOPKINS, T. A. and SELIGER, H. H., *J. Am. chem. Soc.*, **91**, 2178 (1969)
155. SUZUKI, N., SATO, M., NISHIKAWA, K. and GOTO, T., *Tetrahedron Lett.*, 4683 (1969)
156. JOHNSON, F. H., SHIMOMURA, O., SAIGA, Y., GERSHMAN, L. C., REYNOLDS, G. T. and WATERS, J. R., *J. cell. comp. Physiol.*, **60**, 85 (1962)
157. KISHI, Y., GOTO, T., HIRATA, Y., SHIMOMURA, O. and JOHNSON, F. H., *Tetrahedron Lett.*, 3427 (1966); KISHI, Y., GOTO, T., INOUE, S., SUGIURA, S. and KISHIMOTO, H., ibid., 3445
158. JOHNSON, F. H., STACHEL, H. D., TAYLOR, E. C. and SHIMOMURA, O., *Bioluminescence in Progress*, eds, Johnson, F. H. and Haneda, Y., Princeton University Press, 385 (1966)
159. MCCAPRA, F. and CHANG, Y. C., *Chem. Commun.*, 1011 (1967)
160. STONE, H., *Biochem. biophys. Res. Commun.*, **31**, 386 (1968)
161. GOTO, T., INOUE, S., SUGIURA, S., NISHIKAWA, K., ISOBE, M. and ABE, Y., *Tetrahedron Lett.*, 4035 (1968)
162. SHIMOMURA, O., JOHNSON, F. H. and MASUGI, T., *Science, N.Y.*, **164**, 1299 (1969)
163. GOTO, T. and FUKATSU, H., *Tetrahedron Lett.*, 4299 (1969)
164. MCCAPRA, F. and WRIGGLESWORTH, R., *Chem. Commun.*, 1256 (1968)
165. SHIMOMURA, O., JOHNSON, F. H. and HANEDA, Y., *Bioluminescence in Progress*, eds, Johnson, F. H. and Haneda, Y., Princeton University Press, 391 (1966)
166. SHIMOMURA, O. and JOHNSON, F. H., *Biochemistry*, **7**, 1734 (1968)
167. FRACHEBOUD, M. G., SHIMOMURA, O., HILL, R. K. and JOHNSON, F. H., *Tetrahedron Lett.*, 3951 (1969)
168. MAKATSUBO, F., KISHI, Y. and GOTO, T., *Tetrahedron Lett.*, 381 (1970)
169. MCCAPRA, F. and WRIGGLESWORTH, R., *Chem. Commun.*, 91 (1969)

7

SOME ASPECTS OF RECENT WORK ON NITRATION

S. R. Hartshorn and K. Schofield

INTRODUCTION	278
NITRATING SYSTEMS	279
Nitric Acid in Inert Organic Solvents	279
Nitration with Nitronium Salts	280
Nitration via Nitrosation	281
Nitration in Acetic Anhydride	282
NITRATION AND AROMATIC REACTIVITY	293
Limiting Rates of Nitration	293
Methyl- and Halogeno-benzenes	295
Anisole	298
Acetanilide and Related Anilides	300
Methyl Phenethyl Ether	302
Positive Poles	303
Molecular Orbital Treatments of Aromatic Reactivity	308

INTRODUCTION

By the early 1950s the major features of nitration with nitric acid, and with solutions of nitric acid in mineral acids and in inert organic solvents, had been elucidated. There was overwhelming evidence that in all of these systems the nitronium ion was the effective electrophile. The processes involved could be written as follows, in the first case heterolysis of nitric acid being dependent upon auto-protonation and in the second being dependent upon the action of a stronger acid (HX):

$$2HNO_3 \underset{k_2}{\overset{k_1}{\rightleftarrows}} H_2NO_3^+ + NO_3^- \qquad HNO_3 + HX \rightleftarrows H_2NO_3^+ + X^-$$

$$H_2NO_3^+ \underset{k_4}{\overset{k_3}{\rightleftarrows}} NO_2^+ + H_2O$$

$$ArH + NO_2^+ \overset{k_5}{\rightarrow} Products$$

As regards the subsequent process, that of reaction of the nitronium ion with an aromatic molecule, it was concluded that nitration fits the pattern of a two-step reaction now generally accepted for electrophilic substitutions[1].

To name only a few of the workers who contributed to the establishment of the foundations of the subject could be invidious, but it can hardly be disputed that the major contributions came from Ingold and Hughes and their co-workers, from Bennett and Williams, from Westheimer and

Kharasch, and from Melander. Their work has been extensively reviewed[2].

There is evidence for the ability of dinitrogen pentoxide to operate as the nitrating agent when it is used in organic solvents[3], and also that it is the effective electrophile in some reactions in which benzoyl nitrate is used[4]. Nitric acid and acetic anhydride react to give acetyl nitrate and this, usually in acetic anhydride, remains the one important reagent for which the mode of reaction is not yet clarified[2e].

Finally, it was established that with very reactive aromatic substances a process of nitration via nitrosation can become important[5].

In the last ten years a considerable amount of new work on nitration has appeared. Like the earlier work it has had two main themes; the characteristics of the nitrating systems, and the use of nitration for the study of aromatic reactivity. Several reviews of this work have appeared[2d,e,6,7], and for this reason the present article will concentrate on selected aspects of it. Accordingly, some recent work on nitrating systems will be discussed, and also some results which have been obtained from nitration which bear on the problems of aromatic reactivity.

NITRATING SYSTEMS

Nitric Acid in Inert Organic Solvents

The organic solvents most extensively studied by Ingold and his co-workers were nitromethane and acetic acid[2]. Recently, nitration in carbon tetrachloride has been studied[8], and zeroth-order nitration with respect to the concentration of the aromatic compound observed with toluene, the xylenes, and mesitylene. The zeroth-order rate constant depended approximately on the concentration of nitric acid to the fifth power, and the Arrhenius activation energy was about -21 kJ mol^{-1}. Nitronium ion formation was represented as follows[8b]

$$5\ HNO_3 \rightleftarrows \begin{matrix} NO_2^+ \\ \vdots \\ NO_3^-.HNO_3 \end{matrix} + \begin{matrix} H_3O^+ \\ \vdots \\ NO_3^-.HNO_3 \end{matrix}$$

This scheme would lead to a high negative entropy change, and since the concentration of molecular aggregates might decrease with rise in temperature the characteristics of the reaction already mentioned would not be surprising. Carbon tetrachloride differs from other organic solvents which have been used because only low concentrations of nitric acid can be used if homogeneous conditions are to be maintained. The effect of temperature on other zeroth-order processes has not been extensively investigated, but nitromethane appears to give the more familiar type of dependence of rate upon temperature[9,10].

One aspect of the process of nitration in an inert organic solvent implicit in the reaction scheme summarised in the Introduction has not been very fully documented hitherto; the scheme requires that large additions of water to the reaction system shall convert zeroth- into first-order nitrations, because then the bulk reactivity of water surpasses that of the aromatic, $k_4[H_2O] > k_5[ArH]$. (Small quantities of water reduce the zeroth-order rate without altering the kinetic form by means of the reaction $H_2O + H_2NO_3^+ \rightleftarrows$

$H_3O^+ + HNO_3$.) It was reported only that addition of water up to 5% by weight of the *solvent* acetic acid converted the nitrations of toluene and t-butylbenzene from zeroth- to first-order forms[10]. Subsequently, first-order kinetic results were reported for the nitrations at 45°C of toluene and t-butylbenzene in acetic acid containing 10% by weight of water[11].

More recently, zeroth-order nitration has been observed for the reaction of mesitylene with nitric acid ($[HNO_3] = 4.9$ mol l^{-1}) in sulpholan under conditions where the reactions of toluene and benzene were of mixed- and first-order, respectively[12]. In all of the organic solvents examined the zeroth-order rate is dependent upon the concentration of nitric acid to a high power, and sulpholan and acetic acid resemble each other whilst nitromethane is a markedly 'faster' solvent. The superiority of nitromethane in this respect may be related to the fact that solutions in it of sulphuric acid are more acidic than are comparable solutions in the other solvents, and that it is a better solvent for the ionisation of nitric acid[12,13].

The conversion of zeroth-order into first-order nitrations by the addition of water has been observed for reactions occurring in both nitromethane and sulpholan. In the case of sulpholan the addition of about 5% by weight of water (5% by weight of the *total solution*) converted the reaction of toluene and less reactive compounds from the zeroth- to the first-order form. The addition of 7.5% of water achieved the same result with mesitylene and other reactive compounds. In the case of nitromethane as much as 15% of water was needed[12].

Similar experiments with acetic acid as the solvent have recently been carried out[14]. As in the original work[10], benzene, toluene, *m*-xylene, and mesitylene could all be nitrated in zeroth-order processes. Also, the addition of water was observed to convert these zeroth- into first-order processes. Precise comparisons cannot be made because the thresholds of conversion of zeroth- into first-order reactions were not closely studied, and also because comparisons were not made at precisely the same concentrations of nitric acid. However, the behaviour of the various solvents was roughly as would be expected from their relative abilities to maintain zeroth-order reactions[14b].

Nitration with Nitronium Salts

Since Hantzsch first prepared a mixture of nitronium and hydroxonium salts[15], from which later workers separated nitronium perchlorate by fractional crystallisation from nitromethane[16], a number of nitronium salts have become available through the efforts of Olah and his co-workers[6]. The commonest of these is nitronium tetrafluoroborate, now a commercial product.

The low solubilities of nitronium salts in unreactive solvents raises problems over their use in nitrations. Preparatively they have been used neat, in ether, or in sulpholan, and found to give high yields of nitro-compounds. Their use is advantageous when readily hydrolysed functional groups are present[17,18].

Studies of nitrations with nitronium salts have attracted wide attention for two reasons. First, as a consequence of their use in competition reactions between benzene and its derivatives, especially the alkylbenzenes, it was concluded that rate differences almost disappear so that, for example, toluene appears to be very little more reactive than benzene[19]. To account for these observations a new mechanism of nitration was proposed in which the rate-determining step was the formation of a π-complex[19a]. The significance of the experimental work on which this conclusion was based has been questioned[2b,d,20,21], and it is believed that, in particular, the results of Christy, Ridd and Stears[22] on the nitration of dibenzyl prove the experiments, and therefore the conclusions drawn from them, to be invalid owing to the circumstance that under the conditions used the slow process is the mixing of the reactants.

It is important to stress that the idea discussed by Olah and his co-workers of a substitution reaction for which the rate-controlling step has become the formation of a π-complex, rather than the more usual σ-complex, is not excluded on principle (*see* below), but that the evidence proffered is not acceptable.

The second reason why the experiments with nitronium salts attracted attention was that they led Olah and his co-workers to doubt the efficacy of the nitronium ion as the electrophile causing nitration of activated molecules under conditions where the nitronium ion was not pre-formed[13,19a]. Such a conclusion disregards the huge amount of evidence regarding mechanisms of nitration with solutions of nitric acid in mineral acids or in inert organic solvents[2] and must be rejected.

Nitration via Nitrosation

This special nitration process was established by Ingold and his co-workers in the first place for the reactions of phenols and phenolic ethers with solutions of nitric acid in acetic acid, and for aniline derivatives under various conditions[5]. That it was not limited to those structural types, but occurred more generally with compounds of high reactivity, was shown by its occurrence with mesitylene. The importance of this work has not been sufficiently appreciated, and studies of the reactivities of highly active aromatic compounds in nitration have been undertaken either without the necessary care to prevent nitrosation or without recognition of its occurrence.

In a kinetic study of the relative rates of reaction of several compounds in aqueous acetic acid the value for naphthalene with respect to that for benzene was found to be 62. In contrast, that for 1-methylnaphthalene was apparently 20,600, suggesting a quite unreasonable activating power for the methyl group. Despite the very low concentration of nitrous acid present ($[HNO_2] \approx 5 \times 10^{-4}$ mol l^{-1}) the result is almost certainly to be ascribed to nitrosation[14]. Similar observations have been made with 1,6-dimethylnaphthalene, phenol derivatives, and anthanthrene reacting in solutions of nitric acid in aqueous sulpholan[12], and also with a number of very reactive compounds being nitrated with solutions prepared from pure nitric acid and acetic anhydride[23].

Under the latter conditions 1-methylnaphthalene is nitrated smoothly at a rate equal to that of *m*-xylene. This behaviour should be compared with that in acetic acid, where nitration via nitrosation apparently occurs. Thus, changing the solvent from acetic acid to acetic anhydride seems to remove the complications due to nitrosation for this substrate. A similar effect is observed with anisole, and has also been reported more recently with thiophen[24a]. The cause of this effect is not known, but it might be significant that for nitrations in acetic acid much higher concentrations of nitric acid are usually employed than for nitrations in acetic anhydride.

In a variety of nitrating conditions (*see* below) a limiting rate of nitration, usually associated with the nitronium ion, has been established at a particular level of aromatic reactivity. As the reactivity of the substrate is then further increased a point is generally reached at which the rate of nitration suddenly shows a marked increase over the limiting rate; as in the cases mentioned above, this increase can usually be associated with the onset of nitration via nitrosation. Strictly, there is a slight possibility that it may also signify the occurrence of nitration by some electrophile other than the nitronium ion, and less reactive than it[2e,23].

In several studies, undertaken with the aim of using nitration to study relative reactivities in electrophilic substitutions, misleading results may have been obtained because the possibility of nitrosation has been neglected. These include work with naphthalene derivatives[24,25], with polynuclear hydrocarbons and related compounds[26], and with pyrrole[27].

The mechanism for the formation of nitro-products via initial nitrosation can probably be represented by the following scheme,

$$ArH + NOX \rightarrow ArNO + HX$$

$$ArNO + HNO_3 \rightarrow ArNO_2 + HNO_2$$

in which the rate-determining step may be either the nitrosation or the oxidation. It has recently been shown[28a,b], that in carbon tetrachloride dinitrogen tetroxide may function as both the nitrating and the oxidising agent, and that with anisole and *p*-dimethoxybenzene the first step is the slower of the two. However, lack of experimental evidence does not permit any more general conclusions to be reached.

Nitration in Acetic Anhydride

Of all the media which have been used in nitration that prepared by dissolving nitric acid in acetic anhydride has proved to be the most perplexing; the effective electrophile has not been identified, from some substrates higher proportions of *ortho*-substitution are obtained than are given by other reagents, and from some compounds not only nitro-derivatives are produced, but also acetoxylated compounds in significant amounts. The properties and behaviour of solutions of nitric acid in acetic anhydride have been reviewed[2d,e]; this article will concentrate on the kinetic aspects of nitration with these solutions and on the conclusions concerning mechanism which these yield, and also on the phenomenon of acetoxylation.

Kinetic studies Acetic anhydride differs from most of the organic solvents used in nitration in reacting with nitric acid:

$$Ac_2O + HNO_3 \rightleftharpoons AcONO_2 + AcOH$$

A further, much slower reaction gives finally tetranitromethane[2d,e]. Experimental conditions are easily arranged to avoid complications from this latter reaction in studies of nitration.

SOME ASPECTS OF RECENT WORK ON NITRATION

The rates of nitration of activated compounds relative to that of benzene are significantly higher in acetic anhydride than in other organic solvents. Thus, the relative rate for toluene, which has an average value of about 25 with most organic solvents, is 50 with acetic anhydride (*see* below)[29].

The kinetics of nitrations in solutions of acetyl nitrate in acetic anhydride have been studied under conditions where they showed a zeroth-order dependence on the concentration of aromatic[14a,30], and also under first-, order conditions[29,31–34]. In the latter circumstance, in the case of benzene, the rate was found to depend on the concentration of acetyl nitrate to the third power (strictly, upon the third power of the initial concentration of nitric acid, $[HNO_3]_0^3$)[31]. In these experiments the concentration of nitric acid was $[HNO_3]_0 < 1$ mol l^{-1}. A later study gave apparently conflicting

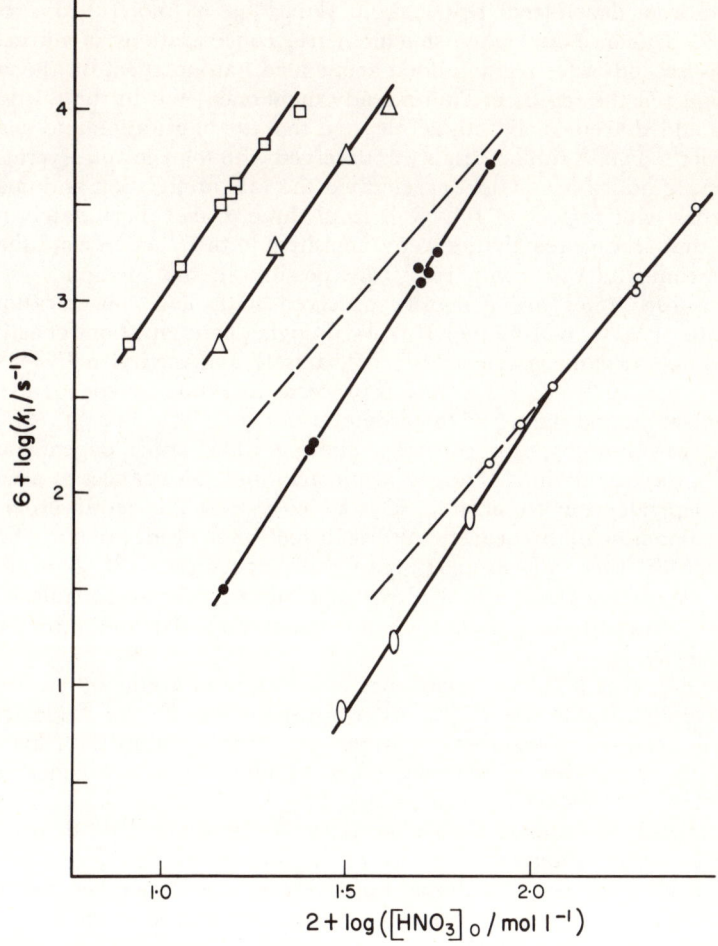

Figure 7.1. The variation of the observed first-order rate constants with the initial concentration of nitric acid for nitration in acetic anhydride at 25°C
○ Benzene; ● toluene; △ *m*-xylene; □ mesitylene (The effect of added acetic acid is shown by the broken lines)

results; in it the concentration of nitric acid was generally $[HNO_3]_0 >$ 1 mol l^{-1} and the kinetic order with respect to it was found to be two[32]. The nitric acid used in this work contained 10% of water, which necessarily introduced a considerable proportion of acetic acid into the medium. However, it was shown that further dilution with acetic acid, up to 50 moles %, had no effect on the rate of nitration, and so it was assumed that the adventitious acetic acid would also have no effect.

More recent results[29,34] have reconciled the earlier observations. The first-order rate constants for the nitration of benzene, toluene, m-xylene and mesitylene were found to vary with the stoichiometric concentration of nitric acid as shown in *Figure 7.1*.

An approximately third-order dependence on $[HNO_3]_0$ was observed with mesitylene, m-xylene and toluene, but with benzene the order changed from a third-order dependence below about $[HNO_3]_0 = 1$ mol l^{-1} to a second-order dependence at higher stoichiometric concentrations of nitric acid. For the second-order region added acetic acid had no effect on the rate of nitration (cf. the results of Paul already mentioned) but in the third-order region added acetic acid both accelerated the rate of nitration and changed the order to two. A similar result was observed with toluene and several other substrates; added acetic acid accelerated the rate of nitration and changed the order with respect to $[HNO_3]_0$ from three to two. Fortunately it was found that relative reactivities were insensitive to the reaction conditions, so that meaningful rate comparisons are possible in this medium.

First-order rates are generally observed with low concentrations of aromatic ($[ArH]_0 \approx 10^{-3}$ mol l^{-1}). With higher concentrations of activated compounds (o- and m-xylene[14a,23,30,33], anisole and mesitylene[14a,33]) when $[ArH]_0 > c.$ 10^{-1} mol l^{-1} nitration proceeds at a rate independent of the concentration and nature of the aromatic.

The zeroth-order rate constants show a third-order dependence on $[HNO_3]_0$ in the absence of added acetic acid, but this changes to a second-order dependence when acetic acid is added[14a,30,33]. The zeroth-order rates of nitration in solutions of acetyl nitrate in acetic anhydride are much greater than zeroth-order rates in inert organic solvents (*Figure 7.2*). Thus, for the same initial concentrations of nitric acid, nitration in acetic anhydride is $c.$ 5×10^5 and 10^4 times faster than nitration in sulpholan and nitromethane, respectively.

The effects of additives upon rates of nitration in acetic anhydride have been studied. In the case of first-order nitrations the effect of acetic acid has already been described above. Nitrate ions depress the rates of first-order reactions; in the case of benzene, added sodium nitrate at a concentration of 10^{-3} mol l^{-1} reduced the rate four-fold[32,34].

As already mentioned, added acetic acid alters the dependence of the observed zeroth-order rate constant on the stoichiometric concentrations of nitric acid. With o-xylene the addition of acetic acid increases the rate in proportion to its concentration, and the rate of nitration follows the expression:

$$\text{rate} = k[HNO_3]_0^2[HOAc]$$

in the presence of sufficient acetic acid (2·2 mol l^{-1})[30]. Similarly, added

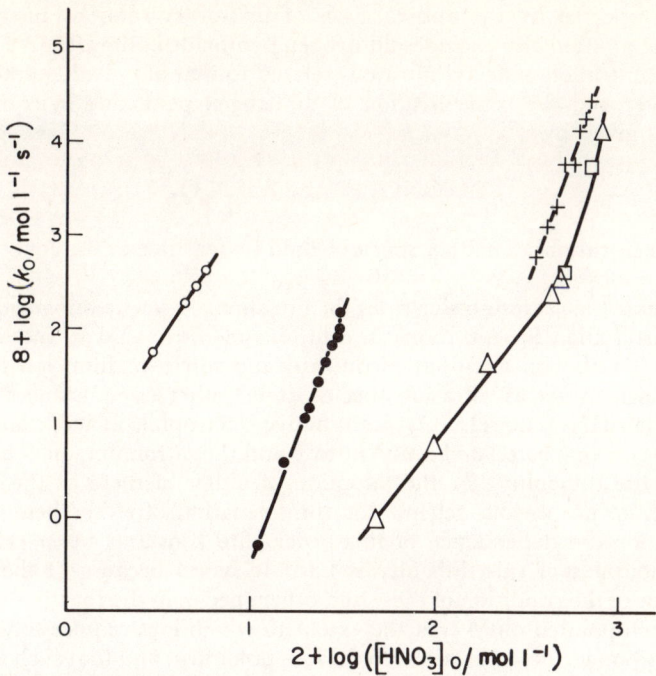

Figure 7.2. The variation of the observed zeroth-order rate constants with the initial concentration of nitric acid for nitration in various organic solvents
○ Acetic anhydride, 25°C; ● carbon tetrachloride, 25°C; + nitromethane, 0°C; △ sulpholan, 25°C; □ acetic acid, 20°C

acetic acid has been found to accelerate the zeroth-order nitration of mesitylene[14a,33]. The zeroth-order nitration of o-xylene was also found to be strongly catalysed by sulphuric acid, without the kinetic form of the reaction being affected[30].

Added lithium nitrate has a pronounced effect on the zeroth-order nitration of o-xylene; the presence of 6×10^{-4} mol l^{-1} of nitrate is sufficient to reduce the rate by a factor of four, and to modify the kinetic form from a zeroth-order dependence on the concentration of aromatic. Small concentrations of sodium nitrate similarly influence the nitration of anisole[33]. It is interesting to recall that the anti-catalytic effect of nitrate upon nitration in inert organic solvents has no influence on the kinetic order[2].

The mechanism of nitration. The species available in solutions of acetyl nitrate in acetic anhydride which might effect nitration could conceivably include the following:

$$HNO_3, \quad H_2NO_3^+, \quad AcONO_2, \quad AcONO_2H^+, \quad NO_2^+, \quad N_2O_5$$

Arguments have been put forward for the involvement of $AcONO_2H^{+[30]}$, $NO_2^{+[2d,32]}$, and $N_2O_5{}^{[35,36]}$. The kinetic results allow some of the species listed to be omitted from further consideration. The observation of nitration at a rate independent of the concentration and nature of the aromatic immediately excludes acetyl nitrate as the reactive species. This conclusion is also supported by the anti-catalysis of first-order rates by nitrate ions, which observation also excludes dinitrogen pentoxide as the effective reagent. The concentration of acetyl nitrate is related to that of the added nitric acid (*see* above) and the concentration of dinitrogen pentoxide in equilibrium with it is given by

$$2AcONO_2 \rightleftharpoons Ac_2O + N_2O_5$$

The concentrations of neither species would be sensitive to the concentration of nitrate ions[2d].

The facts that zeroth-order rates for nitrations in acetic anhydride are so much faster than for nitrations in solutions of nitric acid in inert organic solvents (*see* above), and that nitric acid and nitric acidium ion have not been shown to act as nitrating species under other circumstances[2] suggest that neither HNO_3 nor $H_2NO_3^+$ is the active electrophile in acetic anhydride.

There remain protonated acetyl nitrate and the nitronium ion. The kinetic results raise difficulties for the adoption of either of these as the effective electrophile, no obvious scheme for the formation of either being able to account for the dependence of first-order rate constants upon $[HNO_3]_0$[3]. The seriousness of this difficulty is hard to assess because of the present ignorance of the condition of these and other species in the reaction medium; it has been pointed out[2d] that the extent to which ions require solvation by nitric acid molecules in acetic anhydride is unknown, and that such solvation would influence the apparent kinetic order with respect to $[HNO_3]_0$. The problem of ion-pairs also arises, and the species $AcONO_2H^+NO_3^-$ has been proposed as the electrophile, its formation being supposed to involve a slow proton transfer[30]; the proposal does not account for the anti-catalytic effect of nitrate ions. The similarity in the performance of the electrophile, as regards intermolecular and intramolecular selectivities, to that of the nitronium ion has also been interpreted as evidence that the electrophile is the nitronium ion[2d]. This point is commented on below.

It has not yet proved possible to devise a mechanism which fits all the facts of nitration in acetic anhydride, and, whilst the process gives results recalling those in which the nitronium ion operates, a transition state composed solely of aromatic and nitronium ion seems to be precluded.

Acetoxylation and other side reactions Present knowledge of the acetoxylation which accompanies the nitration of certain compounds with solutions of acetyl nitrate in acetic anhydride is due mainly to the efforts of Fischer and Vaughan and their co-workers[30,37]. *Table 7.1* summarises the results, and shows that with o-xylene acetoxylation is the dominant reaction. For acetoxylation to occur it is not sufficient that a compound be one of high reactivity towards electrophiles, and in the light of what follows it is no longer surprising that acetoxylation does not occur with anisole or naphthalene.

The ratio of acetoxylation to nitration is not changed by the addition of sulphuric acid, acetic acid or lithium nitrate to the reaction solution. Like nitration, acetoxylation can occur by a zeroth-order process. Such observations led to the conclusion that the same electrophile, or intermediate precursor, was responsible for both reactions; it was thought to be protonated acetyl nitrate[30]. However, Ridd pointed out that an alternative

Table 7.1. THE YIELD OF ACETOXYLATION PRODUCT AS A PERCENTAGE OF THE TOTAL PRODUCTS FOR REACTION WITH ACETYL NITRATE IN ACETIC ANHYDRIDE

Compound	Position substituted	Yield, %	Ref.
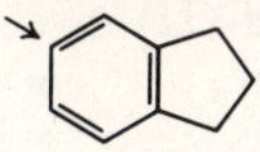 indane	β-	26	38
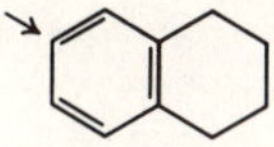 tetralin	β-	28	38
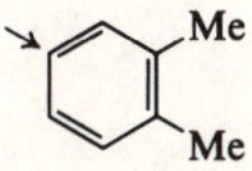 o-xylene	4-	58·6	37c
hemimellitene [a]	4-	10·6	37c
	5-	35·3	
pseudocumene [b]	3-	10·3	37c
	5-	25·6	
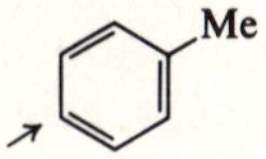 m-xylene (attack shown)	4-	3	37c
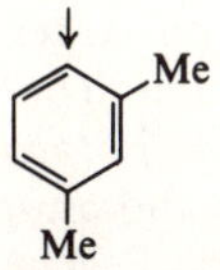 m-xylene	4-	0·7	37c

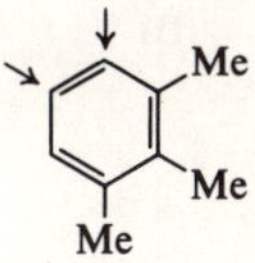

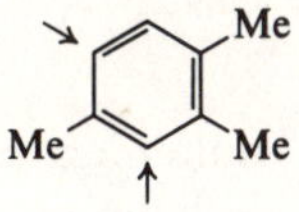

[a] Hemimellitene; [b] pseudocumene

mechanism was possible in which the Wheland intermediate formed in nitration added acetate; subsequent elimination would give the acetoxylated aromatic product[2d].

Recently, support for this idea has been obtained with the isolation of diene intermediates from o-xylene[37d,e], p-xylene[37e] and hemimellitene[37e]. Two intermediates were isolated in each case and shown by spectroscopic studies to be *cis-trans*-isomers[37e]. The adducts from o-xylene are shown below; the figures in parenthesis refer to the τ-values of the aromatic protons measured at 60 MHz in carbon tetrachloride[37d,e]. (One of the adducts was originally assumed to be the 3-nitro-4-acetoxy compound[37d], but this was subsequently shown to be incorrect[37e]).

The 1,4-diene adducts give aryl acetates on decomposition either in aqueous acidic media or when subjected to vapour phase chromatography, and the available evidence suggests that acetates are the only products of decomposition[14a,37e]. Thus the intermediates (I) and (II) give rise to 3,4-dimethylphenyl acetate. The adducts obtained with *p*-xylene (III) give 2,5-dimethylphenyl acetate (IV) on decomposition in acetic acid at 60°C. A similar decomposition in propionic acid containing sodium propionate did not lead to incorporation of propionate, so that the breakdown apparently involves an intramolecular 1,2-acetate shift[37e].

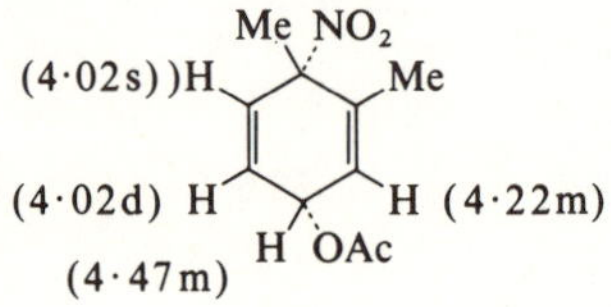

(4·02s))H   Me NO₂   Me   (3·97s) H   Me NO₂   Me
(4·02d) H   H (4·22m)   (3·97d) H   H (4·13m)
         H OAc   H OAc
(4·47m)   (4·28m)
   (I) 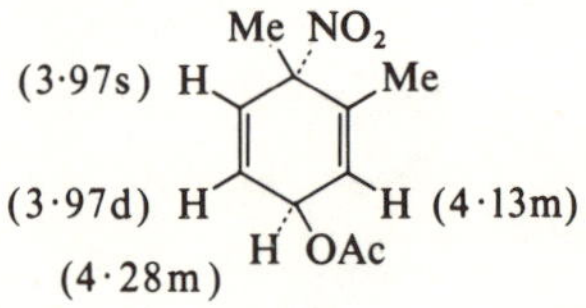   (II)

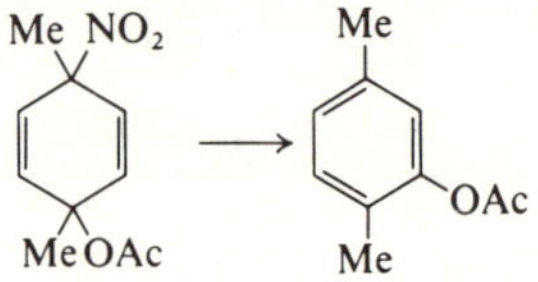

   Me NO₂   Me
              →

   MeOAc   OAc
                Me
   (III)   (IV)

The reactions of several derivatives of benzene containing methyl or alkylene groups with acetyl nitrate in acetic anhydride have been studied by examining the n.m.r. spectra of the solutions in which reaction was occurring[14a,34]. Thus, the reacting solutions of *o*-xylene and tetralin give signals in the region $\tau = 3\cdot8$ to $\tau = 4\cdot5$, characteristic of the diene intermediates (the signals at higher field, of the methyl groups, cannot be used for identification purposes because of the strong solvent signal). With both *o*-xylene and tetralin appreciable amounts of aryl acetates may be isolated after the reaction solution has been decomposed with water (*Table 7.1*). In contrast, the reacting solutions from toluene, mesitylene, and anisole generated no n.m.r. signals in the region 3·8–4·5 τ, in agreement with absence of acetoxylated products from the mesitylene and anisole reactions; toluene gives a very small yield of *p*-tolyl acetate (*Table 7.1*).

Such a study of the reacting solutions also permits the observation of the slow thermal decomposition of the adducts; signals from the latter decay and are replaced by signals from the acetoxylated aromatic products.

It is noteworthy (*Table 7.1*) that all of the compounds which undergo extensive acetoxylation have an alkylated ring position activated towards electrophilic attack by other substituents. The following scheme, illustrated here with *o*-xylene, seems to be consistent with the known facts.

Initial attack by the electrophile at a ring position bearing a hydrogen atom leads to normal nitration (via a Wheland intermediate), but initial attack at a ring position bearing an alkyl or alkylene substituent leads to the formation of a Wheland intermediate of the type (V). If the intermediate (V) is stabilised sufficiently, by additional ring substituents, then attack by a nucleophile to give the diene adduct can compete favourably with the decomposition of (V) by reversal of the first step. Loss of nitrous acid from the adduct leads to the final aryl acetate.

This mechanism can easily account for the almost constant ratio of acetoxylation to nitration observed under different reaction conditions[30]. The ratio of nitro- to acetoxy-product is determined by the relative stabilities (and hence importance in the reaction pathway) of the Wheland intermediates produced by attack of the nitrating agent at ring positions bearing a hydrogen atom and an alkyl group respectively. If a change in the reaction conditions does not significantly alter the discrimination of the nitrating agent amongst the different nuclear positions, then a constant ratio of nitration to acetoxylation should be observed. The observation of zeroth-order kinetics for acetoxylation is also readily explained.

If the additional activating substituent is in the position *para* to the initial point of attack (e.g. as with *p*-xylene), then formation of the adduct involves attachment of acetate at an alkylated ring position. In this case formation of the final aromatic acetate does not involve loss of the alkyl group, but an intramolecular 1,2-acetate shift and loss of nitrous acid. It is thus easy to account for the orientation of acetoxylated products; acetoxylation will always occur in the position *para* to an alkylated ring position. In the case of a *para*-disubstituted substrate a 1,2-acetate shift follows.

The results of *Table 7.1* are consistent with this conclusion. Indan, tetralin, *o*-xylene, toluene and *m*-xylene are each capable of forming only one type of 1,4-diene in which the nitro group is attached to an alkylated ring position, and each produces only one aryl acetate in which the acetoxy group is *para* to an alkylated ring position. With toluene and *m*-xylene the initial Wheland intermediate—analogous to (V)—is not stabilised by additional substituents, and each compound gives only a small percentage of an aryl acetate. Hemimellitene and pseudocumene are both able to form two different 1,4-adducts and both produce two aryl acetates. With hemimellitene the intermediate (VI) will be more stable than the alternative (VII) so that more of the 5-acetate should be produced, as is observed. The results for pseudocumene can be accounted for qualitatively by assuming either that only one adduct is formed and undergoes 1,2-acetate shifts to give both aryl acetates, or that some, or all, of the 5-acetate is produced via a second adduct.

The above assumption that a Wheland intermediate precedes the 1,4-adduct receives some support from the observation that both *cis-* and *trans-*isomers are formed, so that not all the adduct can arise from a concerted 1,4-addition. The assumed intermediacy of a Wheland intermediate is also useful in explaining the formation of side products in some reactions of *p*-disubstituted compounds. A recent study of the reaction of *p*-xylene with acetyl nitrate in acetic anhydride has shown that the following products may be isolated: nitro-*p*-xylene (VIII), 1,4-dimethyl-2,5-dinitrobenzene, *p*-tolylnitromethane (IX), *p*-methylbenzyl acetate (X) and 2,5-dimethylphenyl acetate (XI)[14a,34]. By observing the changes in the n.m.r. spectrum of the reaction solution it was shown that the ring acetate was produced via an intermediate adduct (*see* above). The formation of the dinitro compound is

unusual and cannot be readily explained, but the other products could arise via a Wheland intermediate formed by electrophilic attack at a methylated ring position, as in the following tentative scheme. This modifies an earlier proposal[2e] to take into account the formation of (XI) by a 1,2-shift of acetoxyl[37e].

The results obtained when 1,4-dimethylnaphthalene is treated with acetyl nitrate in acetic anhydride are also interesting in this respect[25]. When the reaction was quenched after 24 h a pale yellow solid with an ultraviolet spectrum similar to that of α-nitronaphthalene was produced. If the mixture was allowed to stand for five days the product was 1-methyl-4-nitromethylnaphthalene. It was suggested that the intermediate was 1,4-dimethyl-5-nitronaphthalene which underwent acid-catalysed rearrangement to the final product. However, it was later pointed out that this was improbable, and an alternative structure (XIII) was suggested for the intermediate, together with a scheme for its formation from an adduct (XII)[39]. A recent study of this reaction by n.m.r. spectroscopy showed, not surprisingly, that (XV) is not produced via 1,4-dimethyl-5-nitronaphthalene[14a]. By analogy with the scheme proposed for the reactions of *p*-xylene (*see* above) it is possible that (XIII) arises by abstraction of a proton from a Wheland intermediate rather than from (XII). Similarly, the further intermediacy of

(XIV) is not required. The observation that 1-methyl-4-nitromethylnaphthalene is formed even when the nitrating system is nitric acid in nitromethane[40a] would not then require separate explanation.

It has been found that polyalkylbenzenes react with fuming nitric acid in solvents such as nitromethane and dichloromethane to give predominantly the substituted benzyl nitrates[40b,c,d]. The mechanism which the authors prefer for these nitro-oxylation reactions involves intramolecular rearrangement of a nitro group through a cyclic transition state. However, in the light

of the above discussion it seems possible to reinterpret the results in terms of a nucleophilic attack subsequent to the formation of a Wheland intermediate at an alkylated ring position. The relevant facts appear to be the following: substituents in the ring affect the rates of both side-chain and ring substitutions to similar extents, which suggests that both processes are mechanistically similar; side-chain substitution is observed with methyl and ethyl groups, but not with t-butyl or with isopropyl groups, when the latter are flanked by other alkyl groups; substitution in the side-chain under nitration conditions produces the nitro compound with some substrates and the nitrate with others. These observations are all consistent with the following scheme, which is analogous to the one proposed above for the reactions of *p*-xylene.

The initially formed Wheland intermediate, if stable enough, can lead to the formation of a methylene cyclohexadiene intermediate. Nucleophilic attack by a nitrite anion at the terminal methylene carbon, departure of the nitro group as an anion and aromatisation will lead to the formation of either a side-chain nitro or nitrite compound. In the presence of nitric acid the latter will be oxidised to the corresponding nitrate. Which nucleophilic atom of the ambident nitrite anion attacks the carbon atom of the terminal methylene group will depend upon the aromatic system being attacked and upon the reaction conditions.

This mechanism is perfectly consistent with all the observed substituent effects[40b,c], if the rate-determining step is the initial attack of the nitrating species to form the Wheland intermediate. Thus, a hydrocarbon undergoes side-chain substitution more readily if the alkylated ring position being attacked by the nitrating species has alkyl groups in the *ortho* positions. The methyl or ethyl group in the *para* position is, of course, essential for the reaction. The effects of other substituents may be illustrated by reference to

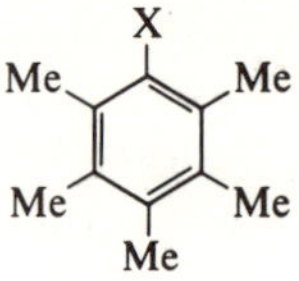

$$
\begin{array}{c}
\text{X} \\
\text{Me} \diagdown \diagup \text{Me} \\
\text{Me} \diagup \diagdown \text{Me} \\
\text{Me}
\end{array}
$$

substituted pentamethylbenzenes. When X was H, Br, and NO_2 the rate of side-chain substitution, relative to that for hexamethylbenzene was reduced, the effects being of similar magnitudes to those found in the ring nitrations of the series C_6H_5X[40b]. This is to be expected if formation of the Wheland intermediate is rate determining and there are negligible steric effects in the polymethylbenzene series.

The side-chain substitutions show high positional selectivity, which may be explained in terms of the normal substituent effects of the group X. When X is electron withdrawing, e.g. NO_2, CO_2H, the initial attack of the nitrating species will be at the position *meta* to substituent X. Substitution will occur into the side-chain *para* to the initial point of attack, which is *ortho* to the substituent X.

If X is *ortho*:*para* directing then nitro-oxylation will give predominantly the *meta* substituted product, derived from initial attack of the nitrating species at the *ortho* position. Initial attack at the *para* position will either lead to no reaction, or to the formation of by-products if the substituent X undergoes a change, e.g. as in the cases of OH, OMe, and NHAc[40c].

The effects of additives on the nitro-oxylation reaction could possibly arise from two sources. First the overall rate, determined by attack of the nitrating species, should be affected by additives if these modify the availability of nitrating species. Second, if added anions are able to compete effectively with the nitrite ion, products other than nitrates should be formed. This seems to be the case when nitrations are performed in acetic acid or, more obviously, in acetic anhydride, when acetates are found in the side-chain products.

Finally, in those cases where both ring nitration and side-chain nitro-oxylation are possible, the ratio of products obtained will depend upon the relative stabilities of the corresponding Wheland intermediates (cf. acetoxylation). For nitro-oxylation there must always be two alkyl groups situated *para* to one another. Thus *m*-xylene and mesitylene give only ring nitro products whereas the more highly substituted alkylbenzenes give predominantly side-chain products.

It will be noticed that side-chain nitration of an alkylbenzene might be represented as the electrophilic attack of a nitrating species on the methylene cyclohexadiene intermediate, or as nucleophilic attack by nitrite on the same intermediate. There is at present no evidence as to which of these processes actually occurs.

NITRATION AND AROMATIC REACTIVITY

Nitration has always played an important part in studies of the connection between structure and reactivity, and of the ways in which substituents and aromatic structures interact[2]. In the last decade much attention has been given to heteroaromatic reactivity, and this work has been reviewed[2e,7]. It necessitated the development of criteria to distinguish between the involvement of a base or its cation when nitration was effected in acid solutions, and further applications of some of these criteria will be mentioned below.

The ways in which substituent effects might vary with reaction conditions will also be considered here. Such variation may be brought about in the following ways:

(a) A change in the interaction between the substituent and the medium.
(b) A change in the effective electrophile with a change of medium.
(c) Changes in the character of a particular electrophile in different media because of changes in its solvation.
(d) A change in the rate-determining step of the reaction mechanism.

The difficulty is, of course, that these factors do not always vary separately, and are not easily unravelled. As regards (d), nitration shows the phenomenon of the limiting rate, and it is useful to discuss this before considering particular substituent effects.

Limiting Rates of Nitration

For nitration at acidities greater than that of 68% sulphuric acid, the introduction of activating substituents into the benzene ring cannot raise the observed rate of reaction more than about 40 times that of benzene. This limiting rate can be convincingly interpreted as the rate of encounter between nitronium ions and aromatic molecules because in this medium the concentration of nitronium ions can be reliably estimated and the appropriate rate constant (Rate = $k[NO_2^+][ArH]$) shown to be close to the calculated encounter rate constant[41]. Thus, the rate-determining process is no longer the attack of the nitronium ion on the aromatic, but the diffusion together of the reactants ('microscopic diffusion control'[21]).

A limiting rate of nitration has also been detected for nitration with solutions of nitric acid in aqueous sulpholan and aqueous nitromethane[12], and with acetyl nitrate in acetic anhydride[14b,23], and regarded analogously as an indication of reaction upon encounter. In these media the limit appears to be less sharp, and the factors which affect it are not well understood. It is not possible to make an estimate of $[NO_2^+]$ independently of the assumption that the limiting rate is the encounter rate.

There is another way in which continued increase in the reactivity of the substrate might fail to effect a continued increase in the rate constant for nitration (or any other reaction), which might for convenience be called the 'Hammond effect'[42]. In a series of reactions in which the electrophile remains constant whilst the reactivity of the aromatic compound is continuously increased, as by the introduction of activating substituents or by annellation, a series of transition states will be involved in which the bonding of the electrophile to the carbon atom being attacked is increasingly less well developed. In this circumstance a situation could be reached where further activating substitution leads to no significant increase in reactivity, and a limiting rate of reaction would be observed. This limiting rate need not necessarily be the same as that already discussed, and could occur sooner than it. It is difficult to imagine that these two possible limiting rates could, however, be much separated, for in the extreme case of the 'Hammond effect' there would be no bonding in the transition state, i.e. reaction would occur with no activation energy, which is what happens when reaction occurs upon encounter. A complementary state of affairs would arise if the aromatic substrate were kept constant whilst the reactivity of the electrophile were continuously increased; this situation has been discussed recently[43].

The two types of limiting situation outlined are not easily distinguished, and for some reactions may, as explained, be indistinguishable. The practical difficulty of measuring encounter rates arises with many substitution processes, and in this respect nitration in sulphuric acid is especially favoured since not only is the concentration of one reagent (NO_2^+) very small, but also that concentration can be fairly well estimated.

Nevertheless, there is an important difference between the two forms of limiting situation which might, in principle, lead to their being distinguished one from the other. The limit which is set by the encounter rate depends on the diffusion through the reaction medium of the reactants; it is conceivable that an aromatic molecule inherently more reactive than another might diffuse more slowly than it, thus leading to a reversal of the situation which would obtain if the 'Hammond effect' were in operation.

The description in terms of an 'early transition state' must lead to a statistical distribution of products amongst activated positions, and in the ultimate limit to a wholly statistical distribution. The description involving the encounter pair will give results which depend on the nature of the latter: if it has the character of a π-complex products will arise in proportions determined by individual positional reactivities[2e,21], and may, or may not, be statistical proportions; if several non-interconverting encounter pairs exist product distribution will necessarily be statistical, a result indistinguishable from that produced in the limit by the 'Hammond effect'.

To determine whether a limiting rate is to be identified with the diffusion rate as, for example, by measuring diffusion rates directly, obviously presents insuperable difficulties for substitutions in general. Equally, there are great difficulties in choosing suitable compounds to study in an attempt to answer this question by examining product distributions.

It might seem possible that steric factors could be responsible for the failure of activating substituents to produce increases in reactivity, and so be responsible for the observed limiting reactivities. It is considered that this explanation is improbable, since nitration at the limiting rate of very reactive compounds with structures quite different from those of the methylbenzenes has been observed[2e,23,41].

It has been argued that the limiting rate found for nitration in sulphuric acid represents reaction at the encounter rate[41]. Analogously, limiting rates detected for nitrations with solution of nitric acid in sulpholan, nitromethane, and of acetyl nitrate in acetic anhydride were, as already mentioned, regarded in the same light[14b]. These reactions are further discussed below, together with the newer results[14] for nitration in acetic acid.

Methyl- and Halogeno-benzenes

Part of the evidence used to support the argument for a common electrophile operating in solutions of nitric acid in inert organic solvents and in solutions of acetyl nitrate in acetic anhydride was the insensitivity of the relative rate of nitration of toluene to the conditions used, and also the similar insensitivity of the isomer proportions formed from toluene[2d]. The methyl substituent can be expected to have only a slight interaction with the solvent so the results apparently suggest that the selectivity of the nitronium ion is not affected to any large extent by the nature of the solvent around it. Some recently determined relative rates are given in *Table 7.2*; the values for toluene are very similar to the earlier ones[2d,e] with the exception of the figure of 50 for reaction in acetic anhydride[29]. This is about double the old value, and the results for the relative rates of toluene now show a larger spread than was previously indicated. However, the spread is still not larger than a factor of three and may be consistent with a variation in the solvation of the nitronium ion in the different solvents. (Difficulties concerning adoption of the nitronium ion as the electrophile operating in solutions of acetyl nitrate in acetic anhydride have already been mentioned.)

The observed relative rates for toluene, *m*-xylene and mesitylene in *Table 7.2* are seen to be very similar for nitrations in the solvents sulpholan, nitromethane and acetic acid, but they are apparently different in acetic anhydride and less markedly so in 57·2% sulphuric acid*. In the last-named solvent, *m*-xylene and mesitylene react at the limiting rate, as already explained. The approach to a limiting rate is largely responsible for the difference between the observed and calculated relative rates (*Table 7.2*). The latter quantities are obtained from the observed relative rate and isomer proportions for the nitrations of toluene, assuming the Additivity Principle to hold. In reactions where a limiting rate is not reached, e.g. molecular chlorination and bromination, the Additivity Principle is obeyed to a high degree of accuracy[48]. The ratios $[k_{\text{rel.}}(\text{calc.})/k_{\text{rel.}}(\text{obs.})]$ at the foot of *Table 7.2* are therefore a measure of the extent to which the occurrence of a limiting

---

* Sulphuric acid of this concentration has been chosen because its viscosity is close to those of the organic solvents.

Table 7.2. RELATIVE RATES OF NITRATION IN DIFFERENT NITRATING SYSTEMS AT 25°C

Compound	H_2SO_4 [a] obs.	H_2SO_4 calc.	Sulpholan [b] obs.	Sulpholan calc.	Nitromethane [c] obs.	Nitromethane calc.	AcOH [d] obs.	AcOH calc.	Ac_2O [e] obs.	Ac_2O calc.
Benzene	1	(1)	1	(1)	1	(1)	1	(1)	1	(1)
Toluene	17	(17)	20	(20)	25	(25)	23	(23)	50	(50)
m-Xylene	95	540	100	750	146	1 200	136	980	870	4 700
Mesitylene	150	18 000	350	29 000	400	56 000	355	43 000	5 000	454 000
Anisole	50[f]	—	—	—	—	—	—	—	1 550	—
Acetanilide	0·016[e]	—	—	—	—	—	—	—	800	—
Methyl phenethyl ether	1·9[e]	—	—	—	—	—	—	—	20	—
Fluorobenzene	0·117[g]	—	0·45[h]	—	—	—	0·22[j]	—	0·165[i]	—
Chlorobenzene	0·064[g]	—	0·14[h]	—	0·03[i]	—	0·10[j]	—	0·035[i]	—

k_{rel} (calc.)/k_{rel} (obs.)

Compound	H_2SO_4	Sulpholan	Nitromethane	AcOH	Ac_2O
Benzene	1	1	1	1	1
Toluene	1	1	1	1	1
m-Xylene	5·7	7·5	8	7	5·5
Mesitylene	118	83	140	120	91

[a] Ref. 41 and 44, 57·2% H_2SO_4, nitrotoluenes (%) o- 60, m- 3, p- 37.
[b] Ref. 12, 7·5% aq. sulpholan, nitrotoluenes (%) o- 61·9, m- 3·5, p- 34·7.
[c] Ref. 12, 15% aq. nitromethane, nitrotoluenes (%) o- 61·5, m- 3·1, p- 35·4.
[d] Ref. 14, aq. acetic acid, nitrotoluenes (%) o- 56·9, m- 2·8, p- 40·3. Acetic anhydride, nitrotoluenes (%) o- 62, m- 3, p- 35.
[e] Ref. 34, for acetanilide the relative rate relates to 75% sulphuric acid.
[f] Ref. 45, 63% H_2SO_4.   [g] Ref. 46, 67·5% H_2SO_4.
[h] Ref. 19, NO_2BF_4.   [i] Ref. 47.   [j] Ref. 10, 20°C.

rate of nitration depresses the apparent reactivity of a compound reacting at or near to the limit; the greater the ratio, the greater is the depression.

On the basis of this comparison the performances of the reagents prepared from nitric acid and inert organic solvents are seen to be very similar. Bearing in mind the relatively high proportions of nitric acid and water which they contain, this is perhaps not very surprising. More important is the similarity brought out between the three such systems on the one hand, and solutions of acetyl nitrate in acetic anhydride and of nitric acid in sulphuric acid on the other. The argument is incomplete, for the way in which the orientation of nitration of toluene may vary with acidity is not known, and further it is not understood exactly how the limiting rate depends on solvent viscosity. Nevertheless, the comparison made shows the performance of the nitronium ion in sulphuric acid to be very similar to its performance in inert aqueous organic solvents, and also to that of the electrophile operating in solutions of acetyl nitrate in acetic anhydride.

The importance of the comparison is two-fold: firstly, since the limiting rate of nitration in sulphuric acid can convincingly be identified as the encounter rate, it suggests that the limiting rates in the other media have the same significance; secondly, it shows that the electrophile operating in acetic anhydride is indeed very similar to a nitronium ion. The difficulties of arguing that it *is* the nitronium ion have already been discussed.

It should be noted that the comparisons made in the lower part of *Table 7.2* are based on the isomer proportions and relative rate for nitration of toluene proper to each solvent. That is, they admit the existence and significance of variations in the behaviour of the electrophiles in the several media, even where these variations are small and where the electrophile is the nitronium ion. Thus, they take account of the factor (c) mentioned at the beginning of this section.

Substrates less reactive than benzene should not be affected by encounter control. Further, in the absence of any strong interaction between the substrate and the solvent, the above results with the methylbenzenes would suggest that there should be no significant variation of the relative rates with reaction conditions. These expectations are borne out by the results obtained with the halogenobenzenes (*Table 7.2*) where a four-fold spread in relative rates is found. This is slightly larger than the spread observed with toluene, but with the more polar substituents in the halogenobenzenes a slightly greater interaction with the solvent might be expected. A small part of the observed differences in reactivity might therefore be due to a slight modification of the substituent effect, although the results show that such an effect can only be of minor importance.

The foregoing discussion enables some generalisations to be made about the importance of the factors (a)–(d) enumerated above. It is apparent that in nitration (c), the change in the solvation of a given electrophile leads to only small changes in the relative rates and the isomer proportions. Factor (d), the occurence of encounter control, will assume importance with the more reactive substrates. The results for the methylbenzenes are useful in affording a measure of the extent to which the encounter rate controls the observed rate of reaction. Thus, any large changes in substituent effects are most probably the result of the operation of factors (a) and/or (b), a variation in the electronic effect of the substituent, or a change in the effective electrophile. This last factor will be of particular importance in the case of reactive substrates which are susceptible to nitrosation.

These points will be illustrated by discussion of the behaviour of three specific substrates. They are chosen also because they are associated with other matters already raised in this article; the unusual $o:p$-ratios observed with some nitrations carried out in acetic anhydride, and the criteria which have been developed for distinguishing between reactions of free bases and their cations in nitrations in sulphuric acid.

Anisole

There have been many reports on the proportions of isomers formed in the nitration of anisole under a variety of conditions[2c,e,49]; a selection of data is included in *Table 7.3*. For nitrations in nitric acid, or with nitric acid in sulphuric acid or acetic acid, typically between 30% and 40% of the *o*-nitro isomer is produced. The proportion of this isomer increases to 70% when nitration is effected with either benzoyl nitrate or acetyl nitrate in acetic anhydride. In both sulphuric acid and acetic anhydride the relative rate (*Table 7.2*) indicates that reaction is occurring at, or close to, the encounter limit.

Table 7.3. PROPORTIONS OF ISOMERS FORMED IN NITRATION UNDER DIFFERENT CONDITIONS

Compound	Conditions	Isomers, %			Ref.
		ortho	meta	para	
Anisole	AcONO$_2$, Ac$_2$O, 25°C	70	—	30	23
	HNO$_3$, sulpholan, 25°C[a]	70	—	30	23
	HNO$_3$, 65% H$_2$SO$_4$, 25°C[a]	60	—	40	23
	HNO$_3$, 65% H$_2$SO$_4$, 25°C[b]	6	—	94	23
	HNO$_3$, AcOH, 20°C	34	—	66	52
	HNO$_3$, H$_2$SO$_4$, 45°C	40	—	60	52
Methyl phenethyl ether	AcONO$_2$, 25°C	62·3	3·7	34·0	35, 36
	AcONO$_2$, Ac$_2$O, 25°C	60	5	33	33
	AcONO$_2$, CH$_3$CN, 0°C	66·0	4·2	29·8	35
	HNO$_3$, CH$_3$NO$_2$, 25°C	41·2	3·0	55·8	35
	HNO$_3$, 65% H$_2$SO$_4$, 25°C	34	9	57	33
	HNO$_3$, 82·9% H$_2$SO$_4$, 25°C	30·8	8·9	60·3	34

[a] Urea present ($c.$ 10^{-2} mol l^{-1}).   [b] Sodium nitrite present ($c.$ 4 × 10^{-2} mol l^{-1}).

Several explanations have been proposed for the high $o:p$-ratio for nitration in acetic anhydride. Halvarson and Melander[50] suggested that either the initial attack of the nitrating species took place at oxygen, followed by an intramolecular rearrangement to the *o*-nitro isomer, or that the orientation observed using acetyl nitrate was the normal one, and the results of experiments carried out using more strongly acidic media had been affected by nitrosation. That nitrosation would affect the isomer distribution in the required direction is clearly shown by the results in Table 7.3. In 65% sulphuric acid in the presence of urea, 60% of the o-nitro isomer is formed, but in the absence of urea and in the presence of added sodium nitrite the amount of the o-nitro isomer falls to 6%. If nitrosation in sulphuric acid were the cause of the change in isomer proportions, then the fact that the rate of reaction is equal to the encounter rate for nitration would be merely a coincidence.

The hypothesis of indirect nuclear nitration assumes that the lone pairs of electrons on the oxygen atom provide a suitable site for the initial attachment of the nitronium ion (or other nitrating species) before migration into the ring. The change in orientation observed is then attributed either to a change in the nitrating agent, which in the case of nitration with acetyl nitrate discriminates in favour of attack at the oxygen atom, or to a modification of the substituent. It is probable that the lone pairs are extensively hydrogen bonded in the more strongly acidic media, and this might decrease the extent of reaction at the oxygen atom and so lead to the change in orientation. As the arguments have been against a change in the actual electrophile, the latter effect would seem more probable. However, it would be necessary then to assume that whilst hydrogen bonding is effective in determining the position of attack, reaction can still occur at the encounter rate.

Another explanation of the change in the $o:p$-ratio emphasises the importance of an electrostatic effect arising from the dipole of the substituent[51]. With anisole the positive end of the dipole is directed away from the ring, resulting in the o-positions being negatively polarised with respect to the p-position, and therefore being more susceptible to electrophilic attack. This electrostatic effect is considered to be more important in acetic anhydride, a solvent of lower dielectric constant, than in mixed acid. However, this effect apparently does not occur in acetic acid which has an even smaller dielectric constant than acetic anhydride (the experiments in acetic acid[52] were performed with $[HNO_3] \approx 3\text{–}6$ mol l^{-1}, so that the dielectric constant of pure acetic acid may not be a fair basis for comparison). This hypothesis was thought to receive some support from the observed variations in the $o:p$-ratio for chlorobenzene[51]. With this substrate the negative end of the dipole is directed away from the ring so that a change from sulphuric acid to acetic anhydride should be accompanied by a decrease in the $o:p$-ratio as was observed, but the results do not agree with those obtained by other workers[2d,e].

More recent results do seem to support the idea, however, and in acetic anhydride the $o:p$-ratio is reduced by the introduction into the solution of a less polar solvent (carbon tetrachloride), and is increased by the addition of a polar solvent (acetonitrile)[53].

Perhaps the simplest account of the situation follows from the discussion of limiting rates (see above). It is to suppose that in situations where anisole gives a predominance of p-nitroanisole nitrosation is occurring, whereas in the other cases o- and p-nitroanisole are being produced in statistical amounts as a consequence of reaction at the encounter rate. It might be necessary to postulate additionally that the situation is modified slightly by substrate–solvent interaction when nitration occurs in sulphuric acid.

Acetanilide and Related Anilides

Acetanilide is another compound known to show large changes in the $o:p$-ratio with change in the reaction conditions[2c,e]. In concentrated sulphuric acid nitration gives predominantly the p-nitro compound, whilst in acetic anhydride the o-nitro compound is the major product[54,55]. These results have again been discussed in terms of a special mechanism operating in acetic anhydride, involving an interaction between the nitrating species and the oxygen or nitrogen atom of the acetylamino group, which is supposed to lead to a high proportion of *ortho*-substitution[2c,35,55].

Some recent results[34] show that whatever may be happening in nitration in acetic anhydride the comparison with nitration in sulphuric acid is irrelevant, since it involves a comparison of unlike processes. The rates of nitration of acetanilide, methane sulphanilide, chloroacetanilide and trifluoroacetanilide have been measured in sulphuric acid and in acetic anhydride. The rates of nitration, relative to that of benzene, for 68·3% sulphuric acid, are 0·016, 1·8, 0·8 and 0·2, and for acetic anhydride are 800, 350, 200 and 6, respectively. Protonation studies and the rate profiles in sulphuric acid (66–81%) indicate that for acetanilide a cation is involved in the nitration whereas with methane sulphanilide and trifluoroacetanilide the free bases are nitrated. The rate profile for chloroacetanilide also indicates that nitration occurs through the equilibrium concentration of free base. This concentration at any given acidity may be calculated[34], and the corrected rate profile, which refers to reaction of the free base species, is parallel to the rate profiles of the other anilides. (The relative rate quoted above for 68·3% sulphuric acid refers to the corrected rate.)

The results for nitration in acetic anhydride, in which the anilides react as the free bases, show that all of the substrates are more reactive than benzene. Acetanilide is nitrated at near to the limiting rate for this medium, and the rates for methane sulphanilide and chloroacetanilide are not far removed from this. In sulphuric acid the anilides which react via the free base species retain the same order of reactivity as observed in acetic anhydride, although they are now less reactive than benzene. Acetanilide, which reacts as the protonated form in sulphuric acid, shows the largest change in reactivity and is the least reactive of the anilides studied in sulphuric acid. The large changes in rates observed between the solvents acetic anhydride and sulphuric acid, even though the reacting species remain the same, can be attributed to a modification of the substituent effect by the strong hydrogen bonding interactions in the latter solvent. Protonation of the acetylamino group, as in acetanilide, produces a greater effect.

Having identified the reacting species, it is now instructive to consider how the isomer proportions depend on the nitration conditions; a selection of data is included in *Table 7.4*. With acetanilide, nitration in 98% sulphuric acid gave 5% of o- and 95% of p-nitroacetanilide, in agreement with earlier work. The proportions of the two isomers changed steadily with decreasing acidity until in 65·8% sulphuric acid (the most dilute acid used) they were 44% of o- and 56% of p-nitroacetanilide. Also in agreement with earlier work, nitration in acetic anhydride gave 77% of o- and 23% of p-nitroacetanilide. These results suggest that one factor involved in determining the isomer proportions is a variation in the electronic effect of the acetylamino group.

SOME ASPECTS OF RECENT WORK ON NITRATION

Table 7.4. THE DEPENDENCE OF ISOMER PROPORTIONS ON NITRATION CONDITIONS AT 25°C

Compound	Conditions[a]	Isomers, % ortho	para
Acetanilide[b]	98% H_2SO_4	5	95
	94·8% H_2SO_4	8	92
	80·9% H_2SO_4	17	83
	72·9% H_2SO_4	32	68
	65·8% H_2SO_4	44	56
	$MeNO_2$	59	41
	Ac_2O	77	23
Chloroacetanilide[c]	98% H_2SO_4	13	87
	94·8% H_2SO_4	20	80
	85·0% H_2SO_4	36	64
	75·5% H_2SO_4	46	54
	68·6% H_2SO_4	54	46
	Ac_2O[d]	80	20
Trifluoroacetanilide[c]	98% H_2SO_4	20	80
	85·3% H_2SO_4	26	74
	80·2% H_2SO_4	30	70
	Ac_2O[e]	34	66
Methane sulphanilide[c]	98% H_2SO_4[f]	30	70
	85·3% H_2SO_4	45	55
	80·2% H_2SO_4	50	50
	Ac_2O	60	40

[a] Quantitative yields of nitro products were obtained except where stated. Ref. 34;
[b] <2% *meta* isomer;   [c] <1% *meta* isomer;   [d] 85% yield;
[e] 82% yield;   [f] 72% yield.

The observed variation in the *o*:*p*-ratio for nitration in sulphuric acid must be attributed to a medium effect on the protonated species. The results obtained with the anilinium ion show that this is not an unreasonable assumption[56,57]. The nitration of aniline in the range of acidity 98–82% sulphuric acid has been shown to involve only the anilinium ion. As the acidity of the medium is decreased the isomer proportions change; at the highest acidity 62% of *m*- and 38% of *p*-nitroaniline are formed, changing to 36% and 59% (with 5% of *o*-nitroaniline) at the lowest acidity. The fact that the cation of acetanilide gives only *o*- and *p*-nitration shows that it is protonated on oxygen rather than on nitrogen.

The results obtained with chloroacetanilide (*Table 7.4*) are very similar to those obtained with acetanilide, although with the former part of the variation in the *o*:*p*-ratio in sulphuric acid is due to protonation of the free base. A solvent effect on the protonated species will only be an important factor at the higher acidities.

Trifluoroacetanilide and methane sulphanilide are such weak bases that even in concentrated sulphuric acid protonation is not complete. The results in *Table 7.4* therefore refer to reactions of the free bases under all conditions. The variations in the isomer proportions in sulphuric acid can thus be attributed to a solvent effect on the free bases.

Of greater interest, however, are the isomer proportions given by these two substrates in acetic anhydride. With trifluoroacetanilide the *o*:*p*-ratio in acetic anhydride is little different from that in 80·2% sulphuric acid, and with methane sulphanilide it is approximately equal to the statistical ratio. The behaviour of acetanilide and chloroacetanilide on the one hand, and of trifluoroacetanilide and methane sulphanilide on the other, can be understood in terms of a special mechanism of nitration in acetic anhydride if it is assumed that this is dependent upon the nucleophilicity of the atom in the substituent which first interacts with the nitrating species. If this is the oxygen atom of the acetylamino group then the special mechanism can operate with acetanilide and chloroacetanilide because this centre is sufficiently nucleophilic, but with the other two anilides the oxygen atom does not provide a suitable site for coordination and so nitration occurs directly in the benzene ring. The reactive methane sulphanilide gives nitro isomers almost in the statistical ratio, whereas the relative unreactive trifluoroacetanilide gives mainly the *para* isomer. The overall behaviour of this family of anilides thus appears to provide fairly clear evidence for the operation in solutions of acetyl nitrate in acetic anhydride of a special mechanism for *ortho*-substitution with appropriate substrates. The electrophile involved remains unidentified.

Methyl Phenethyl Ether

Some results for the nitration of this ether are given in *Table 7.3*. They have been interpreted in terms of three factors: nitronium ion nitration in both mixed acid and acetic anhydride, some degree of protonation of the oxygen atom of the ether in the mixed acids, and the presence of dinitrogen pentoxide in acetyl nitrate solutions in acetic anhydride[35,36]. It was argued that the change in the *o*:*p*-ratio could not be accounted for solely on the basis of a modification of the substrate by protonation, although this factor might be responsible for part of the observed change. It was assumed that in acetic anhydride there were two effective nitrating species, the nitronium ion, giving the same isomer proportions that are observed in mineral acid nitrations, and dinitrogen pentoxide, which leads only to the *o*-isomer by the

following mechanism. One reason that led to the postulation of a special route to the *o*-isomer arose from the use of the benzyltrimethylammonium ion as a model for protonated benzyl methyl ether in nitrations. This led to an estimate for the yield of *o*-nitro isomer which would arise from the free base, and since this fell short of the observed yield benzyl methyl ether, and by analogy methyl phenethyl ether, were considered to require a special mechanism of *ortho* substitution. The use of the benzyltrimethylammonium ion as a model has been questioned[2e,33].

A second point put forward to support the case for a special mechanism for *ortho* nitration of methyl phenethyl ether in acetic anhydride was a comparison of its partial rate factors with those of ethylbenzene[35].

The two compounds, methyl phenethyl ether and ethylbenzene, provide the same sort of comparison as has been made (*see* above) between acetanilide and methane sulphanilide, and recent work shows that for the former pair, as for the latter, the results of nitration in sulphuric acid should be considered separately. In sulphuric acid methyl phenethyl ether is only slightly more reactive than benzene, whilst in acetic anhydride it is more reactive though still less reactive than toluene. The slope of the rate profile in 68–83% sulphuric acid[34] is consistent with nitration via the major species present, which is probably the free base, but the relative reactivity indicates strong interaction with the solvent. This leaves the results for nitration in acetic anhydride to be considered separately, and the comparison with ethylbenzene made by Norman and his co-workers strongly supports their opinion that for the ether there is a special mechanism of *ortho* substitution in acetic anhydride.

Positive Poles

It was mentioned above that nitration of the anilinium ion leads to appreciable amounts of *p*-nitroaniline; indeed, even in 98% sulphuric acid the *p*-position is more reactive than a *m*-position ($f_p = 213 \times 10^{-8}$, $f_m = 173 \times 10^{-8}$)[57]. As the hydrogen atoms of the —NH_3^+ substituent are successively replaced by methyl groups, a smooth change is observed from almost equal *m*- and *p*- to predominant *m*-reactivity. However, even with the —NMe_3^+ substituent 11% of the *p*-nitro compound is still formed[58,59], so that the *m*-position is only about four times more reactive than the *p*-position. Accompanying the change in isomer proportions, the rates of nitration also decrease steadily as the hydrogen atoms of the anilinium cation are replaced by methyl groups.

These results can be understood if it is assumed that in the cases of the protonated cations hydrogen bonding modifies the substituent effect by spreading the positive charge into the medium. However, it appears that both the orientation and the reactivity cannot always be described in terms of a single factor controlling the interaction between the positive pole and the benzene ring. Thus in the following series the relative rate of nitration varies in the order:

$$\text{PhCH}_2\overset{+}{\text{NMe}}_3 > \text{Ph}\overset{+}{\text{NH}}_3 > \text{Ph}\overset{+}{\text{NMe}}_3$$

but the proportions of *p*-nitrations are 15%, 38% and 11% respectively. The unexpectedly high *p* : *m*-ratio produced by the protonated pole may be due to the way in which the charge is spread in the medium, or to release of electrons from it by hyperconjugation*.

These results show positive poles to behave rather differently from previous

---

* Following the arguments of Kirkwood and Westheimer[60], hydrogen bonding of the —NH protons will bring the positive charge nearer to the high dielectric constant of the solvent. This will both increase the distance between the charge on the positive pole and the charges on the aromatic ring, and increase the effective dielectric constants between these charges.

assumptions that they were —I groups which were almost exclusively *m*-directing. Analysis of the way in which these substituents influence the Gibbs' functions of activation at the *m*- and *p*-positions throws more light on their mode of operation[61]. From the partial rate factor, f, the change in Gibbs' function of activation, $\Delta G^{\ddagger}$, at the position to which it refers can be calculated from the equation:

$$\Delta G^{\ddagger}_{m} = -RT \ln f_{m}$$

The difference between the substituent effect at the *p*- and the *m*-positions is then given by the ratio

$$\Delta G^{\ddagger}_{p}/\Delta G^{\ddagger}_{m}$$

which is, to a fair approximation, independent of the nature of the attacking electrophile. Some values of this ratio are given in *Table 7.5*. The results

Table 7.5. THE RELATIVE EFFECTS OF INDUCTIVE SUBSTITUENTS AT THE *meta-* AND *para*-POSITIONS FOR NITRATION[a]

Substituent	$\delta\Delta G^{\ddagger}_{p}/\delta\Delta G^{\ddagger}_{m}$
—$\overset{+}{\text{NMe}_3}$	1·09
—$\overset{+}{\text{NH}_3}$	0·99
—CMe$_3$	3·12
—CH$_3$	4·43

[a] Taken from Ridd, J. H.[61], by courtesy of the Chemical Society.

show that for the positive poles the effects at the *m*- and the *p*-positions are very similar, whereas for the alkyl groups the effect is much larger at the *p*- than at the *m*-position. This contrast suggests that the poles and the alkyl groups exert effects which differ in their modes of action as well as in their direction. With the alkyl substituent the effect can be understood in terms of the π-inductive effect, activating the *p*-position and producing only very

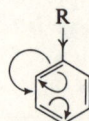

weak activation at the *m*-position; hyperconjugative effects might also be important. With the cations, on the other hand, the *m*- and *p*-positions are deactivated to similar extents because of a direct field effect, which could slightly favour either position depending on the electron distribution in the transition state[61]. This explanation is supported by the observations that the deactivating effects of nitrogen poles fall off less rapidly than those of neutral substituents with distance from the ring, and that, for a given overall deactivating effect, the positive poles give more *para*-substitution than do neutral substituents. It has also been shown recently[62] that the substituent effect of a positive pole depends upon the conformation of the substituent; a result quite consistent with the field effect, but not expected for an inductive effect.

The way in which the substituent effect of a positive pole varies with the distance between the pole and the ring can be seen by considering the results for the series $Ph(CH_2)_n \overset{+}{N}Me_3$ ($n = 0$–3)[63]. The relative rate of nitration does not measure the substituent effect of the pole because of the activating effect of the intervening methylene groups. To remove the latter effect *meta*-substitution was studied and the quantity $x = -\log f_m/f_m^\circ$ (where f_m° is the partial rate factor for nitration at the *m*-position in the corresponding alkylbenzene) was evaluated. The values of log x were plotted against the corresponding values of log r_x, where r_x is the distance between the charge on the pole and the charge on the ring in the Wheland intermediate which was used as a simple model for the transition state. The points were found to lie on the same straight line as those for the corresponding quantities* for the ionisation of the anions $HO_2C \cdot (CH_2)_n \cdot CO_2^-$. Thus,

$$\underset{\underset{NO_2}{H}}{\langle + \rangle} - [CH_2]_n - \overset{+}{N}Me_3 \qquad \longleftarrow r_x \longrightarrow$$

the effect of the positive poles is not transmitted through the methylene chain (σ-inductive effect), but, like the effect of the carboxylate anion upon the second dissociation of a dicarboxylic acid, is a consequence of the field effect. Furthermore, the field effect appears to be accounted for by considering only the electrostatic interactions in the transition state between the pole and the aromatic ring, no allowance being made for interactions between the pole and the ring in the ground state, or between the approaching electrophile and the substrate.

It may not always be safe to neglect the last two factors, as a molecular orbital treatment of the substituent effect of the anilinium ion shows[64]. The Pople perturbation treatment was applied and the electrostatic effect of the pole was allowed to modify the coulomb integral of each carbon atom of the ring by an amount which depended on the distance of the carbon atom from the pole. This produced a ground state in which all three positions carried a positive charge, the charge density being in the order *para* > *ortho* > *meta*. A perturbation caused by the approach of the nitronium ion was then introduced, and this led to marked changes in the charge distribution.

---

* The quantity corresponding to x was defined as $y = -\log(4K_2/K_1)$, where K_1 and K_2 are the first and second dissociation constants respectively. The quantity r_y, analogous to r_x, refers to the minimum distance between the charges on the dianion.

Finally, the force on the nitronium ion approaching each position or the total electrostatic energy was calculated. Now both the *m*- and *p*-positions were seen to be the favoured points of attack. Thus, ground state interactions, transition state interactions, and pole–reagent interactions must all be taken into account.

A slightly different account of the substituent effects of positive poles, in particular of the trimethylammonium group, has recently been presented by Ingold[2b]. Two effects are considered: an inductive effect which enters at the α-carbon atom (XVI), and a field effect which enters mainly at the *o*-carbon atoms (XVII). These combine to deactivate the *o*-positions but

(XVI)   (XVII)

influence *m*- and *p*-positions in opposite ways. In the phenyltrimethylammonium ion the inductive effect is supposed to dominate the field effect. The strong field effect* arises because solvent molecules cannot intervene between the pole and the *o*-positions, at which it exerts its major influence. Ingold suggests the terms 'substrate field effect' and 'reagent field effect' for the ground state and pole–reagent interactions mentioned above. The $\cdot \overset{+}{\text{NMe}}_3$ group is thus supposed to influence substitution by exerting a π-inductive effect modified by a substrate field effect. The discussion is therefore in terms of the ground state of the substrate in contrast to Ridd's use of the transition state.

In a recent series of papers, Ridd and his co-workers have begun an investigation into the relative importance of inductive and field effects in aromatic substitution[65]. The relative substituent effects of —X and —CH$_2$X in nitration were considered in terms of the transmission factor, a, of the methylene group. To minimise complications from conjugative interactions, substitution at the *m*-position was studied, and a was defined by

$$a = \delta \Delta G^{\ddagger}_{\text{CH}_2\text{X}} / \delta \Delta G^{\ddagger}_{\text{X}}$$

where $\qquad \delta \Delta G^{\ddagger}_{\text{CH}_2\text{X}} = -RT \ln (f_{\text{m}}^{\text{CH}_2\text{X}} / f_{\text{m}}^{\text{CH}_3})$

and $\qquad \delta \Delta G^{\ddagger}_{\text{X}} = -RT \ln f_{\text{m}}^{\text{X}}$

Some values of a are shown in *Table 7.6*. These show that a is sensitive to the nature of the substituent and also that there is a clear division between the charged and the dipolar substituents. It was assumed that, in the absence of conjugative interactions, substituent effects could be analysed in terms of separate contributions from field and inductive effects;

$$\delta \Delta G^{\ddagger}_{\text{X}} = F_{\text{X}} + I_{\text{X}}$$

---

* Ingold assumes that in general field effects are of only minor importance. The fact that in this case the π-inductive effect has been modified shows that a strong field effect is in operation.

Table 7.6. TRANSMISSION FACTORS (α)[a]

Substituent (—X)	Class	α
—$\overset{+}{\text{NMe}}_3$	—I	0·59
—$\overset{+}{\text{AsMe}}_3$	—I —M	0·56
—$\overset{+}{(4\text{-}C_5H_5N)}$	—I +M	0·56
—$\overset{+}{\text{NH}}_3$	—I	0·49
—$\overset{+}{\text{PMe}}_3$	—I —M	0·44
—Cl	—I +M	0·37 (0·33)[b]
—CCl$_3$	—I	0·28
—CO$_2$Et	—I —M	0·20
—NO$_2$	—I —M	0·17

[a] From De Sarlo, F. et al.[65c], by courtesy of the Chemical Society;
[b] using different results for PhCH$_2$Cl (see Ref. 65c).

The field effect term (F_X) is the change in the Gibbs' function of activation produced by the electrostatic interaction between the pole or dipole of the substituent and the charge on the aromatic ring or on the electrophile in the transition state. The inductive term (I_X) is the change in the Gibbs' function of activation deriving from the modification of the electronegativity of the α-carbon atom as a result of the difference in the polarity of the C—X and C—H bonds (π-inductive effect). The transmission factor can thus be written as

$$\alpha = (\alpha_F F_X + \alpha_I I_X)/(F_X + I_X)$$

where α_F and α_I are the transmission factors for the field effect and the inductive effect respectively; if the term x is defined as the ratio $x = F_X/(F_X + I_X)$, then

$$\alpha = x\alpha_F + (1 - x)\alpha_I$$

Hence the three terms x, α_F and α_I will, in principle, determine how α varies with the group X. Since α_I is usually regarded as being independent of the substituent, the inconstancy of α (*Table 7.5*), must arise either from a variation of α_F, and/or x.

Further consideration of the effects of hypothetical dipolar substituents —$\overset{+\;-}{\text{AB}}$ on the Gibbs' function of activation led to the conclusion that at least some of the variation in α between charged and neutral substituents comes from a change in α_F. To ignore the inductive effect would, however, leave unexplained the greater reactivity produced at the m-position by $\cdot NO_2$ compared to $\cdot NH_3^+$. With substituents $\cdot CH_2X$ positively charged groups cause greater deactivation than dipolar ones, so that in these cases a_I may be very small. It seems, therefore, that the variation in a can be ascribed to variations in both x and a_F.

Molecular Orbital Treatment of Aromatic Reactivity

This topic has frequently been reviewed and the only comments here will be on the use of the localisation energy as a reactivity index as it has been applied to the results of nitration.

The significance of using the localisation energy, and the definition of this quantity, has been excellently discussed by Dewar and Thompson[66], and in the following the substance of their discussion will be summarised so far as it is relevant to the present purpose. Kinetic theory suggests that only those discussions of reactivity which take note of the transition state can be valid. The problem is then to predict the geometry and energy of the transition state. The problem can be avoided if it be assumed that there is a linear relationship between the Gibbs' function of activation and the Gibbs' function of reaction, and further that for a series of related reactions the entropy of activation is constant; then there would be a linear relation between the potential energy of activation ($\Delta E^\ddagger$) and the potential energy of reaction (ΔE):

$$\Delta E^\ddagger = a' + c\Delta E$$

Consequently

$$-RT \log k = A + c\Delta E$$

This equation is the basis for the use of localisation energies as measures of reactivity. ΔE is the energy difference between the aromatic substrate in the ground state and the 'product' of the reaction, formulated as a Wheland intermediate.

Dewar and Thompson point out that the term localisation energy has been used in two ways:

(a) to denote the change in total π-bond energy when an atom in a conjugated system is removed from conjugation with the rest;

(b) to denote the change in total π-electron energy of a conjugated system when two π-electrons are localised on one atom in it.

In the Hückel M.O. treatment these two definitions are equivalent. They prefer the former definition as being more related to chemical reactivity. The localisation energy is therefore the difference in total π-bond energy between the aromatic substrate and the Wheland intermediate, and it can be used to correlate observed reactivities only if the additional assumption is made that for a given series of related reactions the change of σ-bond energy is constant, i.e.

$$\Delta E = \text{constant} + \Delta E_\pi$$

Under these conditions the localisation energy is related to the rate of reaction by the relation:

$$-RT \log k = A' + C\Delta E_\pi$$

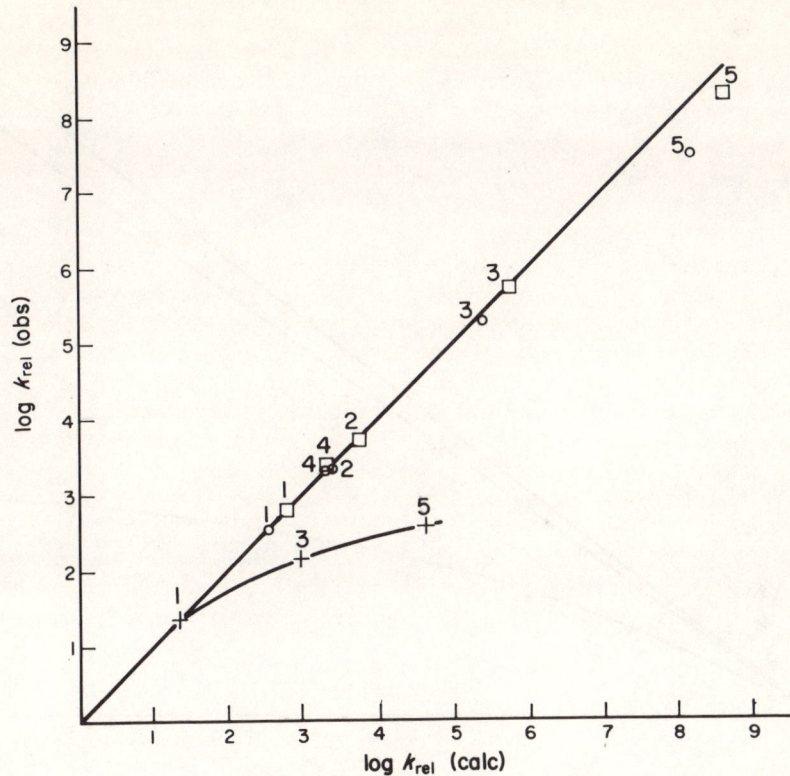

Figure 7.3. The relation between observed and calculated rate constants relative to those of benzene in acetic acid at 25°C
○ Chlorination (Ref. 48); □ bromination (Ref. 48); + nitration (*Table 7.2*)
1 Toluene; 2 o-xylene; 3 m-xylene; 4 p-xylene; 5 mesitylene

Benzene being the reference compound in aromatic substitution, then:

$$\log f = C(\Delta E_\pi^\circ - \Delta E_\pi)/RT$$

where ΔE_π° is the localisation energy for reaction with benzene. Thus the logarithms of the partial rate factors should be linearly related to $(\Delta E_\pi^\circ - \Delta E_\pi)$.

A linear correlation has recently been reported between cation localisation energies calculated for benzene, toluene and the xylenes and the logarithms of the partial rate factors for nitration in acetic acid[67]. The particular theoretical model used took into account both the inductive and hyperconjugative effects of the methyl group, but nothing in the treatment assumes additivity in the effects of the substituents or relates them particularly to nitration. In an earlier section it has been shown that with the methylbenzenes the additivity principle is not obeyed in nitration, but rather a limiting reactivity is approached. Therefore, the calculated localisation energies should not correlate data for nitration and at the same time lead to satisfactory correlations with results for other electrophilic substitutions, such as molecular

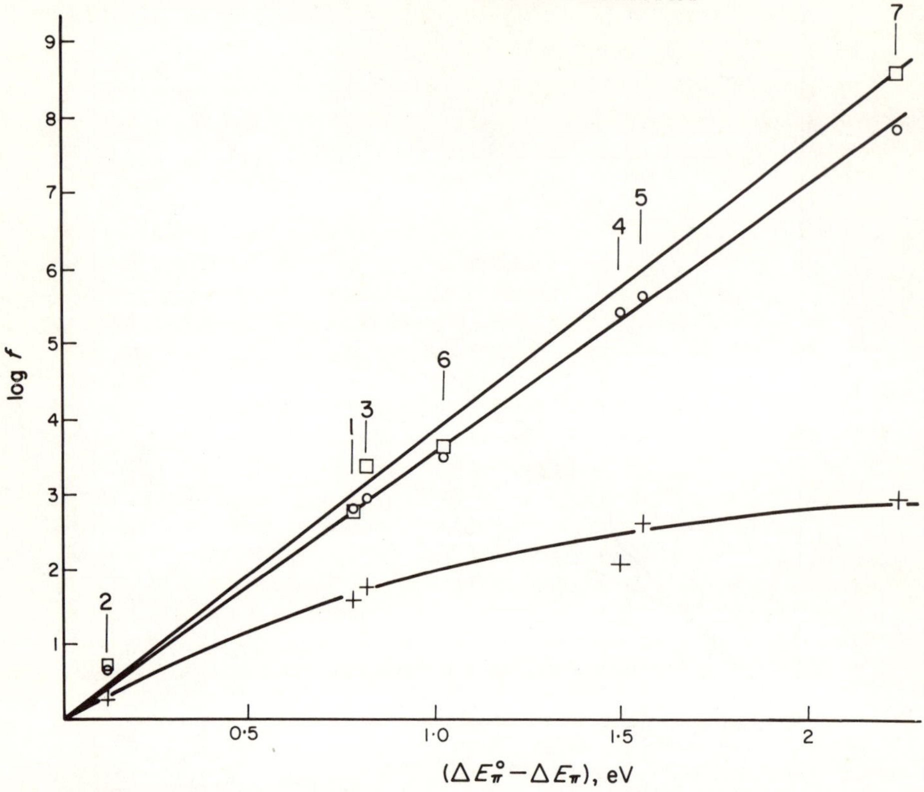

Figure 7.4. The relation between partial rate factors and calculated localisation energies (Refs. 14b, 67)
○ Chlorination (Rel. rates, Ref 48; toluene *o-* 59·78%, *m-* 0·48%, *p-* 39·74%; *m*-xylene 2-23%, 4-77%)
□ Bromination (Rel. rates, Ref. 48; toluene *o-* 32·9%, *m-* 0·3%, *p-* 66·8%)
+ Nitration (Rel. rates, Table 7.2; toluene *o-* 56·9%, *m-* 2·8%, *p-* 40·3%; *m*-xylene 2-14%, 4-86%)
1 Toluene, 2 posn; 2 toluene, 3 posn; 3 toluene, 4 posn; 4 *m*-xylene, 2 posn; 5 *m*-xylene, 2 posn; 5 *m*-xylene, 4 posn; 6 *p*-xylene; 7 mesitylene

chlorination or bromination, for which the substituent effects are to a high degree additive. The contrast between the substituent effects in nitration and the halogenations is clearly illustrated in *Figure 7.3*, where logarithms of the observed rates relative to that of benzene are plotted against the logarithms of the relative rates, calculated assuming additivity. The points for the halogenations fall on or close to a straight line of unit slope, whereas the points for nitration fall on a smooth curve. In *Figure 7.4* the logarithms of the partial rate factors for these three reactions are plotted against $(\Delta E_\pi^\circ - \Delta E_\pi)$. The results for halogenation do give good linear plots, that for chlorination being somewhat better than for the bromination. In contrast, the results for nitration seem to be best fitted to a smooth curve[14b].

Linearity between logarithms of partial rate factors and localisation energies will be found only if in a series of substituted benzenes continued substitution influences a particular position in such a way that the change in potential energy of the transition state for substitution at that position is throughout proportional to the change in potential energy of the corresponding Wheland intermediate. If this situation does not hold, so that with continued substitution the transition state comes more and more to resemble the reactants rather than the Wheland intermediate, linearity cannot be expected. Evidently, nitration is a process of the latter kind, where the differential stabilisation of transition state and intermediate is noticeable at relatively low substrate reactivities.

The results above refer to nitrations in acetic acid but, as already discussed, there is evidence that limiting reactivities are reached in nitrations under other conditions. In acetic anhydride, for example, the limiting reactivity appears to be at about 5 000 times that of benzene[14b]. It seems unlikely therefore that the very large partial rate factors ($c.$ 10^6) that have been reported[26] for certain polycyclic compounds refer to nitration, but they could refer to nitration via nitrosation. However, a good linear correlation is obtained when these data are plotted against reactivity numbers[26d] (which are similar to localisation energies); the significance of such a correlation is not clear.

The results obtained from nitrations have been used extensively for testing theoretical treatments of chemical reactivity. The above discussion indicates that great care should be exercised when using data which refer to substrates more reactive than benzene. These comments do not apply, of course, to those substrates less reactive than benzene; these compounds still provide valuable data for testing theoretical treatments of reactivity–structure relationships.

Manuscript received April 1971.

REFERENCES

1. BERLINER, E., *Prog. Phys. Org. Chem.*, Vol. 2, 253 (1964); ZOLLINGER, H., *Adv. Phys. Org. Chem.*, Vol. 2, 163 (1964)
2. (a) GILLESPIE, R. J. and MILLEN, D. J., *Q. Rev.*, **2**, 277 (1948); (b) INGOLD, C. K., *Structure and Mechanism in Organic Chemistry*, Bell, London (1953, 2nd edn 1969); (c) DE LA MARE, P. B. D. and RIDD, J. H., *Aromatic Substitution: Nitration and Halogenation*, Butterworths, London (1959); (d) RIDD, J. H., *Studies on Chemical Structure and Reactivity*, Methuen, London, 133 (1966); (e) HOGGETT, J. G., MOODIE, R. B., PENTON, J. R. and SCHOFIELD, K., *Nitration and Aromatic Reactivity*, Cambridge University Press (1971)
3. GOLD, V., HUGHES, E. D., INGOLD, C. K. and WILLIAMS, G. H., *J. chem. Soc.*, 2452 (1950)
4. GOLD, V., HUGHES, E. D. and INGOLD, C. K., *J. chem. Soc.*, 2467 (1950)
5. (a) BUNTON, C. A., HUGHES, E. D., INGOLD, C. K., JACOBS, D. I. H., JONES, M. H., MINKOFF, G. J. and REED, R. I., *J. chem. Soc.*, 2628 (1950); (b) GLAZER, J., HUGHES, E. D., INGOLD, C. K., JAMES, A. T., JONES, G. T. and ROBERTS, E., *J. chem. Soc.*, 2657 (1950)
6. OLAH, G. A. and KUHN, S. J., *Friedel-Crafts and Related Reactions*, Interscience, New York, Vol. 3, Chap. 43 (1964)
7. (a) KATRITZKY, A. R., *Quaderni de 'La Ricerca Scientifica'*, 53 (1968); (b) RIDD, J. H., *Z. Chemie*, **8**, 201 (1968); (c) AKSEL'ROD, ZH. I. and BEREZOVSKII, V. M., *Russ. chem. Rev.*, **39**, 627 (1970)
8. (a) BONNER, T. G., HANCOCK, R. A. and ROLLE, F. R., *Tetrahedron Lett.*, 1665 (1968); (b) COOMBES, R. G., *J. chem. Soc. B*, 1256 (1969)
9. BENFORD, G. A. and INGOLD, C. K., *J. chem. Soc.*, 929 (1938)
10. HUGHES, E. D., INGOLD, C. K. and REED, R. I., *J. chem. Soc.*, 2400 (1950)
11. COHN, H., HUGHES, E. D., JONES, M. H. and PEELING, M. G., *Nature*, **169**, 291 (1952)
12. HOGGETT, J. G., MOODIE, R. B. and SCHOFIELD, K., *J. chem. Soc. B*, 1 (1969)
13. OLAH, G. A., KUHN, S. J., FLOOD, S. H. and EVANS, J. C., *J. Am. chem. Soc.*, **84**, 3687 (1962)

14. (a) THOMPSON, M. J., unpublished work; (b) HARTSHORN, S. R., MOODIE, R. B., SCHOFIELD, K. and THOMPSON, M. J., *J. chem. Soc. B*, 2447 (1971)
15. HANTZSCH, A., *Ber. dt. chem. Ges.*, **58**, 941 (1925); *Z. phys. Chem.*, **149**, 161 (1930)
16. GODDARD, D. R., HUGHES, E. D. and INGOLD, C. K., *J. chem. Soc.*, 2559 (1950)
17. OLAH, G. A., KUHN, S. J. and MLINKÓ, A., *J. chem. Soc.*, 4257 (1956)
18. KUHN, S. J. and OLAH, G. A., *J. Am. chem. Soc.*, **83**, 4564 (1961)
19. OLAH, G. A., KUHN, S. J. and FLOOD, S. H., (a) *J. Am. chem. Soc.*, **83**, 4571 (1961); (b) *J. Am. chem. Soc.*, **83**, 4581 (1961)
20. TOLGYESI, W. S., *Can. J. Chem.*, **43**, 343 (1965)
21. RIDD, J. H., *Accts Chem. Res.*, **4**, 248 (1971)
22. CHRISTY, P. F., RIDD, J. H. and STEARS, N. D., *J. chem. Soc. B*, 797 (1970)
23. HOGGETT, J. G., MOODIE, R. B. and SCHOFIELD, K., *Chem. Commun.*, 605 (1969)
24. (a) BUTLER, A. R. and HENDRY, J. B., *J. chem. Soc. B*, 102 (1971); (b) ALCORN, P. G. E. and WELLS, P. R., *Aust. J. Chem.*, **18**, 1377 (1965); (c) ALCORN, P. G. E. and WELLS, P. R., *Aust. J. Chem.*, **18**, 1391 (1965)
25. DAVIES, A. and WARREN, K. D., *J. chem. Soc. B*, 873 (1969)
26. DEWAR, M. J. S. (a) with MOLE, T., *J. chem. Soc.*, 1441 (1956); (b) with MOLE, T., URCH, D. S. and WARFORD, E. W. T., *J. chem. Soc.*, 3572 (1956); (c) with MOLE, T. and WARFORD, E. W. T., *J. chem. Soc.*, 3576 (1956); (d) with MOLE, T. and WARFORD, E. W. T., *J. chem. Soc.*, 3581 (1956); (e) with LOGAN, R. H., *J. Am. chem. Soc.*, **90**, 1924 (1968)
27. COOKSEY, A. R., MORGAN, K. J. and MORREY, D. P., *Tetrahedron*, **26**, 5101 (1970)
28. (a) BONNER, T. G. and HANCOCK, R. A., *Chem. Commun.*, 780 (1967); (b) BONNER, T. G., HANCOCK, R. A., YOUSIF, G. and (in part) ROLLE, F. R., *J. chem. Soc. B*, 1237 (1969)
29. HARTSHORN, S. R., MOODIE, R. B. and SCHOFIELD, K., *J. chem. Soc. B*, 1256 (1971)
30. FISCHER, A., READ, A. J. and VAUGHAN, J., *J. chem. Soc.*, 3691 (1964)
31. COHN, F. H. and WIBAUT, J. P., *Recl Trav. chim. Pays-Bas Belg.*, **54**, 409 (1935)
32. PAUL, M. A., *J. Am. chem. Soc.*, **80**, 5329 (1958)
33. HOGGETT, J. G., Ph.D. Thesis, University of Exeter (1969)
34. HARTSHORN, S. R., MOODIE, R. B. and SCHOFIELD, K., *J. chem. Soc. B.*, 2454 (1971)
35. NORMAN, R. O. C. and RADDA, G. K., *J. chem. Soc.*, 3030 (1961)
36. KNOWLES, J. R. and NORMAN, R. O. C., *J. chem. Soc.*, 3888 (1961)
37. (a) FISCHER, A., PACKER, J., VAUGHAN, J. and WRIGHT, G. J., *Proc. chem. Soc.*, 369 (1961); (b) FISCHER, A., PACKER, J., VAUGHAN, J. and WRIGHT, G. J., *J. chem. Soc.*, 3687 (1964); (c) FISCHER, A., VAUGHAN, J. and WRIGHT, G. J., *J. chem. Soc. B*, 368 (1967); (d) BLACKSTOCK, D. J., FISCHER, A., RICHARDS, K. E., VAUGHAN, J. and WRIGHT, G. J., *Chem. Commun.*, 641 (1970); (e) BLACKSTOCK, D. J., CRETNEY, J. R., FISCHER, A., HARTSHORN, M. P., RICHARDS, K. E., VAUGHAN, J. and WRIGHT, G. J., *Tetrahedron Lett.*, **32**, 2793 (1970)
38. VAUGHAN, J., WELCH, G. J. and WRIGHT, G. J., *Tetrahedron*, **21**, 1665 (1965)
39. ROBINSON, R., *J. chem. Soc. B*, 1289 (1970)
40. (a) ROBINSON, R. and THOMPSON, H. W., *J. chem. Soc.*, 2015 (1932); (b) NAKAMURA, K., *Bull. chem. Soc. Japan*, **44**, 133 (1971); (c) SUZUKI, H. and NAKAMURA, K., ibid., **44**, 227 (1971); (d) Idem, ibid., **44**, 303 (1971)
41. COOMBES, R. G., MOODIE, R. B. and SCHOFIELD, K., *J. chem. Soc. B*, 800 (1968)
42. HAMMOND, G. S., *J. Am. chem. Soc.*, **77**, 334 (1955)
43. OLAH, G. A., TASHIRO, M. and KOBAYASHI, S., *J. Am. chem. Soc.*, **92**, 6369 (1970)
44. COOMBES, R. G., unpublished work
45. DENO, N. C. and STEIN, R., *J. Am. chem. Soc.*, **78**, 578 (1956)
46. COOMBES, R. G., CROUT, D. H. G., HOGGETT, J. G., MOODIE, R. B. and SCHOFIELD, K., *J. chem. Soc. B*, 347 (1970)
47. BIRD, M. L. and INGOLD, C. K., *J. chem. Soc.*, 918 (1938)
48. BACIOCCHI, E. and ILLUMINATI, G., *Prog. Phys. Org. Chem.*, Vol. 5, 1 (1967)
49. GRIFFITHS, P. H., WALKEY, W. A. and WATSON, H. B., *J. chem. Soc.*, 631 (1934)
50. HALVARSON, K. and MELANDER, L., *Ark. Kemi*, **11**, 77 (1957)
51. PAUL, M. A., *J. Am. chem. Soc.*, **80**, 5332 (1958)
52. BUNTON, C. A., MINKOFF, G. J. and REED, R. I., *J. chem. Soc.*, 1416 (1947)
53. SPARKS, A. K., *J. org. Chem.*, **31**, 2299 (1966)
54. ARNALL, F. and LEWIS, T., *Chemy Ind.*, 159T (1929)
55. LYNCH, B. M., CHEN, C. M. and WIGFIELD, Y.-Y., *Can. J. Chem.*, **46**, 1141 (1969)
56. BRICKMAN, M. and RIDD, J. H., *J. chem. Soc.*, 6845 (1965)
57. HARTSHORN, S. R. and RIDD, J. H., *J. chem. Soc. B*, 1063 (1968)

58. BRICKMAN, M., UTLEY, J. H. P. and RIDD, J. H., *J. chem. Soc.*, 6851 (1965)
59. GASTAMINZA, A., MODRO, T. A., RIDD, J. H. and UTLEY, J. H. P., *J. chem. Soc. B*, 534 (1968)
60. KIRKWOOD, J. G. and WESTHEIMER, F. H., *J. chem. Phys.*, **6**, 506 (1938)
61. RIDD, J. H., *Spec. Publ. chem. Soc.*, **21**, 149 (1967)
62. MOSSA, G., RICCI, A. and RIDD, J. H., *Chem. Commun.*, 332 (1971)
63. MODRO, T. A. and RIDD, J. H., *J. chem. Soc. B*, 528 (1968)
64. CHANDRA, A. K. and COULSON, C. A., *J. chem. Soc.*, 2210 (1965)
65. (a) DE SARLO, F. and RIDD, J. H., *J. chem. Soc. B*, 712 (1971); (b) GRYNKIEWICZ, G. and RIDD, J. H., *J. chem. Soc. B*, 716 (1971); (c) DE SARLO, F., GRYNKIEWICZ, G., RICCI, A. and RIDD, J. H., *J. chem. Soc. B*, 719 (1971)
66. DEWAR, M. J. S. and THOMPSON, C. C., *J. Am. chem. Soc.*, **87**, 4414 (1965)
67. CLARK, D. T. and FAIRWEATHER, D. J., *Tetrahedron*, **25**, 4083 (1969)

8

HOMOGENEOUSLY CATALYSED HYDROGENATION

F. J. McQuillin

General	314
Catalytically Active Complexes	320
Homogeneous Hydrogenation of Olefinic and Acetylenic Groups	322
The Olefinic Group	322
Conjugated Olefins	327
Unsaturated Aldehydes	329
The Acetylenic Bond	329
The Stereochemistry of Hydrogenation of Olefins and Acetylenes	329
Olefin Isomerisation	329
Hydrogenation of Heteroenoid Groups	331
Hydrogenation of the Carbonyl Group	331
Halogen Compounds	332
Hydrogenolysis of Vinyl Cyclopropanes	332
Catalysed Deuteration	333
Induced Asymmetry via Homogeneous Hydrogenation	334
Conclusion	335

GENERAL

Although the long-established metallic catalysts for hydrogenation remain generally very convenient and effective, a number of transition metal complexes are now available which catalyse hydrogenation homogeneously, and at rates comparable with those of the better-known metallic catalysts. Moreover, certain of these complexes are much more selective than the metallic catalysts for the types of groups which may be hydrogenated readily, and this property may, in particular cases, offer a considerable practical advantage.

The development of soluble complexes which exhibit hydrogenation catalysis has, in addition, very much clarified our understanding of the mechanism of hydrogen activation and of the process of hydrogen transfer to an olefinic acceptor. One clear essential for catalysis is that the complex should react with molecular hydrogen to form a reactive metal hydride, but activation of the olefin through co-ordination to the metal is also generally necessary. In this way hydrogen transfer occurs within a hydrido–metal olefin complex, and the catalytic sequence may be represented schematically as follows.

(a) Hydride formation, which may lead to a mono- or di-hydride:

$$L_xM + H_2 \longrightarrow L_xMH \quad \text{or} \quad L_xMH_2$$
$$\qquad\qquad\qquad\quad (I) \qquad\qquad\quad (II)$$

(b) Hydrogen transfer:

(i) From a monohydride L_xMH:

$$L_xMH + \text{olefin} \longrightarrow L_xM\overset{H}{\underset{(III)}{|}}\!\!\!\!\!\!\!\!\underset{}{\diagdown\!\!\!\diagup} \longrightarrow L_xM\underset{(IV)}{\overset{H}{\diagdown}}$$

$$\xrightarrow{H_2} L_xMH + \text{alkane}$$

(L_x = ligand groups)

Or, (ii) From a dihydride L_xMH_2:

$$L_xMH_2 + \text{olefin} \longrightarrow L_xM\overset{H}{\underset{H}{|}}\!\!\!\!\!\!\!\!\underset{(V)}{\diagdown\!\!\!\diagup} \longrightarrow L_xM\underset{(VI)}{\overset{H\ H}{\diagdown}}$$

$$\longrightarrow L_xM + \text{alkane}$$

(L_xM = metal complex, bearing ligands L)

For the better known and most active homogeneous catalysts the hydrido–metal complex has been fully characterised, and examples are given below of catalysis by both mono- (I), and dihydrido- (II) species. Olefin co-ordination has also generally been demonstrated, and there is, in many instances, direct or indirect evidence for the metal alkyl intermediates (IV) or (VI).

For the sequence of reactions (a), and (b) above to constitute a rapid catalytic process, however, there must be a suitable balance of bond energies so that hydride formation, olefin co-ordination and hydrogen transfer may occur with small activation energy. Thus, catalytic activity depends critically on the choice of metal M, and very substantially on the nature of the ligand groups L. Clearly, the co-ordinating power of the olefin is also important, and both the olefin and the hydride contribute as ligands to the reactivity of the complex. It will be seen from the examples given below that the ligands L may include a halogen, and, most frequently, also π-acceptor groups such as R_3P, CO, or CN, which stabilise the hydride and serve to protect the metal complex against reduction to the free metal.

The reaction sequence of equations (a), and (b) (i) or (b) (ii) represents a cyclical increase and decrease in the co-ordination number, or formal oxidation state, of the metal in the complex. The importance of this underlying principle in the catalytic process was clearly recognised by Vaska[1] who observed the reversible addition of hydrogen to bis(triphenylphosphine)-chlorocarbonyliridium.

$$\text{IrCl(CO)(PPh}_3)_2 + \text{H}_2 \longrightarrow \text{IrH}_2\text{Cl(CO)(PPh}_3)_2$$
$$\text{(VII)} \qquad\qquad\qquad \text{(VIII)}$$

However, the iridium dihydride (VIII) is a hexa-coordinated complex, and the additional co-ordination of an olefin which is necessary for catalytic hydrogenation requires a ligand displacement. The importance of ligand displacement, which does not readily occur with $\text{IrH}_2\text{Cl(CO)(PPh}_3)_2$, was established in Wilkinson's brilliant study[2] of the related rhodium complex $\text{RhH}_2(\text{Cl})(\text{PPh}_3)_2$. This dihydridorhodium complex is readily formed on shaking tris(triphenylphosphine)chlororhodium in solution, e.g. in benzene, with hydrogen:

$$\text{RhCl(PPh}_3)_3 + \text{H}_2 \longrightarrow \text{RhH}_2\text{Cl(PPh}_3)_2$$
$$\text{(IX)} \qquad\qquad\qquad \text{(X)}$$

Addition of hydrogen in this case, however, is accompanied by loss of a triphenylphosphine ligand group so that the hydrido complex (X) carries one ligand fewer than the iridium complex (VIII). However, what appears to be phosphine dissociation* is in fact solvent displacement of phosphine, and in the dihydrido complex $\text{RhH}_2(\text{Cl})(\text{PPh}_3)_2$, in which the rhodium is formally penta-coordinated, the sixth site is, in solution, occupied by a solvent molecule. Although the dihydride may be isolated as a solvent-free crystalline solid, solvates such as $[\text{RhH}_2(\text{Cl})(\text{PPh}_3)_2]_2 \cdot \text{CH}_2\text{Cl}_2$ are readily obtained[2]. Thus the structure of the dihydride in solution is represented as (XI), and co-ordination with an olefin molecule as in (XII) occurs by solvent displacement. Within the resultant complex (XII) the olefin and hydride

* The parent complex was reported[2] to dissociate in solution: $\text{RhCl(PPh}_3)_3 \rightleftarrows \text{RhCl(PPh}_3)_2 + \text{PPh}_3$, but the evidence for this appears to be associated with an oxygen-induced reaction[3].

ligands are *cis*-related, and hydrogen transfer occurs readily with the formation of the alkane, and with release of $\text{RhCl(PPh}_3)_2$ from which the dihydride is re-formed; the catalytic cycle is then repeated.

The related tris(triphenylphosphine)chloroiridium also forms a dihydride:

$$\text{Ir(PPh}_3)_3\text{Cl} + \text{H}_2 \longrightarrow \text{IrH}_2(\text{PPh}_3)_3\text{Cl}$$

In this case, however, the dihydride is formed without displacement of triphenylphosphine, and the six co-ordinated dihydride is not active in hydrogenation catalysis[4]. The bis(triphenylphosphine) derivative, which may be obtained by displacement from the bis(cyclo-octene) complex,

$$[\text{Ir}(\text{C}_8\text{H}_{14})_2\text{Cl}]_2 + 4\,\text{PPh}_3 \longrightarrow [\text{Ir}(\text{PPh}_3)_2\text{Cl}]_2$$

is, however, found to be highly active in the catalytic hydrogenation of hex-l-ene[5a]. Similar observations have been made with the trihydrides $(Ph_3P)_2IrH_3$ and $(Ph_3P)_3IrH_3$, the former being active for the hydrogenation of hex-l-ene, for example, and the latter being inactive under the same conditions[5b].

The mode of operation of the rhodium complex $Rh(PPh_3)_3Cl$, which has so far proved the most widely useful catalyst for homogeneous hydrogenation, emphasises the importance of the relative co-ordinating strength of the olefin, and also that the decrease in olefin co-ordinating strength with increasing alkyl substitution of the olefinic bond is an important factor underlying the selectivity exhibited by this and related homogeneous catalysts discussed below.

In the dihydridorhodium olefin complex (XII) hydrogenation may be represented as an effectively simultaneous *cis*-addition of two hydrogen ligands to the olefin[2]. However, simultaneous addition is not necessary for effective catalysis, and hydrogen transfer may be stepwise. Indeed, the hydride ligands in (XII) are not in entirely equivalent environments in the complex, and hydrogen transfer to the two termini of an unsymmetrical olefin would in any case be expected to occur at appreciably different rates[6a]. Stepwise hydrogen transfer is also relevant to the problem of olefin isomerisation which may accompany hydrogenation, and the activity as hydrogenation catalysts of monohydrido complexes[7] such as $RhH(CO)(PPh_3)_3$ and $RuH(Cl)(PPh_3)_3$ necessarily involves sequential hydrogen transfer.

In the case of a monohydrido complex, reaction with a co-ordinated olefin yields a metal–alkyl intermediate which with more hydrogen undergoes hydrogenolysis forming the alkane and re-forming the parent monohydride. For hydridocarbonyltris(triphenylphosphine)rhodium the reaction sequence may be represented as a phosphine dissociation step:

$$RhH(CO)(PPh_3)_3 \longrightarrow RhH(CO)(PPh_3)_2$$

followed by olefin co-ordination and hydrogen transfer to form the metal–alkyl intermediate (XV) which undergoes hydrogenolysis.

$\longrightarrow (Ph_3P)_2Rh(H)CO$ + alkane

The mode of formation of the various catalytically active hydridometal complexes is of interest. The dihydrides $RhCl(PPh_3)_2(H_2)$ and $IrCl(CO)(PPh_3)_2(H_2)$ are obtained by direct addition of hydrogen, e.g.:

$$RhCl(PPh_3)_3 \xrightarrow{H_2} RhCl(PPh_3)_2(H_2) + PPh_3$$

This process amounts to an insertion of the metal into the H—H bond of molecular hydrogen, and in the above example the formal charge state of the rhodium is increased from Rh^I to Rh^{III}. Thus hydride formation constitutes an oxidative addition, and in the subsequent transfer of hydrogen to an olefin acceptor the metal returns to the state Rh^I.

Monohydride complexes such as $RhH(CO)(PPh_3)_3$ or $RuH(Cl)(PPh_3)_3$, on the other hand, are obtained by various methods of reductive displacement of halogen. Hydridochlorotris(triphenylphosphine)ruthenium may be prepared from the corresponding dichloride by reaction with hydrogen in benzene solution:

$$RuCl_2(PPh_3)_3 + H_2 \longrightarrow RuH(Cl)(PPh_3) + HCl$$

but a suitably basic co-solvent such as ethanol, or an added base, e.g. triethylamine, is also necessary. Sodium borohydride is, however, a more convenient reducing agent[8]. Hydridocarbonyltris(triphenylphosphine)-rhodium is prepared similarly by reduction of the chloride with sodium borohydride[9].

$$RhCl(CO)(PPh_3)_3 \longrightarrow RhH(CO)(PPh_3)_3$$

Reaction of dichlorotris(triphenylphosphine)ruthenium with hydrogen involves heterolysis of the hydrogen molecule. The hydridopentacyanocobalt ion, a monohydride long recognised as a catalyst for homogeneous hydrogenation[10-12], is formed, on the other hand, by homolysis of hydrogen.

$$2\ Co(CN)_5^{3-} + H_2 \longrightarrow 2\ HCo(CN)_5^{3-}$$

The catalytic activity of the hydridopentacyanocobalt complex is, however, largely confined to the hydrogenation of conjugated olefins such as acrylonitrile or butadiene, which readily undergo hydride addition. Buta-1,3-diene, for example, gives a butenyl cobalt intermediate which has been characterised[13].

$$HCo(CN)_5^{3-} + CH_2{=}CH{-}CH{=}CH_2 \longrightarrow CH_2{=}CH\overset{\underset{\displaystyle|}{Me}}{C}H{-}Co(CN)_5^{3-}$$

and is hydrogenolysed by reaction with a second equivalent of hydridopentacyanocobalt to yield but-1-ene.

$$CH_2{=}CH{-}\overset{\underset{\displaystyle|}{Me}}{C}HCo(CN)_5^{3-} + HCo(CN)_5^{3-} \longrightarrow CH_2{=}CH{-}\overset{\underset{\displaystyle|}{Me}}{C}H_2 + 2\ Co(CN)_5^{3-}$$
$$(XVII)$$

The hydrogenation product formed is, however, dependent on the CN^-/Co ratio, and with a low concentration of cyanide ion in the medium *cis*- and *trans*-but-2-enes are obtained instead of but-1-ene. This change arises from the stepwise nature of the hydrogenation[12-14], and because the butenyl cobalt intermediate (XVII) undergoes dissociation of CN^- at low cyanide concentrations.

This dissociation arises from displacement of a CN^- ligand by the olefinic bond of the butenyl residue yielding the cobalt π-butenyl derivative (XVIII), which may clearly arise as two stereoisomers (XVIIIa) and (XVIIIb), giving on hydrogenolysis *trans-* and *cis-*but-2-enes.

This reaction sequence illustrates one mechanism by which during hydrogenation an olefinic bond may be translocated to yield a product different from that expected from direct 1,2-addition of hydrogen. However,

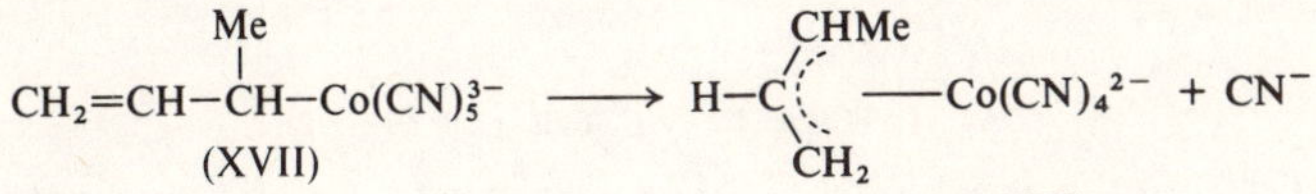

$$CH_2=CH-\underset{Me}{CH}-Co(CN)_5^{3-} \longrightarrow H-C\underset{CH_2}{\overset{CHMe}{\diagup}}-Co(CN)_4^{2-} + CN^-$$
(XVII) (XVIII)

(XVIII) + $HCo(CN)_5^{3-}$ + CN^- → $CH_3CH=CHCH_3$ + 2 $Co(CN)_5^{3-}$

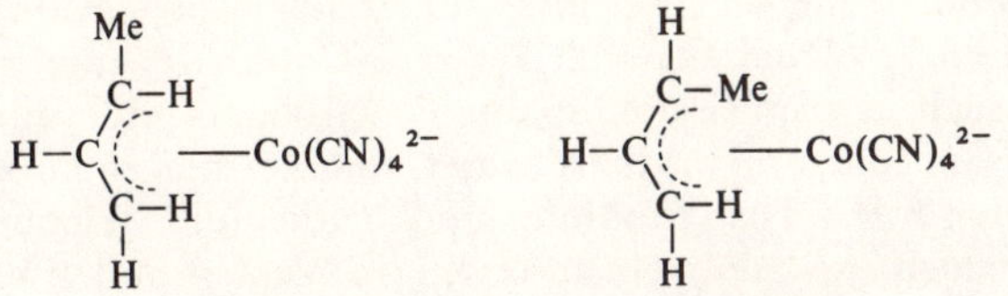

$$H-C\underset{\underset{H}{C-H}}{\overset{\overset{Me}{C-H}}{\diagup}}-Co(CN)_4^{2-} \qquad H-C\underset{\underset{H}{C-H}}{\overset{\overset{H}{C-Me}}{\diagup}}-Co(CN)_4^{2-}$$

(XVIIIa) (XVIIIb)

isomerisation via a σ-allyl → π-allyl transformation is much less common than isomerisation by hydride abstraction from an intermediate metal alkyl. An example of this type arises during the hydrogenation of *cis*-pent-2-ene by means of $Rh(hal)(PPh_3)_3$, (hal=Cl, Br, I) in benzene–alcohol solution. Product analysis during the reaction showed the presence of *trans*-pent-2-ene and small amounts of pent-1-ene in addition to *cis*-pent-2-ene and pentane[6a]. The isomerised olefins are considered to arise by a reversal of the addition step.

$RhH_2(Cl)(PPh_3)_2 + C_2H_5CH=CHCH_3 \longrightarrow$

$$(Ph_3P)_2Rh(H)(Cl)-CH\overset{CH_2C_2H_5}{\underset{CH_3}{\diagdown}}$$

(XIX)

Re-formation of $RhH_2(Cl)(PPh_3)_2$ and olefin from the alkyl intermediate (XIX) may occur in three ways. Hydride abstraction from the CH_2 group may give either *cis-* or *trans*-pent-2-ene, whilst abstraction of hydride from the CH_3 group will lead to pent-1-ene.

Clearly, the factor determining the importance of isomerisation is the relative rate of the reversion

$$L_xM-alkyl \longrightarrow L_xM-H + \text{isomerised alkene}$$

in comparison with the rate of hydrogenolysis of the metal alkyl:

$$L_xM(H)alkyl \longrightarrow L_xM + alkane,$$
$$\text{or } L_xM-alkyl + H_2 \longrightarrow L_xMH + alkane$$

The forward reaction is usually rapid and isomerisation is therefore generally unimportant. However, it may become a more serious competitor with direct hydrogenation when the iso-alkene is appreciably stabilised by substitution, as in the case of damsin discussed below, or by conjugation as may arise when the reaction is started from a 1,4-diene.

$$L_xMH + RCH=CHCH_2CH=CHR'$$
$$\longrightarrow L_xM-CH(CH_2R)CH_2CH=CHR'$$
$$\longrightarrow L_xMH + RCH_2CH=CH-CH=CHR'$$

CATALYTICALLY ACTIVE COMPLEXES

The metal complexes which have been shown to be effective catalysts for homogeneous hydrogenation may be considered under four headings.

(a) Transition metal hydrides of the type $RhH_2(Cl)(PPh_3)_2$, $RhH(CO)(PPh_3)_3$, $RuH(Cl)(PPh_3)_3$ or $HCo(CN)_5^{3-}$ are stabilised by π-acceptors such as triphenylphosphine[2,7], carbonyl[7], or cyanide[10] ligands. The pentacyanocobalt hydride is water soluble, and hydrogenates conjugated olefins[10-12]. The rhodium and ruthenium triphenylphosphine complexes, which are relatively insoluble in alcohol and similar solvents, are effective in benzene or benzene–alcohol solution for the hydrogenation of olefins[2,7,8].

(b) Cationic complexes of the type[15] $[MH_2(PPh_3)_2S]^+X^-$ (where M = Ir or Rh, S = co-ordinated solvent such as acetone or alcohol, and $X^- = ClO_4^-$ or Ph_4B^-), which are more soluble in protic solvents than the rhodium and ruthenium complexes in (a), have been shown to hydrogenate olefins and acetylenes[15] and ketones[16]. These cationic complexes are obtained as shown.

$$[M(diene)Cl]_2 + Ph_3P \text{ in protic solvent} \longrightarrow [M(diene)(PPh_3)_2]^+$$
$$\xrightarrow{H_2} [MH_2(PPh_3)_2S]^+$$

where S = Me_2CO, EtOH, dioxan, dimethylformamide, M = Ir or Rh, and diene = a chelating diene such as cyclo-octa-1,5-diene.

The bis(acetonitrile) complexes $[M(diene)(CH_3CN)_2]^+BF_4^-$, M = Rh or Ir, prepared by diene displacement from $[M(diene)_2]^+BF_4^-$, also exhibit activity as homogeneous catalysts[17] for the hydrogenation of olefins.

A related group of cationic complexes showing catalytic activity may be derived[18] from dimeric bridged acetates such as $Rh_2(OAc)_4$ or $Ru_2(OAc)_4Cl$ which have strong metal–metal bonds. In strongly acid solution these acetates give binuclear cationic species, as in the following example.

$$Rh_2(OAc)_4 + 4\ HBF_4 \longrightarrow Rh_2^{4+} + 4\ BF_4^- + 4\ HOAc$$

The binuclear rhodium cation when combined with triphenylphosphine (2 molar equivalents) in methanol gives a cationic complex which is capable of catalysing the hydrogenation of hex-1-yne, hex-1-ene, hexa-1,5-diene, and of hept-2-ene and cyclohexene at a slower rate. The corresponding ruthenium derivative shows similar activity.

(c) Catalytic activity has been observed also for various transition metal salts in the presence of ligands other than phosphine or carbonyl derivatives.

Following an observation of Cramer[19], the combination of chloroplatinic acid and stannous chloride was shown to act as a homogeneous catalyst for the hydrogenation of various alkynes and alkenes in alcohol solution[20]. The rhodium pyridine complex 1,2,6-py_3RhCl_3 was also shown to exhibit catalysis in alcohol solution[21]. James and Rempel, however, noted the particular importance of the solvent used, and demonstrated an enhanced catalytic activity for rhodium trichloride when dissolved in dimethylacetamide[22]. In this solvent the Rh^{III} salt is reduced by hydrogen to an Rh^I derivative which was found to be active for the hydrogenation of maleic acid.

Many of these simpler systems are, however, of limited activity and also rather unstable towards reduction to the metal. A rhodium catalyst which appears to overcome these limitations makes use of py_3RhCl_3 activated by means of sodium borohydride in dimethylformamide solution. A catalytically active complex $py_2(dmf)RhCl_2(BH_4)$ has been isolated[23], and this catalyst system has been shown to be effective for the hydrogenation of a range of alkenes and alkynes and also groups[24] such as —CH=N—, —N=N—, or $ArNO_2$.

(d) Various catalyst systems which have been the subject of industrial study include iron pentacarbonyl[25] using cyclohexane as solvent at 180°C, chromium carbonyl arene derivatives[26] such as $Cr(CO)_3 \cdot PhCO_2Me$ at 160°C under pressure in a hydrocarbon solvent, and various transition metal acetylacetonates, e.g. $Cr(C_5H_7O_2)_3$, in combination with an aluminium alkyl, e.g. Al(iso-Bu)$_3$, in a hydrocarbon solvent[27]. A variety of other catalytic systems have been described[28].

The laboratory preparation of the more useful catalytically active complexes may be outlined.

(i) *RhCl(PPh₃)₃* A solution of rhodium trichloride ($RhCl_3 \cdot 3H_2O$) in ethanol with 6M excess of freshly recrystallised triphenylphosphine is refluxed for 0·5 h, filtered and cooled. The product exists as red and yellow crystalline forms. Corresponding bromo- and iodo- derivatives are obtained by metathesis using LiBr or LiI in hot ethanol[2].

(ii) *RhH(CO)(PPh₃)₂* Rhodium carbonyl chloride, $[Rh(CO)_2Cl]_2$, prepared from $RhCl_3 \cdot 3H_2O$ and carbon monoxide[29], is treated with triphenylphosphine in benzene solution to yield the phosphine carbonyl chloride[30], $RhCl(CO)(PPh_3)_2$, which is reduced using sodium borohydride in hot ethanol[9].

(iii) *RuH(Cl)(PPh₃)₃* The dichloride $RuCl_2(PPh_3)_3$ in benzene is reduced by means of sodium borohydride in a little water by being heated under reflux[8].

(iv) *Rhpy₂(dmf)(Cl₂)(BH₄)* This is conveniently prepared and used in solution. Py_3RhCl_3 dissolved in dimethylformamide is treated with 1 molar equivalent of sodium borohydride under hydrogen[23].

(v) *HCo(CN)₅³⁻* This material is prepared and used in solution; for example, cobalt chloride in water or alcohol is treated with rather more than 5 equivalents of potassium or lithium cyanide and shaken under hydrogen[9-13,31].

The Olefinic Group

For the group of complexes $Rhhal(PPh_3)_3$, Wilkinson[2] has established the following generalisations.

(a) As the halogen is varied the rate of hydrogenation increases in the order: hal = Cl < Br < I.

(b) The rate of hydrogenation is increased by electron donor substituents in the phosphine ligands[32]; i.e. the rate falls in the order

$$R_3P = (p\text{-}MeOC_6H_4)_3P > (p\text{-}MeC_6H_4)_3P > Ph_3P > PhEt_2P >$$
$$> (p\text{-}FC_6H_4)_3P > (p\text{-}ClC_6H_4)_3P$$

Horner reports[33] similar conclusions, and Stern[34] and his co-workers have shown that phosphines of the type: Ph_2RP and PhR_2P, where

R = [piperidinyl] or [morpholinyl]

are effective ligands.

(c) Terminal olefins, for example hex-1-ene, hept-1-ene or dodec-1-ene, react rapidly as do non-conjugated dienes such as hexa-1,5-diene or octa-1,7-diene. Of a pair of *cis*- and *trans*-isomeric olefins, for example *cis*- and *trans*-hex-2-ene, or *cis*- and *trans*-4-methylpent-2-ene, the *cis*-isomer reacts the more rapidly. Tri- and tetra-substituted olefins, for example 1-methylcyclohexene or 2,3-dimethylbut-2-ene, react relatively very slowly. Conjugated dienes, for example penta-1,3-diene, and chelating dienes such as cyclo-octa-1,5-diene are resistant to hydrogenation. However, styrene, acrylonitrile, acrylamide and allyl alcohol are readily hydrogenated[2].

The enthalpy and entropy of activation for the hydrogenation of various olefins have been determined[2].

Harmon and his co-workers have described[35] the hydrogenation of the following substances by means of $RhCl(PPh_3)_3$ and hydrogen in benzene at 40–60°C and slightly elevated pressures.

(i) $Ar.CH=C(R).CO_2H$, $Ar = Ph$, $R = H$, Me, Ph, or $Ar = p\text{-}MeOC_6H_4$, $R = H$;

(ii) $PhCH=CHCN$, but not $PhCH=C(Ph)CN$ or $PhCH=C(Ph)COMe$;

(iii) $ArCH=CHNO_2 \longrightarrow ArCH_2CH_2NO_2$, $Ar = p\text{-}NO_2C_6H_4$ or $3,4\text{-}(MeO)_2C_6H_3$;

(iv) itaconic and citraconic acids.

Using the $py_3RhCl_3/NaBH_4$ catalyst in dimethylformamide, Jardine and McQuillin[36] found that the rates of hydrogenation of a series of cycloalkenes fall with the heat of hydrogenation, viz: norbornadiene > cyclohexene > cyclopentene > cycloheptene > cyclo-octene, and within this series a plot of the log rate *v.* heat of hydrogenation was found to be almost linear.

In a study of the hydrogenation of various cyclohexenes by means $RhCl(PPh_3)_3$, Hussey and Takeuchi[37] found the relative rate sequence shown.

       10   4·5   3   1

This influence of substitution on reaction rate is the basis of selective hydrogenation of differently substituted olefins. Thus in a series described by Sims, Honwad and Selman[38], 1,4,5,8-tetrahydro- (XX) and 1,2,3,4,5,8-hexahydronaphthalene (XXI) were hydrogenated using $RhCl(PPh_3)_3$ to give mainly (XXII), accompanied by some 20% of (XXIII). Similarly, methyl-1,4,5,8-tetrahydronaphthoate (XXIV) gave (XXV) in 96% yield accompanied by a trace of (XXVI). The principal products in these examples are formed by preferential hydrogenation of the least substituted olefinic bonds.

However, formation of (XXIII) may be ascribed to isomerisation to a conjugated diene before hydrogenation, whilst (XXVI) is the result of aromatisation by conjugation of all three olefinic bonds in (XXIV).

The incidence of olefin isomerisation is discussed more fully below, but with $RhCl(PPh_3)_3$ isomerisation is generally only a minor side reaction as in these examples.

The catalyst hydridocarbonylbis(triphenylphosphine)rhodium $RhH(CO)(PPh_3)_2$ is found[7] to be highly specific for the hydrogenation of the terminal $-CH=CH_2$ group which, however, reacts at about half the rate found for catalytic hydrogenation with $RhCl(PPh_3)_3$. The ruthenium complex[8] $RuH(Cl)(PPh_3)_3$ also shows high specificity towards terminal olefins which, moreover, are hydrogenated with this catalyst some five times as rapidly as with $RhCl(PPh_3)_3$. Examples of easily hydrogenated olefins include pent-1-ene, hex-1-ene, hept-1-ene and dec-1-ene, whilst disubstituted olefins such as cis- or trans-hex-2-ene, hept-3-ene or cyclohexene react at only 10^{-3} or 10^{-4} of the rate for a terminal $-CH=CH_2$ group.

In a survey of the use of the complex $RhCl(PPh_3)_3$ for the hydrogenation of a wider range of chemical types, Birch and Walker[39] reported selective hydrogenation of the vinyl group of linalool (XXVII) and of the isopropenyl group of carvone (XXVIII). Also, 1-methoxycyclohexa-1, 4-diene (XXIX) was hydrogenated to a dihydro-derivative yielding cyclohexanone on hydrolysis, that is there is selective hydrogenation of the less substituted double bond to give (XXX). However, (XXXI), a closely similar cyclohexadiene, was resistant to hydrogenation, and cholesterol (XXXII) also could not be hydrogenated. In ergosterol, (XXXIII) the 7,8-olefinic group reacted selectively.

Djerassi and Gutzwiller[41] who examined the hydrogenation of a range of steroids by means of $RhCl(PPh_3)_3$ found that olefinic bonds at the 1,2-, 2,3-, and 3,4-position are hydrogenated whilst the more substituted 4,5-, 8,14-, 11,12- and 14,15- olefins are resistant. Similarly it was found that 3-keto-Δ^1-steroids are hydrogenated whilst 3-keto-Δ^4- or 3-keto-Δ^5-steroids are largely unaffected. Dienones such as androsta-1,4-dien-3,17-dione (XXXIV) or androsta-4,6-dien-3,17-dione (XXXV) were found to be selectively hydrogenated at the less substituted olefin bond. However, a small amount of the fully hydrogenated product is also formed by slow hydrogenation of the 4,5-olefinic bond. Birch and Walker[39] also observed a very slow hydrogenation of a 3-keto-Δ^4-steroid. Testosterone (XXXVII) was hydrogenated over a period of days to give androstan-17β-ol-3-one (XXXVIII).

Differentiation between the 1,2- and 4,5-olefinic bonds in the hydrogenation of steroids is, however, not invariably observed. With the 11α-acetoxy steroid 1,4-dien-3-one (XXXIX), Wieland and Anner[40] found both olefinic

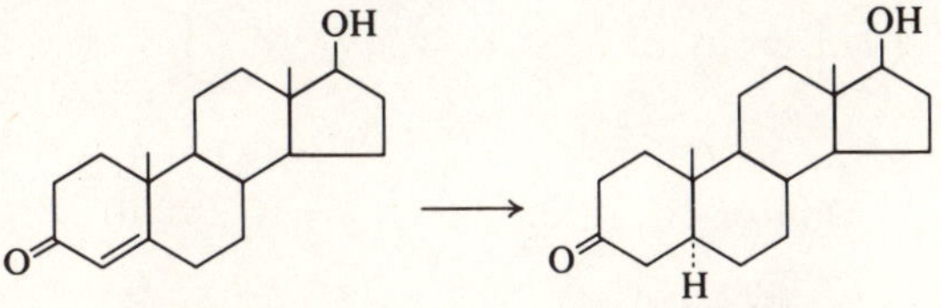

(XXXVII) (XXXVIII)

bonds to react at very similar rates when hydrogenated using $RhCl(PPh_3)_3$ in benzene. After hydrogenation for six hours the product contained the 1,2-dihydro- and 1,2,4,5-tetrahydro-derivatives as well as unchanged (XXXIX).

Using the $py_3RhCl_3/NaBH_4$ catalyst in dimethylformamide, Jardine and McQuillin were able to hydrogenate a series of 3-keto-Δ^4-steroids without

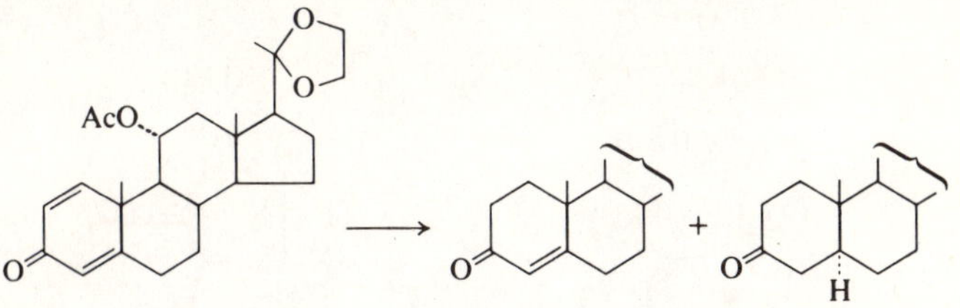

(XXXIX)

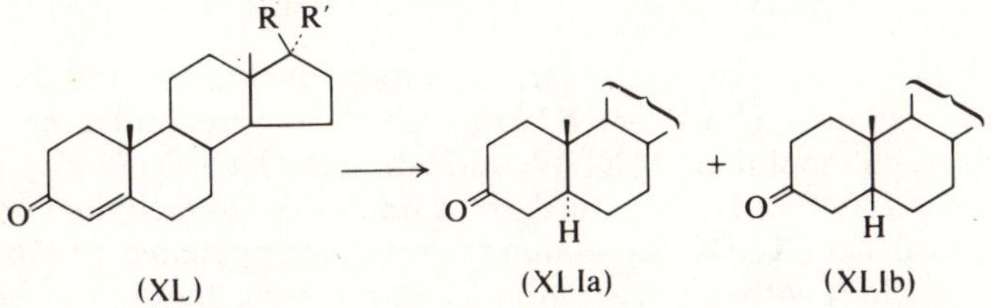

(XL) (XLIa) (XLIb)

difficulty. In a study of the proportions of 5αH, (XLIa), and 5βH-, (XLIb), hydrogenation product obtained from variously substituted examples (XL) the following results were obtained:

R	R'		
COMe	H	78%	22%
HO	H	78	22
HO	Me	25	75
iso-C_8H_{17}	H	20	80

The authors draw attention to the influence of a remote substituent on the stereoselectivity of these reactions. The results are attributed to conformational transmission of compressions arising from the steric bulk of the substituents R and R′. In the examples studied the proportions of products (XLIa) and (XLIb) were found to be appreciably different from those obtained by heterogeneous hydrogenation.

A number of instances of selective hydrogenation by means of RhCl(PPh$_3$)$_3$

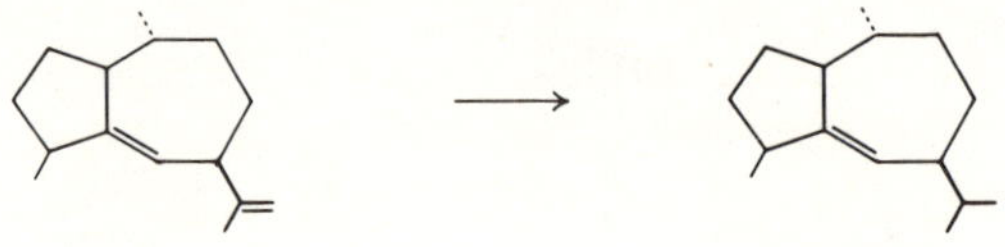

(XLII) → (XLIII)

(XLIV) → (XLV)

(XLVI) → (XLVII)

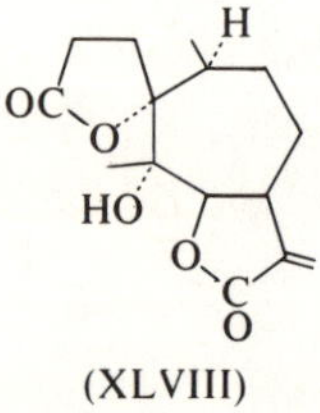

(XLVIII)

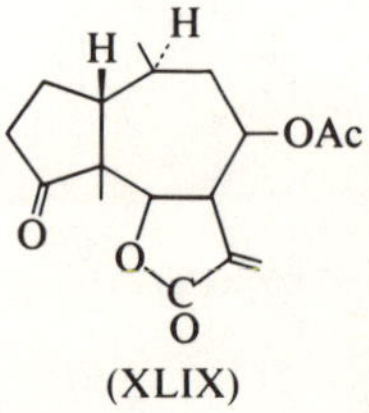

(XLIX)

have also been observed in the terpene group. Biellmann and Liesenfelt[43] report the example (XLII → (XLIII). Similarly, eremophilone (XLIV) gives dihydroeremophilone[44] (XLV), and santonin (XLVI) gives 1,2-dihydro santonin[38] (XLVII). In two further examples, psilostachyine (XLVIII) and confertiflorin (XLIX) were found[45] to be hydrogenated to the dihydro derivatives, whilst with the structurally similar terpene damsin (L), isomerisation to iso-damsin (LI) rather than hydrogenation was observed. However,

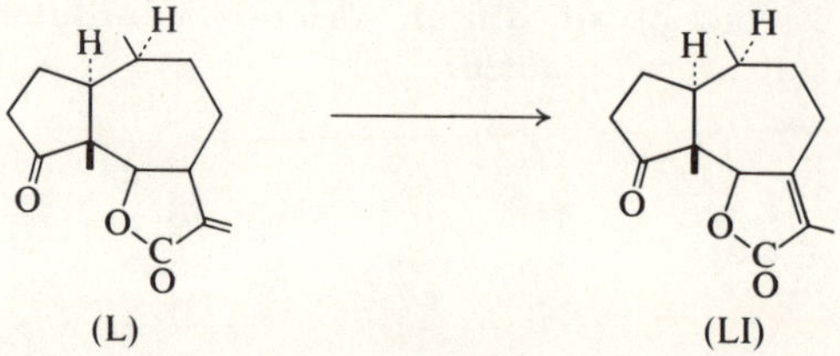

(L) → (LI)

isomerisation rather than hydrogenation by means of RhCl(PPh$_3$)$_3$ is relatively very rare, and this soluble catalyst offers considerable advantage over the available heterogeneous methods in this respect[45].

The hydrogenation of (LII) selectively at the enedione grouping to give (LIII) provides an illustration in synthetic chemistry[46] of the value of homogeneous hydrogenation by means of RhCl(PPh$_3$)$_3$.

Amongst various examples of olefins containing a hetero atom, Birch and

Walker[39] have described the hydrogenation of dihydropyran (LIV) using RhCl(PPh$_3$)$_3$, and the selective hydrogenation of thebaine (LVI) to give the 8,14-dihydro derivative (LVII). Heterogeneous hydrogenation of thebaine yields also tetrahydrothebaine.

Hydrogenation of the olefinic bond in the thiophen derivatives (LVIII)–(LXI) illustrates the utility of RhCl(PPh$_3$)$_3$ with substances containing a heteroatom and the insensitiveness of this catalyst to inactivation by sulphur compounds[47].

Conjugated Olefins

Although Wilkinson[2] and his co-workers found simple conjugated dienes to be relatively resistant to hydrogenation with $RhCl(PPh_3)_3$, the examples discussed above indicate that some conjugated olefins, such as ergosterol and cholestadienones, may be successfully hydrogenated using this or related catalysts.

The hydrogenation of conjugated olefins by means of the hydridopentacyanocobalt catalyst has already been mentioned. For butadiene, penta-1,3-diene, and isoprene the nature of the hydrogenation product was found[12] to depend on the ratio CN^-/Co^{II} used, as shown below. The reason

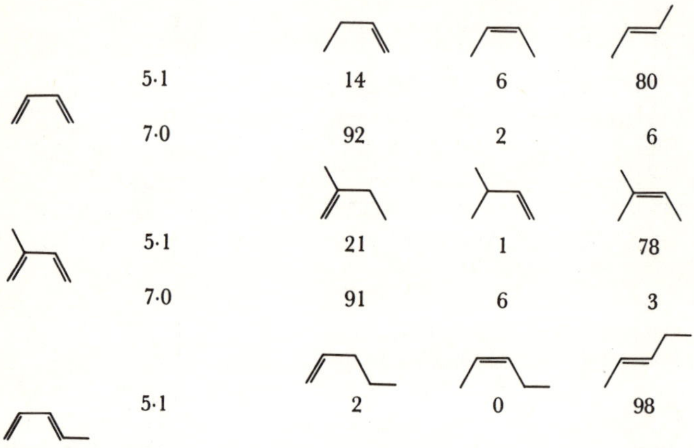

	CN^-/Co^{II}	Products, %		
butadiene	5.1	14	6	80
	7.0	92	2	6
isoprene	5.1	21	1	78
	7.0	91	6	3
penta-1,3-diene	5.1	2	0	98
	7.0	21	12	67

(After Kwiatek, J. and Seyler, J. K., *Homogeneous Catalysis, Advances in Chemistry Series*, No. 70, p. 213–214 (1968) by courtesy of the Authors and the American Chemical Society)

for this dependence of the reaction sequence on the cyanide concentration has already been discussed (page 318). However, the instance of penta-1,3-diene indicates that the extent of isomerisation is also structure dependent, since in this case a high cyanide concentration fails to suppress the rearrangement.

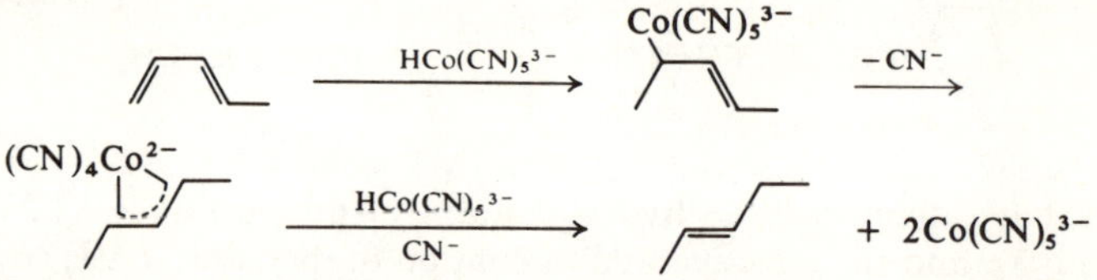

Hydrogenation with $HCo(CN)_5^{3-}$ was also found[12] to be quite sensitive to steric inhibition; 2,5-dimethylhexa-2,4-diene could not be hydrogenated. Similarly, 1,1-diphenylethylene was found to be resistant[9], although styrene and α-methylstyrene could be hydrogenated. Indene was reported to be

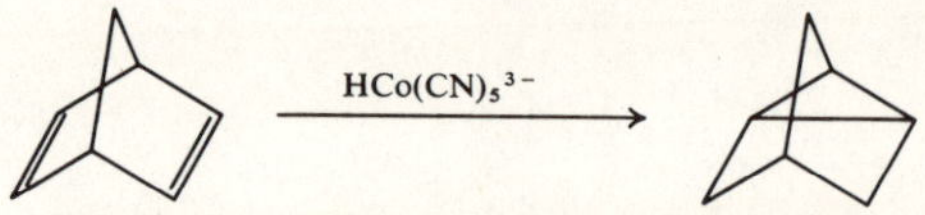

(LXII) $\xrightarrow{HCo(CN)_5^{3-}}$ (LXIII)

resistant[10]. However, cyclohexa-1,3-diene and 3-methylenecyclohexene could be hydrogenated[12], and other conjugated compounds found to react[10] included cinnamic, sorbic, acrylic and α-methacrylic acids.

It is interesting to note that the formally unconjugated norbornadiene (LXII) could be hydrogenated[12] by means of $HCo(CN)_5^{3-}$, but yielded mainly nortricyclene (LXIII); the reaction must involve a conjugative addition process.

Unsaturated Aldehydes

The hydrogenation of unsaturated aldehydes may present a particular problem owing to the ease of carbon monoxide abstraction from the aldehyde group[48]. Thus with heptanal and tris(triphenylphosphine)chlororhodium the following reaction occurs:

$$RhCl(PPh_3)_3 + C_6H_{13}CHO \longrightarrow RhCO(Cl)(PPh_3)_2 + C_6H_{14} + \text{some } C_6H_{12}$$

This process also leads to inactivation of the catalyst since the carbonyl complex $RhCO(Cl)(PPh_3)_2$ does not readily activate hydrogen.

The difficulty may, however, be mitigated if the unsaturated aldehyde is added in dilute solution to $RhCl(PPh_3)_3$ presaturated with hydrogen. A high catalyst ratio and higher hydrogen pressures are also helpful. In this way[48] propenal was hydrogenated rapidly and in satisfactory yield to propanal, but but-2-enal, and 2-methylpent-2-enal were found to react more slowly. Also $PhCH=C(Me)CHO$ has been reported[33] to be resistant to hydrogenation with this catalyst.

The Acetylenic Bond

The acetylenic bond is reported[2] to be hydrogenated by means of $RhCl(PPh_3)_3$, or by using the $py_3RhCl_3/NaBH_4$ catalyst[23], but relatively few examples are described.

THE STEREOCHEMISTRY OF HYDROGENATION OF OLEFINS AND ACETYLENES

For the catalyst $RhCl(PPh_3)_3$ Wilkinson[2] established *cis* hydrogen addition to maleic and fumaric acids which in deuterium were shown to give *meso*- and (±)-1,2-dideuterosuccinic acids respectively. Using $RuCl_2(PPh_3)_3$ in benzene with hydrogen, Jardine and McQuillin[49] demonstrated *cis* hydrogenation of diphenylacetylene to give *cis*-stilbene. Abley and McQuillin also observed[50] *cis* hydrogen addition using their $py_3RhCl_3/NaBH_4$/dimethylformamide catalyst in the cases of dimethyl acetylenedicarboxylate, and butyn-1,4-diol, and with deuterium maleic and fumaric acids gave the expected *meso*- and (±)-dideuterosuccinic acids. An exception was found, however, in the hydrogenation of diphenylacetylene with this catalyst which gave *trans*-stilbene.

OLEFINIC ISOMERISATION

The isomerisation of damsin (L) to isodamsin (LI) noted[45] above (page 326) is an exception to the general experience that hydrogenation with $RhCl(PPh_3)_3$ is singularly free from olefin isomerisation. This instance is, however, important in pointing to the intervention of a rhodium alkyl intermediate.

This matter has been discussed by Bond[6a] who observed isomerisation of *cis*- and *trans*-pent-2-enes during hydrogenation by means of $RhCl(PPh_3)_3$ in benzene–ethanol solution: *cis*-pent-2-ene⟶*trans*-pent-2-ene + pent-1-ene. Isomerisation is observed also with $RhH(CO)(PPh_3)_3$. Competitive hydrogenation/isomerisation has also been studied by Abley and McQuillin[51] using oct-1-ene and the much less active complexes $MX_2(PPh_3)_2$ where M = Pt, Pd, Ni; X = Cl, Br, I, CN.

From a very careful study of deuterium addition to various dimethylcyclohexenes catalysed by $RhCl(PPh_3)_3$, Hussey and Takeuchi[37] obtained the following data:

$[^2H_0]$: 1·9%		1·5%
$[^2H_1]$: 5·3		5·3
$[^2H_2]$: 84·4		92·5
$[^2H_3]$: 7·8		0·8

The different extent of side reaction in the pathways leading to the *cis*- and *trans*-isomers has been noted in other instances[37].

In the absence of added hydrogen $RhCl(PPh_3)_3$ has been shown to be an effective catalyst for isomerisation of 1,4-dienes

—CH=CH—CH$_2$—CH=CH— ⟶ —CH=CH—CH=CH—CH$_2$—

but only in refluxing chloroform or benzene solution. Birch and Subba-Rao[52] report the following examples:

The extent to which isomerisation may intervene during hydrogenation with $RhCl(PPh_3)_3$ has, however, been shown to depend on the experimental conditions. Augustine and Van Peppen[53] have reported that whilst there is no significant isomerisation of hept-1-ene using $RhCl(PPh_3)_3$ in benzene, use of benzene–ethanol as solvent leads to formation of *cis*- and *trans*-hept-2-ene at an early stage in hydrogenation. Olefin isomerisation is much reduced if the complex is first equilibrated with hydrogen, whilst the presence of oxygen appears to potentiate isomerisation.

HYDROGENATION OF HETEROENOID GROUPS

Using the $py_3RhCl_3/NaBH_4$ catalyst in dimethylformamide, Jardine and McQuillin found[24] that azobenzene is rapidly hydrogenated to hydrazobenzene and then very slowly to aniline:

$$PhN=NPh \longrightarrow PhNHNHPh \longrightarrow 2PhNH_2$$

Benzalaniline was also hydrogenated, but not cleaved by hydrogenolysis: $PhCH=NPh \longrightarrow PhCH_2NHPh$. Nitrobenzene gave aniline.

Of particular interest, however, was the observed hydrogenation of pyridine and quinoline to give piperidine and 1,2,3,4-tetrahydroquinoline respectively[24].

Styrene epoxide has been shown to be hydrogenated to 2-phenylethanol by means of the hydridopentacyanocobalt catalyst, and cyclohexene epoxide similarly gave cyclohexanol[10].

HYDROGENATION OF THE CARBONYL GROUP

Whereas $RhCl(PPh_3)_3$ and related rhodium and ruthenium complexes do not appear to catalyse the hydrogenation of a carbonyl group, various cationic complexes such as $[RhH_2(PPh_3)_2 \text{ solvent}]^+$ have been found to be effective for the carbonyl group as well as for olefinic and acetylenic bonds[54]. With this catalyst in hydrogen, acetone gave isopropanol, and in deuterium

t-Bu   H t-Bu   H t-Bu
(LXIV)   (LXV) 86%   (LXVI) 14%

Me_2CDOD. Methyl ethyl ketone, cyclohexanone, acetophenone and 2,2,4,4-tetramethyl cyclobuta-1,3-dione were also hydrogenated to the alcohols, but benzophenone was found to be resistant. 4-t-Butylcyclohexanone (LXIV) gave both *cis*- and *trans*-4-t-butylcyclohexanols, the *trans*-isomer (LXV) with equatorial OH being predominant.

Water appears to be an essential component of this catalyst system and it has been proposed[54] that water may have a role in the lysis of an alkoxy rhodium intermediate.

Using hydridopentacyanocobalt, Kwiatek, Mador and Seyler[10] noted the hydrogenation of benzil, but only to the benzoin stage:

$$PhCOCOPh \longrightarrow PhCH(OH)COPh$$

A less well defined catalyst system developed by Henbest[55] is interesting in that it shows marked selectivity towards formation of the axial alcohol in the hydrogenation of cyclohexanones and ketosteroids. However, this catalyst does not use molecular hydrogen. Reduction is carried out with sodium chloroiridate or chloriridic acid in isopropanol under reflux with

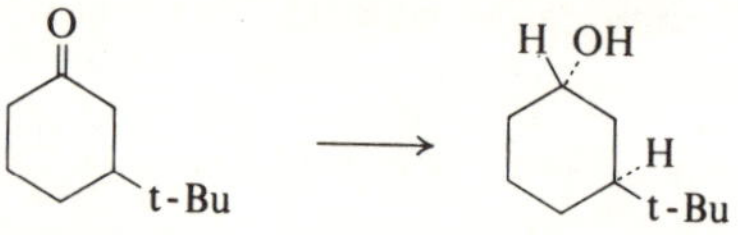

(LXVII) → (LXVIII)

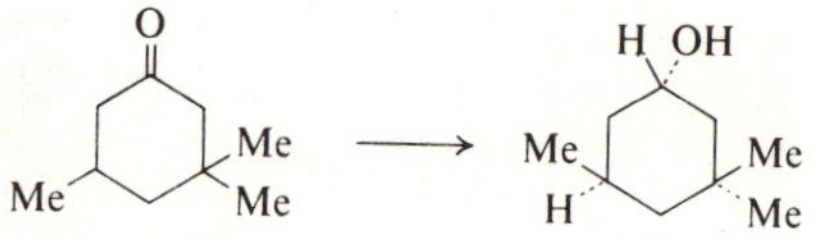

(LXIX) → (LXX)

trimethyl phosphite which is hydrolysed to phosphorous acid and presumably acts as a hydride source. Reaction is slow and may require 24 hours or longer, but the axial cyclohexanol is obtained with a purity of better than 90%. Similarly 5αH- and 5βH-3-ketosteroids gave essentially pure axial steroid alcohols. This stereospecificity suggests hydride transfer from a bulky reagent, and, in agreement, the more hindered 2-ketosteroid group was reduced only slowly, a 4-keto-group was resistant, and the 17-keto-group less reactive than a keto-group in position 3.

HALOGEN COMPOUNDS

Hydrogenolysis of halides has not been extensively studied. Several allylic halides are hydrogenated by means of hydridopentacyanocobalt, and the following cases are reported[11].

$$Ph_2CHCl \longrightarrow Ph_2CH-CHPh_2$$
$$Ph_3CCl \longrightarrow Ph_3C\text{-dimer}$$

Birch and Walker[39] found benzyl bromide to be resistant to hydrogenolysis using the $RhCl(PPh_3)_3$ catalyst, but cinnamyl chloride showed some hydrogenolysis as well as hydrogenation.

HYDROGENATION OF VINYLCYCLOPROPANES

Heathcock and Poulter[56] report that hydrogenation of vinylcyclopropanes using $RhCl(PPh_3)_3$ in benzene solution is accompanied, to a greater or lesser degree, by hydrogenolysis of the cyclopropane:

HOMOGENEOUSLY CATALYSED HYDROGENATION

[reaction scheme: methylenecyclopropane → methylcyclopropane 85% + butene 14%]

[reaction scheme: vinylcyclopropane → ethylcyclopropane 70% + pentene 27%]

[reaction scheme: isopropylidenecyclopropane → isopropylcyclopropane 97%]

[reaction scheme: bicyclic → bicyclic 82% + methylcyclohexane 17%]

It is suggested by these authors[56] that the hydrogenolysis represents an alternative mode of reaction of an alkyl rhodium intermediate, e.g.

$$\triangleright\!\!-\!\!\text{Rh(H)}_2\text{L}_3 \longrightarrow \triangleright\!\!-\!\!\overset{\text{CH}_3}{\underset{\text{Rh(H)L}_3}{|}}$$

$$\triangleright\!\!-\!\! + \text{RhL}_3 \quad \text{or} \quad \text{L}_3\text{Rh(H)}\!\!-\!\!\sim\!\!\sim$$

$$\sim\!\!\sim + \text{RhL}_3$$

CATALYSED DEUTERATION

The *cis* hydrogen addition mechanism and the generally unimportant isomerisation or exchange reactions observed with $RhCl(PPh_3)_3$ makes this a very suitable catalyst for specific *cis* deuterium addition. This principle was established by Wilkinson in the demonstration[2] that maleic and fumaric acids give respectively *meso*- and $(\pm)$-1,2-dideuterosuccinic acids. Zech, Jones and Djerassi[57] have applied the method to the preparation of (LXXII) and related substances required for mass spectroscopic studies.

The catalyst $RhCl(PPh_3)_3$ was also employed[58] to prepare dideuterocyclo-octatriene (LXXIV) required for mechanistic studies. Djerassi and

[structure LXXI: R = H or Me] → [structure LXXII]

[structure LXXIII: cyclooctatriene] → [structure LXXIV: dideuterocyclooctatriene]

Gutzwiller[41] have demonstrated the stereochemistry of deuterium addition to cholest-1-en-3-one (LXXV) and cholesta-1,4-diene-3-one (LXXVI), using the $RhCl(PPh_3)_3$ catalyst.

However, whilst deuteration in benzene solution generally leads to the dideutero derivative in good yield, use of benzene–alcohol as solvent with prolonged reaction times is accompanied by exchange processes which result in partially deuterated products[59].

The chromium carbonyl developed by Frankel permits specific deuteration in a different manner. This catalyst effects 1,4-hydrogen addition to conjugated 1,3-dienes, or to 1,4-dienes which are brought into conjugation by the catalyst[26]. Using the chromium carbonyl methyl benzoate complex, $(PhCO_2Me)Cr(CO)_3$, Frankel, Selke and Glass[60] found specific 1,4-deuterium addition to methyl sorbate (LXXVII).

$$MeCH=CHCH=CHCO_2Me \longrightarrow \underset{\underset{CH=CH}{\diagup \diagdown}}{MeCHD \quad\quad CHDCO_2Me}$$

(LXXVII) (LXXVIII)

The chromium carbonyl appears to form a 1,4-adduct with the diene which undergoes hydrogenolysis or deuterolysis. The product, e.g. (LXXVIII), is necessarily the *cis*-isomer.

INDUCED ASYMMETRY VIA HOMOGENEOUS HYDROGENATION

Homogeneous hydrogenation which involves hydrogen transfer within an olefin–metal complex opens up the possibility of asymmetric hydrogenation by the introduction of an asymmetric ligand proximate to the reaction site. This possibility has been realised. Knowles and Sabacky[61a] prepared $[(-)\text{-}Me(Pr_i)PhP]_3RhCl_3$ which was reduced in hydrogen in the presence of triethylamine, and the derived complex used for catalysis of the hydrogenation:

$$\underset{(LXXIX)}{R-\underset{\underset{CH_2}{\parallel}}{\overset{\overset{CO_2H}{\mid}}{C}}} \longrightarrow \underset{(LXXX)}{R-CH(Me)CO_2H}$$

This gave, in the case (LXXIX, R = Ph), hydratropic acid, (LXXX, R = Ph) of 15% optical purity and, in the case R = CH_2COOH, a product (LXXX, R = CH_2COOH) showing 3% optical purity. Use of a phosphine ligand PhP($CH_2\overset{*}{C}$HMeEt)$_2$ which is asymmetric at a more remote carbon centre (C*) was less successful, atropic acid (LXXIX, R = Ph) giving hydratropic acid of only 1% optical purity. Similar results are reported by Horner, Siegel and Büthe[61b].

A much more effective and convenient method of asymmetric hydrogenation has, however, been described by Abley and McQuillin[62a]. The use of py$_3$RhCl$_3$ activated with NaBH$_4$ in the presence of a series of optically active amides for the hydrogenation (LXXXI) → (LXXXII) gave the results listed below[62b].

$$Ph-\underset{\underset{Me}{|}}{C}=CH-CO_2Me \longrightarrow Ph-CH(Me)CH_2CO_2Me$$
$$(LXXXI) \qquad\qquad\qquad (LXXXII)$$

	Aysmmetric amide	Product (LXXXII) $[\alpha]_D$
(a)	(+)-PhCH(Me)NHCHO	+ 33°
(a)	(−)- "	− 28°
(b)	(+)- "	+ 32°
(c)	(+)-[PhCHMeNHCO]$_2$	+ 26°
(a)	(−)-CH$_3$CH(OH)CONMe$_2$	− 9°
(c)	(+)-N-acetylglucosamine	− 8°
(b)	(−)-N-formylbornylamine	+ 26°
(b)	(−)-N-formylisobornylamine	+ 16°

(a) Using the amide as solvent, (b) using the amide as a 5% solution in ethyl digol, (c) using the amide as a 5% solution in diethyleneglycol monoethyl ether/water (10:1).

Since pure methyl 3-phenylbutanoate, (LXXXII), shows an optical rotation of (+) or (−) 58 degrees these results represent, in the best cases, an induced asymmetry of some 60%. The results are interesting not only for this high level of induced asymmetry, but also because they demonstrate the feasibility of asymmetric induction over a sequence of five intervening atoms in the rhodium complex.

More successful hydrogenations using rhodium complexes of suitable chiral phosphines have recently been reported. The degree of induced asymmetry appears, however, to depend rather critically on the substance hydrogenated[63].

CONCLUSION

The development of homogeneous hydrogenation catalysts has made it possible to express formerly rather obscure processes in precise molecular terms. The active complexes at present available make possible not only highly selective hydrogenation and deuterium addition, but also hydrogenation of carbonyl and other hetero-groups, and in addition asymmetric hydrogenation with a useful level of induced asymmetry.

Manuscript received March 1971.

REFERENCES

1. VASKA, L., *Inorg. Nucl. Chem. Lett.*, **1**, 89 (1965); VASKA, L. and RHODES, R. E., *J. Am. chem. Soc.*, **87**, 4970 (1965); cf. CHOCK, P. B. and HALPERN, J., *J. Am. chem. Soc.*, **88**, 3511 (1966)
2. OSBORN, J. A., JARDINE, F. H., YOUNG, J. F. and WILKINSON, G., *J. chem. Soc. A*, 1711 (1966)
3. LEHMANN, D. D., SCHRIVER, D. F. and WHARF, I., *Chem. Commun.*, 1486 (1970)
4. BENNETT, M. A. and MILNER, D. L., *J. Am. chem. Soc.*, **91**, 6983 (1969)
5. (a) VAN GAAL, H., CUPPERS, H. G. A. M. and VAN DER ENT, A., *Chem. Commun.*, 1964 (1970); (b) GUSTINIANI, M., DOLCETTI, G., NICOLINI, M. and BELLUCO, U., *J. chem. Soc. A*, 1961 (1969)
6. (a) cf. BOND, G. C. and HILLYARD, R. A., *Discuss. Faraday Soc.*, **46**, 20 (1968); (b) STROHMEIER, W. and REHDER-STIRAWEISS, W., *J. organomet. Chem.*, **26**, C22 (1971)
7. O'CONNOR, C. and WILKINSON, G., *J. chem. Soc. A*, 2665 (1968)
8. HALLAM, P. S., MCGARVEY, B. R. and WILKINSON, G., *J. chem. Soc. A*, 3143 (1968)
9. YAGUPSKY, G., EVANS, D. and WILKINSON, G., *J. chem. Soc. A*, 2660 (1968)
10. KWIATEK, J., MADOR, I. L. and SEYLER, J. K., *J. Am. chem. Soc.*, **84**, 304 (1962)
11. KWIATEK, J. and SEYLER, J. K., *J. organomet. Chem.*, **2**, 421 (1965)
12. KWIATEK, J. and SEYLER, J. K., *Adv. Chem. Ser.*, **70**, 207 (1968)
13. HALPERN, J. and WONG, L.-Y., *J. Am. chem. Soc.*, **90**, 6665 (1968)
14. BURNETT, M. G., CONNOLLEY, P. J. and KEMBALL, C., *J. chem. Soc. A*, 991 (1968)
15. SHAPLEY, J. R., SCHROCK, R. R. and OSBORN, J. A., *J. Am. chem. Soc.*, **91**, 2816 (1969)
16. SCHROCK, R. R. and OSBORN, J. A., *Chem. Commun.*, 567 (1970)
17. GREEN, M., KUC, T. A. and TAYLOR, S. H., *Chem. Commun.*, 1553 (1970)
18. LEGZDINS, P., REMPEL, G. L. and WILKINSON, G., *Chem. Commun.*, 825 (1969); HUI, B. C. and REMPEL, G. L., *Chem. Commun.*, 1195 (1970); LEGZDINS, P., MITCHELL, R. W., REMPEL, G. L., RUDDICK, J. D. and WILKINSON, G., *J. chem. Soc. A*, 3322 (1970)
19. CRAMER, R. D., JENNER, E. L., LINDSAY, R. U. and STOLBERG, V. G., *J. Am. chem. Soc.*, **85**, 1691 (1963)
20. VAN BEKKUM, H., VAN GOGH, J. and VAN MINNEN-PATHUIS, G., *J. Catalysis*, **7**, 292 (1967); VAN'T HOF, L. P. and LINSEN, B. G., *J. Catalysis*, **7**, 295 (1967)
21. OSBORN, J. A., WILKINSON, G. and YOUNG, J. F., *Chem. Commun.*, 17 (1965)
22. JAMES, B. R. and REMPEL, G. L., *Can. J. Chem.*, **44**, 233 (1966); JAMES, B. R. and REMPEL, G. L., *Discuss. Faraday Soc.*, **46**, 48 (1968)
23. JARDINE, I. and MCQUILLIN, F. J., *Chem. Commun.*, 477 (1969)
24. JARDINE, I. and MCQUILLIN, F. J., *Chem. Commun.*, 626 (1970)
25. FRANKEL, E. N., MOUNTS, T. L., BUTTERFIELD, R. O. and DUTTON, H. J., *Adv. Chem. Ser.*, **70**, 177 (1968)
26. FRANKEL, E. N. and BUTTERFIELD, R. O., *J. org. Chem.*, **34**, 3830 (1969); FRANKEL, E. N., SELTE, E. and GLASS, C. A., *J. Am. chem. Soc.*, **90**, 2447 (1968)
27. SLOAN, M. P., MATLACK, A. S. and BRESLOW, D. S., *J. Am. chem. Soc.*, **85**, 4014 (1963)
28. (a) BURNETT, M. G., *Chem. Commun.*, 507 (1965); (b) HALPERN, J., HARROD, J. F. and JAMES, B. R., *J. Am. chem. Soc.*, **83**, 753 (1967); (c) TAKEGARNI, Y., UENO, T. and FUJII, T., *Bull. chem. Soc. Japan*, **38**, 1279 (1965); (d) SCHWAB, G.-M. and MANDRE, G., *J. Catalysis*, **12**, 103 (1968); (e) STERN, R. and SAJUS, L., *Tetrahedron Lett.*, 6313 (1968); (f) HIDAI, M., KUSE, T., HIKITA, T., UCHIDA, Y. and MISONA, A., *Tetrahedron Lett.*, 1715 (1970); (g) OGATA, I., IWATA, R. and IKEDA, Y., *Tetrahedron Lett.*, 3011 (1970)
29. MCCLEVERTY, J. H. and WILKINSON, G., *Inorg. Synth.*, **8**, 214 (1966)
30. EVANS, D., OSBORN, J. A. and WILKINSON, G., *Inorg. Synth.*, **11**, 99 (1968)
31. GRIFFITH, W. P. and WILKINSON, G., *J. chem. Soc.*, 2759 (1959); KING, N. K. and WINFIELD, M. E., *J. Am. chem. Soc.*, **83**, 3366 (1961)
32. O'CONNOR, C. and WILKINSON, G., *Tetrahedron Lett.*, 1375 (1969)
33. HORNER, L., BÜTHE, H. and SIEGEL, H., *Tetrahedron Lett.*, 4023 (1968)
34. STERN, R., CHEVALLIER, Y. and SAJUS, L., *C. r. Acad. Sci. C*, **264**, 1740 (1967); cf. *Tetrahedron Lett.*, 1197 (1969)

35. HARMON, R. E., PARSONS, J. L., COOKE, D. W., GUPTA, S. K. and SCHOOLENBERG, J., *J. org. Chem.*, **34**, 3684 (1969)
36. JARDINE, I. and MCQUILLIN, F. J., *Chem. Commun.*, 502 (1969)
37. HUSSEY, A. S. and TAKEUCHI, Y., *J. org. Chem.*, **35**, 643 (1970); cf. *J. Am. chem. Soc.*, **91**, 672 (1969)
38. SIMS, J. J., HONWAD, V. K. and SELMAN, L. H., *Tetrahedron Lett.*, 87 (1969); cf. PIERS, F. and CHENG, K. F., *Can. J. Chem.*, **46**, 377 (1968)
39. BIRCH, A. J. and WALKER, K. A. M., *J. chem. Soc. C*, 1894 (1966)
40. WIELAND, P. and ANNER, G., *Helv. chim. Acta*, **51**, 1698 (1968)
41. DJERASSI, C. and GUTZWILLER, J., *J. Am. chem. Soc.*, **88**, 4537 (1966)
42. JARDINE, I. and MCQUILLIN, F. J., *Chem. Commun.*, 503 (1969)
43. BIELLMANN, J.-F. and LIESENFELT, H., *Bull. Soc. chim. Fr.*, 4029 (1966)
44. BROWN, M. and PISZKIEWIZ, L. W., *J. org. Chem.*, **32**, 2013 (1967)
45. BIELLMANN, J.-F. and JUNG, M. J., *J. Am. chem. Soc.*, **90**, 1673 (1968)
46. HOFFSOMMER, R. D., TAUB, D. and WENDLER, N. L., *J. org. Chem.*, **32**, 3074 (1967)
47. HORNFELT, A. B., GRONOWITZ, J. S. and GRONOWITZ, S., *Acta chem. scand.*, **22**, 2725 (1968); cf. BIRCH, A. J. and WALKER, K. A. M., *Tetrahedron Lett.*, 1935 (1967)
48. JARDINE, F. H. and WILKINSON, G., *J. chem. Soc. C*, 270 (1967)
49. JARDINE, I. and MCQUILLIN, F. J., *Tetrahedron Lett.*, 4871 (1966)
50. ABLEY, P. and MCQUILLIN, F. J., *Chem. Commun.*, 1503 (1969)
51. ABLEY, P. and MCQUILLIN, F. J., *Discuss. Faraday Soc.*, **46**, 31 (1968)
52. BIRCH, A. J. and SUBBA-RAO, G. S. R., *Tetrahedron Lett.*, 3797 (1968)
53. AUGUSTINE, R. L. and VAN PEPPEN, J. F., *Chem. Commun.*, 495, 497, 571 (1970)
54. SCHROCK, R. R. and OSBORN, J. A., *Chem. Commun.*, 567 (1970)
55. HADDAD, Y. M. Y., HENBEST, H. B., HUSBANDS, J. and MITCHELL, T. R. B., *Proc. chem. Soc.*, 361 (1964); cf. BROWN, P. A. and KIRK, D. N., *J. chem. Soc. C*, 1653 (1969); ORR, J. C., MERCEREAU, M. and SANFORD, A., *Chem. Commun.*, 162 (1970)
56. HEATHCOCK, C. H. and POULTER, S. R., *Tetrahedron Lett.*, 2755 (1969)
57. ZECH, B., JONES, G. and DJERASSI, C., *Chem. Ber.*, **100**, 3204 (1967)
58. ST. JACQUES, M. and PRUD'HOMME, R., *Tetrahedron Lett.*, 4833 (1970)
59. VOLTER, W. and DJERASSI, C., *Chem. Ber.*, **101**, 58 (1968)
60. FRANKEL, E. N., SELKE, E. and GLASS, C. A., *J. Am. chem. Soc.*, **90**, 2447 (1968)
61. (a) KNOWLES, W. S. and SABACKY, M. J., *Chem. Commun.*, 1445 (1968); (b) HORNER, L., SIEGEL, H. and BÜTHE, H., *Angew. Chem. (Int. Edn)*, **7**, 942 (1968)
62. (a) ABLEY, P. and MCQUILLIN, F. J., *Chem. Commun.*, 477 (1969); (b) ABLEY, P. and MCQUILLIN, F. J., *J. chem. Soc. C*, 844 (1971)
63. DANG, T. P. and KAGAN, H. B., *Chem. Commun.*, 481 (1971); MORRISON, J. D., BURNETT, R. E., AGUIAR, A. M., MORROW, C. J. and PHILLIPS, C., *J. Am. chem. Soc.*, **93**, 1301 (1971)

INDEX

Abscissin II, 35
β-Acoradiene, 42
Actinocin, 197, 200, 201
Actinomycins, 197
   structure, 197
   synthetic studies
      aniso-series, 200
      iso-series, 197
Adouetines, 210
Alamethicin
   structure, 187
Albomycins, 168
Albonoursin
   synthesis, 131
Aldol condensation, 24
Alumichrome
   H^1-n.m.r. studies, 167
Amanita fungi
   peptides from, 204
Amatoxins, 205
Anhydropenicillins, 114
Antanamide, 207
   strucutre, 207
   synthesis, 207
Aplysin, 35
Aranotins, 133
Aspochracin, 140
(−)-Atisirene, 53

Bacitracin, 203
Beauvericin, 160
Bioluminescence
   Cypridina Hilgendorfii, 269
   Cypridina luciferin, 270
   Firefly luciferin, 267
   Latia luciferin, 256
   Latia neritoides, 271
Bisabolol, 32
Bis(triphenylphosphine)-
   chlorocarbonyliridium, 315

Calcitonin, 215
(±)-Campherenone, 43
Camphor, 30
Cannabichromene, 86
Cannabicyclol, 85
Cannabidiol, 80
Cannabidiolic acid, 85
Cannabidiolic acid tetrahydrocannabitriol ester, 88

Cannabidivarin, 88
Cannabielsoic acids, 88
Cannabigerol, 80
Cannabigerol monomethyl ether, 88
Cannabigerolic acid, 85
Cannabinoids, 78
   biogenesis, 97
   isomerisation and cyclisation, 92
Cannabinol, 80
Cannabipinol, 86
Cannabis Sativa L., 78
Cannabivarin, 99
Capreomycidine, 192
Capreomycin, 192
Carba-oxytocins, 214
Catalytically active complexes, 320
Cation transport by
   enniatins, 160
   model system, 147
   valinomycin, 183
Cationic complexes, 320
Ceanothines, 210
Cedrene, 41
Cedrol, 41
Cephams, 115
Chaetocin, 134
α-Chamigrene, 35
Charas, 78
Chemiluminescence, 231
   autoxidation, Russell mechanism, 243
   bis-2,3-dinitrophenyl oxalate, 258
   dioxetanedione, 258
   1,2-dioxetanes, 251
   electrochemically generated, 238
   electron transfer reactions, 239
      acridan oxidation, 241
      radical anions, oxidation, 239
      reduction reactions, 242
   Grignard reagent, 240
   hydrazides,
      cyclic, 248
      linear, 250
   monoperoxyoxalic acid, 258
   nitriles, 245
   peroxide reactions, 247
      acridan esters, 262
      acridine esters and related compounds, 259
      active esters and related compounds, 257
      alkylidene peroxides, 245
      aromatic endoperoxides, 247

INDEX

Chemiluminescence—*continued*
    dioxetane, 251
    hydrazide, 248
    imine peroxides, 255
    indolenyl hydroperoxides, 250
    indolenyl peroxides, 255
    lucigenin, 263
    oxalic acid derivatives, 257
    quantum yield, 234
    sensitised, 234
    singlet oxygen, 243
    sources, 245
    tetrakisdimethylaminoethylene, 257
    weak, 243
$\Delta^{8,14}$-cholesten-3α-ol, 62
Δ^{11}-cholesten-3α-ol, 62
Chromium carbonyl arenes, 321
Conformation of cyclic peptides
    cyclohexapeptides, 148–150
    cyclopentapeptides, 146–147
    cyclotetrapeptides, 142–144
    cyclotripeptides, 140
    2, 5-dioxopiperazines, 135
    enniatins, 161–162
    ferrichromes, 166
    gramicidin S, 174, 176
    valinomycin, 183-184
Cyclic peptides from Amanita fungi, 205–206
Cyclisation
    antanamide peptides, 209
    cysteine peptides, 216
    diastereoisomeric peptides, 144, 145
    gramicidin S peptides, 169
    hexapeptides, 159
    pentapeptides, 146, 170
Cyclodecapeptides
    gramicidin S, 169–180
    tyrocidines, 180–181
Cyclodiastereoisomerism, 152
Cyclodimerisation, 156
    of pentapeptides, 169
    of tripeptides, 159
    stereospecificity of, 145, 159
Cycloenantioisomerism, 152
Cycloheptamycin, 191
Cyclohexapeptides
    cyclodimerisation, 156
    spectroscopic studies, 150, 154
    stereochemistry of, 148, 149, 152
Cyclooctenyl intermediate, 46
Cyclopentapeptides, 146
    cyclisation studies, 146, 170
Cyclotetrapeptides, 141, 144
    cyclisation studies, 144
Cyclotripeptides, 140
Cysteine peptides
    oxidation studies, 216

Dehydroisocalamanediol, 39
16,17-Dehydroprogesterone, 59

Depsipeptides
    enniatins, 159
    monamycins, 189
    valinomycin, 182
Deuteration, 333
    cholest-1-en-3-one, 334
    cholesta-1,4-diene-3-one, 334
    conjugated 1,3-dienes, 334
    dideuterocyclooctatriene, 333
    fumaric acid, 333
    maleic acid, 333
    methyl sorbate, 334
Dihydrolanosterol, 58
Dihydroparthenolide, 38
Dimethylbiacridene, 253, 265
2,5-Dioxopiperazines, 131
    conformation studies, 135
    metabolites, 131
    preparation, 134
    spectroscopic studies, 137, 139
Direct field effect, 304
($\pm$)-Drimenin, 35
Drimenol, 32

Echinomycin, 196
Echinulin, 131
Elemane-type compounds, 38
β-Elemene, 36
Enantio-enniatin B, 160
Energy transfer, 250, 251, 252, 265
Enniatins, 159
    analogues, 159
    cation binding, 159
    conformation, 159, 162
    structure–activity relationships, 159
Enzyme models, 155
Enzyme substrates, 155
($\pm$)-Epicampherenone, 43
($\pm$)-Epi-β-santalene, 45
Etamycin, 195
Eudesmols, 36
Evolidine, 168

trans, trans-Farnesic acid, 34
Farnesiferol A, 33
Farnesiferol C, 33
trans, trans-Farnesyl acetate monoepoxide, 32
Fernadiene, 68
Fern-8-ene, 68
Ferrichrome
    H^1-n.m.r. studies, 166
    synthesis, 165
Ferrichrome A
    X-ray studies, 166
Ferrichromes
    structure, 163
($\pm$)-Fichtelite, 51
Fungisporin, 144

INDEX

Geranylgeraniol, 45, 51
Gliotoxin, 132
Gramicidin S, 169
   analogues, 170, 173
   conformation, 176, 179
   cyclodimerisation, 169
   structure–activity relationships, 170, 171
   synthesis, 169

Hashish, 78
(−)-Hibaene, 55
Homogeneous hydrogenation, 314
   acetylenic bond, 329
   conjugated olefins, 327
   olefins, 322
   unsaturated aldehydes, 329
Hopenone-1, 59
Hydridocarbonyltris(triphenylphosphine)-rhodium, 317
Hydridochlorotris(triphenylphosphine)-ruthenium, 318
Hydridopentacyanocobalt, 318
Hydrogenation
   acetone, 331
   acetophenone, 331
   azobenzene, 331
   benzalaniline, 331
   benzil, 332
   4-t-butylcyclohexanone, 331
   carbonyl group, 331
   confertiflorin, 326
   cyclohexa-1,3-diene, 329
   cyclohexanone, 331
   cyclohexene epoxide, 331
   damsin, 326
   dihydropyran, 327
   dimethylcyclohexenes, 330
   diphenylacetylene, 329
   eremophilone, 326
   ergosterol, 324
   heteroenoid group, 331
   indene, 328
   3-keto-Δ^4-steroids, 325
   5αH-3-ketosteroid, 332
   5βH-3-ketosteroid, 332
   linalool, 324
   metal-alkyl intermediate, 317
   1-methoxycyclohexa-1,4-diene, 324
   methyl ethyl ketone, 331
   2-methylpent-2-enal, 329
   α-methylstyrene, 328
   naphthalene, 323
   nitrobenzene, 331
   cis-pent-2-ene, 330
   propenal, 329
   psilostachyne, 326
   pyridine, 331
   quinoline, 331
   santonin, 326
   selective, 324
   styrene, 328
   styrene epoxide, 331
   terminal olefins, 322
   2,2,4,4-tetramethyl cyclobuta-1, 3-dione, 331
   thebaine, 327
   thiophen derivatives, 327
Hydrogenolysis
   vinyl cyclopropanes, 332
14α-Hydroxyhibanes, 46
16-Hydroxyisokaurene, 53
Hymenocardine, 211

Imidazopyridinone, 270
Immunoglobulins, 221
Induced assymetry, 334
π-Inductive effect, 304
Infra-red studies on cyclic peptides
   cyclo-hexapeptides, 154
   2,5-dioxopiperazines, 139
Insulin, 217
   chain recombination studies, 218
   synthesis, 218
   synthesis of a model, 219
(−)-Isoatisirene, 53
(±)-Isocampherenol, 43
Iso-cannabichromene, 92
(+)-Isokaurene, 55
(−)-Ioskaurene, 53, 54
Isotetrahydrocannabinols, 86

(−)-Kaurene, 53, 54
12-Keto-3α-cholestanol, 62

Lanosterol, 56
Lasiodines, 211
β-Levantenolide, 51
Limonene, 29
Localisation energy, 308
Lophine, 250, 255
Luminescence
   electron transfer, 236
Luminol(5-amino-2,3-dihydro-1,4-phthalazinedione), 248

(±)-Malabaricanediol, 57
Malformin, 213
Marihuana, 78
Mayurone, 41
Metal hydride, 314
Metal ion complexes of cyclic peptides
   enniatins, 159
   ferrichromes, 163–166
   model system, 147
   valinomycin, 183

INDEX

7-Methoxymethylbicycloheptenone, 14
Methyl angolensate, 71, 72
Methyl-1,4,5,8-tetrahydronaphthoate, 323
Mexicanolide, 74
Monamycins, 189
Monogynol, 55
Mycobacillin
  structural studies, 186

Neurohypophyseal hormones, 213
  structure–activity relationships, 213
Nitration
  acetoxylation and other side reactions, 286
  limiting rates of, 293
Nitration in
  acetic anhydride, 282
  carbon tetrachloride, 279
  dichloromethane, 291
  nitromethane, 280, 291
  sulpholan, 280
Nitration of
  acetanilide, 300
  anisole, 298
  1,4-dimethylnaphthalene, 290
  methyl phenethyl ether, 302
  polyalkylbenzenes, 291
  o-xylene, 288
  p-xylene, 290
Nitronium salts, 280
Nitro-oxylation, 291
Nitrosation, 281
Norphalloin
  synthesis, 207

Olefin isomerisation, 329
o.r.d. and c.d. studies of cyclic peptides
  cyclo-hexapeptides, 154
  2,5-dioxopiperazines, 137
  enniatins, 161
  gramicidin S analogues, 173
Oxidative studies on cysteine peptides, 216
Oxytocin, 213

Pandamine, 211
β-Patchoulene, 38
Penicillin, 102
  conformation, 104
  deaminations, 123
  epimerisations, 116
    mechanism, 118
    thermodynamics, 117
  mode of action, 104
  nomenclature, 102
  rearrangements, 124
    1,2-bond cleavages, 124

    1,5-bond cleavages, 125
    4,7-bond cleavages, 125
  structure–activity relationships, 105
  sulphenic acids, 114
  sulphoxides, 109
  transformation, to cephalosporin, 108
Pentanorlanosterol, 58
Peptide alkaloids, 210
Peptide antibiotics containing 3-hydroxy-picolinic acid, 193
Phallotoxins, 205–206
Phosphine ligands, 322
Pimara-8,15-dienes, 46
β-Pinene, 30
Piperazic acid derivatives, 189
Positive poles, 303
Preisocalamanediol, 39
Presqualene alcohol, 67
Presqualene pyrophosphate, 66
Pristinamycins, 195
Pro-insulin, 220
  C-peptide, 221
Prostaglandins, 1
  biological activity, 2
  biosynthesis, 3
    arachidonic scid, 3
    bishomo-γ-linolenic acid, 3
  metabolism, 3
    15-hydroxyl, oxidation, 4
    β-oxidation, 4
  non-mammalian source, 22
  primary, 2
Prostaglandin-A_1, 1
Prostaglandin-A_2, 1
  15-epimer, 22
Prostaglandin-A_3, 1
Prostaglandin-B_1
  isomer, 12
Prostaglandin-E_1, 1, 12, 16
  15-dehydro-, 9
  13,14-dihydro-, 23
  11-epimer, 8, 9
  15-epimer, 8
  15-epimer esters, 13
  indane route, synthesis, 19
  8-isomer, 13
  15-methyl, 13
  methyl ester, 12
  3-oxa-, 13
  synthesis of, 5
  1,1,1-trichlorethyl ester, 12
  5,6-tritiated analogues, 16
Prostaglandin-E_2, 13, 16, 23
  15-epimer, 13
Prostaglandin-E_3
  methyl ester, 13
Prostaglandin-F_1
  bicyclo[3,1,0]heptane route, synthesis, 10
  isomers, 12

INDEX

Prostaglandin-$F_{1\alpha}$, 1, 13, 16
  separation of C-15 epimers, 16
  5,6-tritiated analogues, 16
Prostaglandin-$F_{1\beta}$,
  11-desoxy, 25
Prostaglandin-$F_{2\alpha}$, 1, 14, 16, 23
  bicyclo[2,2,1]heptane route, synthesis, 14
  tetrahydropyranyl ethers, 14
Prostaglandin-$F_{3\alpha}$, 1
Prostanoic acid, 1
Pyridomycins, 195

Quinomycins, 196
Quinoxaline antibiotics, 196

Ribonuclease, 217
Rosadiene, 48
Rosenonolactone, 49

($\pm$)-α-Santalene, 45
($\pm$)-β-Santalene, 45
Schiff bases, 256
Secopenicillins, 116
Serratamolides, 144
Sporidesmins, 133
S-Protecting groups, 219
Squalene, 63, 65, 66
Squalene monoepoxide, 56
Staphylomycin, 195
Stendomycidine, 193
Stendomycin, 193
Stereochemistry of hydrogenation of
  olefins and acetylenes, 329
Δ^{11}-Steroids, 63
Δ^{16}-Steroids, 63

Telomycin, 193
α-Terpineol, 29
Tetrahydrocannabinols
  carbocyclic and heterocyclic analogues, 97
  interconversion of cis- and trans-, 92
  mass spectrometry, 97
  metabolites, 97
  p.m.r. analysis, 97
  stability studies, 88
Δ^{1}-Tetrahydrocannabinol, 79
$\Delta^{1(6)}$-Tetrahydrocannabinol, 80
Δ^{8}-trans-iso-Tetrahydrocannabinol, 94
Δ^{1}-Tetrahydrocannabinol acid A, 84
Δ^{1}-Tetrahydrocannabinol acid B, 84
Tetrahydrocannabivarin, 99
Tetranortriterpenes, 68
Thiostrepton, 201
  X-ray studies, 203
Threonine-4-oxytocin, 214
Thujopsadiene, 41
Thyrocalcitonin,
  see calcitonin
Triostins, 196
2,4,5-Triphenylimidazole, 255
Tris(triphenylphosphine)chlororhodium, 316
Tryptathionine, 206
Tuberactinomycin, 192
Tyrocidines, 180
  aggregation studies, 181
  structure, 180
  synthesis, 181

Valinomycin, 182–184
Verticillin, 134
Vetiver (zizaane) sesquiterpenes, 39
Viomycidine, 192
Viomycin, 191

Zinc tetraphenylporphine, 242
Zizyphine, 211

QD
245
P7
v.8
1973

JUL 9 1973